DATE DUE

TRANSFORMATIONS IN TELECOMMUNICATIONS AND MEDIA

ELEMENTS AND ISSUES FOR CONSIDERATION

MEDIA AND COMMUNICATIONS - TECHNOLOGIES, POLICIES AND CHALLENGES

Additional books in this series can be found on Nova's website under the Series tab.

Additional E-books in this series can be found on Nova's website under the E-book tab.

MEDIA AND COMMUNICATIONS - TECHNOLOGIES, POLICIES AND CHALLENGES

TRANSFORMATIONS IN TELECOMMUNICATIONS AND MEDIA

ELEMENTS AND ISSUES FOR CONSIDERATION

IRWIN CAVAZOS
EDITOR

nova
publishers

New York

For permission to use material from this book please contact us:
Telephone 631-231-7269; Fax 631-231-8175
Web Site: http://www.novapublishers.com

NOTICE TO THE READER

The Publisher has taken reasonable care in the preparation of this book, but makes no expressed or implied warranty of any kind and assumes no responsibility for any errors or omissions. No liability is assumed for incidental or consequential damages in connection with or arising out of information contained in this book. The Publisher shall not be liable for any special, consequential, or exemplary damages resulting, in whole or in part, from the readers' use of, or reliance upon, this material. Any parts of this book based on government reports are so indicated and copyright is claimed for those parts to the extent applicable to compilations of such works.

Independent verification should be sought for any data, advice or recommendations contained in this book. In addition, no responsibility is assumed by the publisher for any injury and/or damage to persons or property arising from any methods, products, instructions, ideas or otherwise contained in this publication.

This publication is designed to provide accurate and authoritative information with regard to the subject matter covered herein. It is sold with the clear understanding that the Publisher is not engaged in rendering legal or any other professional services. If legal or any other expert assistance is required, the services of a competent person should be sought. FROM A DECLARATION OF PARTICIPANTS JOINTLY ADOPTED BY A COMMITTEE OF THE AMERICAN BAR ASSOCIATION AND A COMMITTEE OF PUBLISHERS.

Additional color graphics may be available in the e-book version of this book.

Library of Congress Cataloging-in-Publication Data

ISBN: 978-1-62948-413-6

Published by Nova Science Publishers, Inc. † New York

CONTENTS

PREFACE

The passage of the 1996 Telecommunications Act resulted in a major revision of the Communications Act of 1934 to address the emergence of competition in what were previously considered to be monopolistic markets. Since its passage, however, the advancement of broadband technology to supply data, voice, and video; the growing convergence of the telecommunications and media sectors; and the growth in demand for usable radio frequency spectrum has led to a consensus that the laws that govern these sectors have become inadequate to address this rapidly changing environment and have, according to a growing number of policymakers, made it necessary to consider revising the current regulatory framework.

Chapter 1 - This report provides an overview of selected topics that, while far from a definitive list, provide a broad overview of issues that are central to the telecommunications/ media convergence debate. The issues covered in this report include broadband deployment, broadband regulation and access, broadcast media ownership rules, funding for the Corporation for Public Broadcasting, emergency communications, legal issues regarding facilities siting, Federal Communications Commission oversight and reform, Internet governance and the domain name system, reauthorization of statutory copyright and communications provisions in the Satellite Television Extension and Localism Act, spectrum policy and wireless broadband deployment, and Universal Service Fund reform.

Rather than addressing the specific legislative, regulatory, and industry activities, this report provides an overview of these major issues.

Chapter 2 - The "digital divide" is a term that has been used to characterize a gap between "information haves and have-nots," or in other words, between those Americans who use or have access to telecommunications and information technologies and those who do not. One important subset of the digital divide debate concerns high-speed Internet access and advanced telecommunications services, also known as broadband. Broadband is provided by a series of technologies (e.g., cable, telephone wire, fiber, satellite, wireless) that give users the ability to send and receive data at volumes and speeds far greater than traditional "dial-up" Internet access over telephone lines.

Broadband technologies are currently being deployed primarily by the private sector throughout the United States. While the numbers of new broadband subscribers continue to grow, studies and data suggest that the rate of broadband deployment in urban/suburban and high income areas is outpacing deployment in rural and low-income areas. Some policymakers, believing that disparities in broadband access across American society could have adverse economic and social consequences on those left behind, assert that the federal government should play a more active role to avoid a "digital divide" in broadband access.

With the conclusion of the grant and loan awards established by the American Recovery and Reinvestment Act of 2009 (P.L. 111-5), there remain two ongoing federal vehicles which direct federal money to fund broadband infrastructure: the broadband and telecommunications programs at the Rural Utilities Service (RUS) of the U.S. Department of Agriculture and the Universal Service Fund (USF) programs under the Federal Communications Commission (FCC). Although the USF's High Cost Program does not explicitly fund broadband infrastructure, subsidies are used, in many cases, to upgrade existing telephone networks so that they are capable of delivering high-speed services. Additionally, subsidies provided by USF's Schools and Libraries Program and Rural Health Care Program are used for a variety of telecommunications services, including broadband access. Currently the USF is undergoing a major transition to the Connect America Fund, which is targeted to the deployment, adoption, and utilization of both fixed and mobile broadband.

To the extent that the 113th Congress may consider various options for further encouraging broadband deployment and adoption, a key issue is how to strike a balance between providing federal assistance for unserved and underserved areas where the private sector may not be providing acceptable levels of broadband service, while at the same time minimizing any deleterious effects that government intervention in the marketplace may have on competition and private sector investment.

Chapter 3 - As congressional policymakers continue to debate telecommunications reform, a major point of contention is the question of whether action is needed to ensure unfettered access to the Internet. The move to place restrictions on the owners of the networks that compose and provide access to the Internet, to ensure equal access and non-discriminatory treatment, is referred to as "net neutrality." While there is no single accepted definition of "net neutrality," most agree that any such definition should include the general principles that owners of the networks that compose and provide access to the Internet should not control how consumers lawfully use that network, and they should not be able to discriminate against content provider access to that network.

A major focus in the debate is concern over whether it is necessary for policymakers to take steps to ensure access to the Internet for content, services, and applications providers, as well as consumers, and if so, what these steps should be. Some policymakers contend that more specific regulatory guidelines may be necessary to protect the marketplace from potential abuses which could threaten the net neutrality concept. Others contend that existing laws and policies are sufficient to deal with potential anti-competitive behavior and that additional regulations would have negative effects on the expansion and future development of the Internet. The December 21, 2010, adoption, and November 20, 2011, implementation, by the Federal Communications Commission (FCC) of its Open Internet Order has focused attention on the issue. Although most concede that networks have always needed and will continue to need some management, the use of prioritization tools, such as deep packet inspection, as well as the initiation of metered/usagebased billing practices have further fueled the debate.

A consensus on the net neutrality issue has remained elusive and support for the FCC's Open Internet Order has been mixed. While some Members of Congress support the action, and in some cases would have supported an even stronger approach, others feel that the FCC has overstepped its authority and that the regulation of the Internet is not only unnecessary, but harmful. Internet regulation and the FCC's authority to implement such regulations has been a topic of legislation (H.R. 96, H.R. 166, S. 74, H.R. 2434, H.R. 1, H.R. 3630, H.J.Res. 37, and S.J.Res. 6) and hearings in the 112th Congress. The House, on April 8, 2011, passed (240-179) H.J.Res. 37, to state disapproval of and remove the force and effect of the FCC's Open Internet Order. However, an identical resolution of disapproval (S.J.Res. 6) failed to pass the Senate on November 10, 2011, by a 52-46 vote. Attempts to prohibit implementation through the appropriations process, by the withholding of FCC funds for such purposes, have also been unsuccessful. It is anticipated that the issue of Internet access will be of continued interest to policymakers.

The net neutrality issue has also been narrowly addressed within the context of the American Recovery and Reinvestment Act of 2009 (ARRA, P.L. 111-5). Provisions required the National Telecommunications and Information Administration (NTIA), in consultation with the FCC, to establish "nondiscrimination and network interconnection obligations" as a requirement for grant participants in the Broadband Technology Opportunities Program (BTOP). These obligations were released, July 1, 2009, in conjunction with the issuance of a notice of funds availability soliciting applications. Recipients of these awards have been selected and continued congressional oversight is expected.

Chapter 4 - The convergence of wireless telecommunications technology with the Internet Protocol (IP) is fostering new generations of mobile technologies. This transformation has created new demands for advanced communications infrastructure and radio frequency spectrum capacity that can support high-speed, content-rich uses. Furthermore, a number of services, in addition to consumer and business communications, rely at least in part on wireless links to broadband (highspeed/high-capacity) infrastructure such as the Internet and IP-enabled networks. Policies to provide additional spectrum for mobile broadband services are generally viewed as drivers that would stimulate technological innovation and economic growth.

The Middle Class Tax Relief and Job Creation Act of 2012 (P.L. 112-96, signed February 22, 2012) contained provisions in Title VI that expedite the availability of spectrum for commercial use. The provisions in Title VI —also known as the Public Safety and Spectrum Act, or the Spectrum Act—included expediting auctions of licenses for spectrum designated for mobile broadband; authorizing incentive auctions, which would permit television broadcasters to receive compensation for steps they might take to release some of their airwaves for mobile broadband; requiring that specified federal holdings be auctioned or reassigned for commercial use; and providing for the availability of spectrum for unlicensed use. The act also included provisions to apply future spectrum license auction revenues toward deficit reduction; to establish a planning and governance structure to deploy public safety broadband networks, using some auction proceeds for that purpose; and to assign additional spectrum resources for public safety communications.

Increasing the amount of spectrum available to support new mobile technologies is one step toward meeting future demand for mobile services. This report discusses some of the commercial and federal spectrum policy changes required by the act. It also summarizes new policy directions for spectrum management under consideration in the 113th Congress, such as the encouragement of new technologies that use spectrum more efficiently.

Chapter 5 - The Federal Communications Commission's (FCC's) broadcast media ownership rules, which place restrictions on the number of media outlets that a single entity can own or control in a local market or nationally, are intended to foster the three long-standing goals of U.S. media policy— competition, localism, and diversity of voices. The FCC is statutorily required to review these rules every four years to determine whether they continue to serve the public interest or should be modified or eliminated. One part of these rules, the FCC's attribution rules, identify criteria for determining when an entity holds sufficient ownership or control of a broadcast station that such ownership or control should be attributed to the entity for the purposes of applying the media ownership rules.

In December 2011, the FCC proposed a number of rule changes, which it has not yet adopted. It proposed eliminating its Radio/Television Cross-Ownership rule because it is no longer needed to foster the goals of diversity of voices and localism. It also proposed modifying its Newspaper/Broadcast Cross-Ownership rule to allow certain types of combinations in the 20 largest markets. It proposed a technical change in its Local Television Ownership Rule, but otherwise would continue to prohibit ownership of two stations in a local market unless one is not among the four highest-ranked stations in the market and, after the combination, there would still be eight independently owned and operating commercial full-power television stations. The FCC proposed that its Local Radio Ownership and Dual Network rules be retained as is. The FCC also sought public comment on how to define the criteria for an entity to be eligible for programs intended to promote the diversity of media ownership, and, in particular, to promote ownership by women and minorities.

In recent years, many television stations have entered into sharing arrangements with other stations in their local market to jointly sell advertising and/or produce local news programming, typically with one station managing that shared operation and perhaps providing most or all of the staffing and other resources. The FCC sought public comment on how, for the purposes of the media ownership rules, to attribute control of a broadcast television station that has entered into such a sharing arrangement. Currently, the only sharing agreement-related attribution rule for television stations covers local marketing agreements in which one station both purchases blocks of time from another station in the same market and sells the advertising for the purchased time— that is, the broker station provides both the programming and the advertising—for at least 15% of the brokered station's broadcasting time. The FCC has enforced this as a bright-line rule. As long as (1) the block of time covered by an agreement does not exceed 15% of the brokered station's programming time, and (2) the agreement contains a certification and perhaps other language indicating that the licensee of the brokered station maintains ultimate control over station finances, personnel, and programming, the agreement will not trigger the attribution rule. Other evidence is considered immaterial. As a result, in many cases the FCC has not deemed a station to have control over another station in the same market even if such control is considered to exist, and must be reported, under generally accepted accounting practices. Such agreements create what is known in the industry as "virtual duopolies."

In late 2012, the FCC released—and made available for public comment—a report on broadcast ownership by gender, ethnicity, and race, and invited the public to comment on how its proposed ownership rule changes might affect female and minority ownership. It delayed adoption of new broadcast ownership rules until those public comments could be analyzed. It is expected to adopt new rules early in 2013.

Chapter 6 - The Corporation for Public Broadcasting (CPB) receives virtually all of its funding through federal appropriations; overall, about 15% of all public television and radio broadcasting funding comes from the federal appropriations that CPB distributes. CPB's appropriation is allocated through a distribution formula established in its authorizing legislation and has historically received two-year advanced appropriations. Congressional policymakers are increasingly interested in the federal role in supporting CPB due to concerns over the federal debt, the role of the federal government funding for public radio and television, and whether public broadcasting provides a balanced and nuanced approach to covering news of national interest.

It is also important to note that many congressional policymakers defend the federal role of funding public broadcasting. They contend that it provides news and information to large segments of the population that seek to understand complex policy issues in depth, and in particular for children's television broadcasting, has a significant and positive impact on early learning and education for children.

On June 20, 2012, the Corporation for Public Broadcasting released a report, Alternative Sources of Funding for Public Broadcasting Stations. The report was undertaken in response to the conference report accompanying the Military Construction and Veterans Affairs and Related Appropriations Act of 2012 (incorporated into the Consolidated Appropriations Act, FY2012, H.R. 2055, P.L. 112-74). The CPB engaged the consulting firm of Booz & Company to explore possible alternatives to the federal appropriation to CPB. Among its findings, the report stated that ending federal funding for public broadcasting would severely diminish, if not destroy, public broadcasting service in the United States.

On September 28, 2012, President Obama signed a Continuing Resolution (CR) of federal funding for FY2013 into law (H.J.Res. 117, P.L. 112-175). It maintains CPB's advanced appropriations for FY2013 at $445 million from October 1, 2012, through October 1, 2013.

Chapter 7 - Digital and Internet protocol technologies have spawned a number of online video distributors (OVDs) whose "over-the-top" video services are in some ways akin to, and in some ways different from, traditional cable and satellite video programming distribution services. However, most of the statutory and regulatory framework for video predates the commercial Internet and was developed within a policy debate that could not consider digital technology and online services. As a result, many statutory provisions apply only to cable companies or satellite carriers, or only to "multichannel video programming distributors" (MVPDs)—a category that includes cable and satellite operators, but as currently interpreted by the Federal Communications Commission excludes online video distributors.

Congress has begun to consider this issue. At both the June 27, 2012, House Energy and Commerce Subcommittee on Communications and Technology hearing on "The Future of Video" and the July 24, 2012, Senate Commerce, Science, and Transportation Committee hearing on "The Cable Act at 20," questions were posed about which of the existing statutory provisions and regulatory rules, if any, should be applied to the new service providers, which provisions and rules should be modified in light of the new technologies and new market realities, and even whether changed circumstances are so great that major statutory reform is needed.

Statutory provisions and regulatory rules affecting media and communications typically are shaped by negotiations among the many stakeholders present at the time the statutes and regulations are being developed.

The resulting framework creates obligations, prohibitions, privileges, and even rights for the various stakeholders. The industry players construct business models based on these. But statutes and regulations that are tailored to existing technologies may create impediments to the deployment of new technologies, especially if they create privileges or rights that are technology-specific.

Some observers have raised concerns that the current statutory and regulatory framework no longer fosters the long-standing U.S. media policy goals of competition, diversity of voices, localism, and innovation because it does not extend to online video distributors the privileges and rights—and also the obligations and prohibitions—that are applicable to traditional video distributors.

For example, competition and innovation may be harmed if online video distributors are denied the access to programming that MVPDs enjoy through the program access and retransmission consent rules or if they are denied the guaranteed low cost compulsory copyright license that cable companies and satellite carriers enjoy. At the same time, localism may be harmed if online video distributors that rebroadcast broadcast television signals are not required to carry their subscribers' local broadcast stations and are not required to black out distant broadcast signals that duplicate the network and syndicated programming on local stations.

Chapter 8 - The Internet is often described as a "network of networks" because it is not a single physical entity, but hundreds of thousands of interconnected networks linking hundreds of millions of computers around the world. As such, the Internet is international, decentralized, and comprised of networks and infrastructure largely owned and operated by private sector entities. As the Internet grows and becomes more pervasive in all aspects of modern society, the question of how it should be governed becomes more pressing.

Currently, an important aspect of the Internet is governed by a private sector, international organization called the Internet Corporation for Assigned Names and Numbers (ICANN), which manages and oversees some of the critical technical underpinnings of the Internet such as the domain name system and Internet Protocol (IP) addressing. ICANN makes its policy decisions using a multistakeholder model of governance, in which a "bottom-up" collaborative process is open to all constituencies of Internet stakeholders.

National governments have recognized an increasing stake in ICANN policy decisions, especially in cases where Internet policy intersects with national laws addressing such issues as intellectual property, privacy, law enforcement, and cybersecurity. Some governments around the world are advocating increased intergovernmental influence over the way the Internet is governed. For example, specific proposals have been advanced that would create an Internet governance entity within the United Nations (U.N.). Other governments (including the United States), as well as many other Internet stakeholders, oppose these proposals and argue that ICANN's multistakeholder model, while not perfect and needing improvement, is the most appropriate way to govern the Internet.

Currently, the U.S. government, through the National Telecommunications and Information Administration (NTIA) at the Department of Commerce, enjoys a unique influence over ICANN, largely by virtue of its legacy relationship with the Internet and the domain name system. A key issue for the 113th Congress is whether and how the U.S. government should continue to maximize U.S. influence over ICANN's multistakeholder Internet governance process, while at the same time effectively resisting proposals for an increased role by international governmental institutions such as the U.N. An ongoing concern is to what extent will future intergovernmental telecommunications conferences (such as the December 2012 World Conference on International Telecommunications or WCIT) constitute an opportunity for some nations to increase intergovernmental control over the Internet, and how effectively will NTIA and other government agencies (such as the State Department) work to counteract that threat? H.R. 1580, introduced on April 16, 2013, states that "[I]t is the policy of the United States to preserve and advance the successful multistakeholder model that governs the Internet."

The ongoing debate over Internet governance will likely have a significant impact on how other aspects of the Internet may be governed in the future, especially in such areas as intellectual property, privacy, law enforcement, Internet free speech, and cybersecurity. Looking forward, the institutional nature of Internet governance could have far-reaching implications on important policy decisions that will likely shape the future evolution of the Internet.

Chapter 9 - Navigating the Internet requires using addresses and corresponding names that identify the location of individual computers. The Domain Name System (DNS) is the distributed set of databases residing in computers around the world that contain address numbers mapped to corresponding domain names, making it possible to send and receive messages and to access information from computers anywhere on the Internet. Many of the technical, operational, and management decisions regarding the DNS can have significant impacts on Internet-related policy issues such as intellectual property, privacy, Internet freedom, e-commerce, and cybersecurity.

The DNS is managed and operated by a not-for-profit public benefit corporation called the Internet Corporation for Assigned Names and Numbers (ICANN). Because the Internet evolved from a network infrastructure created by the Department of Defense, the U.S. government originally owned and operated (primarily through private contractors) the key components of network architecture that enable the domain name system to function. A 1998 Memorandum of Understanding (MOU) between ICANN and the Department of Commerce (DOC) initiated a process intended to transition technical DNS coordination and management functions to a private-sector not-for-profit entity. While the DOC played no role in the internal governance or day-to-day operations of the DNS, ICANN remained accountable to the U.S. government through the MOU, which was superseded in 2006 by a Joint Project Agreement (JPA). On September 30, 2009, the JPA between ICANN and DOC expired and was replaced by an Affirmation of Commitments (AoC), which provides for review panels to periodically assess ICANN processes and activities.

Additionally, a contract between DOC and ICANN authorizes the Internet Assigned Numbers Authority (IANA) to perform various technical functions such as allocating IP address blocks, editing the root zone file, and coordinating the assignment of unique protocol numbers. With the current contract due to expire on September 30, 2012, NTIA announced on July 2, 2012, the award of the new IANA contract to ICANN for up to seven years.

With the expiration of the ICANN-DOC Joint Project Agreement on September 30, 2009, the announcement of the new AoC, the renewal of the IANA contract, and the rollout of the new generic top level domain (gTLD) program, the 113th Congress and the Administration are likely to continue assessing the appropriate federal role with respect to ICANN and the DNS, and examine to what extent ICANN is positioned to ensure Internet stability and security, competition, private and bottom-up policymaking and coordination, and fair representation of the global Internet community. Controversies over the new gTLDs and the addition of the .xxx domain have led some governments to criticize the ICANN policymaking process and to suggest various ways to increase governmental influence over that process.

How these and other issues are ultimately addressed and resolved could have profound impacts on the continuing evolution of ICANN, the DNS, and the Internet.

Chapter 10 - Since September 11, 2001, when communications failures contributed to the tragedies of the day, Congress has passed several laws intended to create a nationwide emergency communications capability. Yet the United States has continued to strive for a solution that assures seamless communications among first responders and emergency personnel at the scene of a major disaster. To address this problem, Congress included provisions in the Middle Class Tax Relief and Job Creation Act of 2012 (P.L. 112-96) for planning, building, and managing a new, nationwide, broadband network for public safety communications, and assigned additional spectrum to accommodate the new network. In addition, the act has designated federal appropriations of over $7 billion for the network and other public safety needs. These funds will be provided through new revenue from the auction of spectrum licenses. The cost of construction of a nationwide network for public safety is estimated by experts to be in the tens of billions of dollars over the long term, with similarly large sums needed for maintenance and operation. In expectation that public-private partnerships to build the new network will reduce costs to the public sector, the law has provided requirements and guidelines for shared use.

The act has mandated that technical standards developed for the new network incorporate commercial standards for Long Term Evolution (LTE). LTE is a fourth-generation wireless technology that bases its operating standards on the Internet Protocol (IP). IP-enabled networks and wireless devices provide higher capacity and transmission speeds than earlier generations of technology. LTE represents the convergence of wireless technology with the Internet, bringing the capacity and resiliency of packet-switched networks to emergency communications. It is generally believed that the use of LTE and IP standards will greatly enhance communications for emergency response and recovery.

There are many challenges for public safety leaders and policy makers in establishing IP-enabled technologies as the baseline for the development of future solutions for response and recovery. One of the immediate challenges in developing standards is the need for a clear policy on the use of spectrum for commercial and public safety LTE. Because public safety planning has lagged behind commercial efforts to build LTE networks, the work on design and development of technical requirements is incomplete. Many experts are concerned that these delays may place public safety officials at a disadvantage in negotiating with potential partners, increase costs, and add further delays in moving forward to build a nationwide broadband network. Requirements in the act for standards development may be insufficient to overcome current technical obstacles for desired network features such as roaming between public safety and commercial networks.

In addition to monitoring progress in building the new broadband network for public safety, Congress may want to consider reviewing the role of commercial networks in emergency response and recovery. Once commercial communications lines are compromised because of infrastructure failures, interdependent public safety networks are threatened and the ability to communicate vital information to the public is diminished. New policy initiatives may be needed to identify critical gaps in communications infrastructure and the means to fund the investments needed to close these gaps.

Chapter 11 - The Federal Communications Commission (FCC) is an independent federal agency with its five members appointed by the President, subject to confirmation by the Senate. It was established by the Communications Act of 1934 (1934 Act) and is charged with regulating interstate and international communications by radio, television, wire, satellite, and cable. The mission of the FCC is to ensure that the American people have available—at reasonable cost and without discrimination—rapid, efficient, nation- and world-wide communication services, whether by radio, television, wire, satellite, or cable.

Although the FCC has restructured over the past few years to better reflect the industry, it is still required to adhere to the statutory requirements of its governing legislation, the Communications Act of 1934. The 1934 Act requires the FCC to regulate the various industry sectors differently. Some policymakers have been critical of the FCC and the manner in which it regulates various sectors of the telecommunications industry—telephone, cable television, radio and television broadcasting, and some aspects of the Internet. These policymakers, including some in Congress, have long called for varying degrees and types of reform to the FCC. Most proposals fall into two categories: (1) procedural changes made within the FCC or through congressional action that would affect the agency's operations or (2) substantive policy changes requiring congressional action that would affect how the agency regulates different services and industry sectors. Nine bills have been introduced during the 112th Congress that would change the operation of the FCC.

For FY2014, the FCC has requested a budget of $359,299,000. The FCC's budget is derived from regulatory fees collected by the agency rather than through a direct appropriation. The fees, often referred to as "Section (9) fees," are collected from license holders and certain other entities (e.g., cable television systems) and deposited into an FCC account. The law gives the FCC authority to review the regulatory fees and to adjust the fees to reflect changes in its appropriation from year to year. It may also add, delete, or reclassify services under certain circumstances.

In the 113th Congress, three hearings have been held on FCC oversight, reform, and management and four bills have been introduced that would affect the manner in which the FCC conducts its business.

Chapter 12 - The National Telecommunications and Information Administration (NTIA), a bureau of the Department of Commerce, is the executive branch's principal advisory office on domestic and international telecommunications and information policies. Its mandate is to provide greater access for all Americans to telecommunications services, support U.S. efforts to open foreign markets, advise on international telecommunications negotiations, and fund research for new technologies and their applications.

NTIA also manages the distribution of funds for several key grant programs. Its role in managing radio frequency spectrum allocated for federal use includes addressing policies for sharing, and monitoring and resolving questions regarding usage, including causes of interference. It is responsible for identifying federal spectrum that can be transferred to commercial use through the auction of spectrum licenses, conducted by the Federal Communications Commission. Many of the NTIA's responsibilities are shared with other agencies.

With the passage of the Middle Class Tax Relief and Job Creation Act of 2012 (P.L. 112-96), in February 2012, Congress has given the NTIA new responsibilities in spectrum management and the support of public safety initiatives. The 113th Congress may wish to review the NTIA's performance in meeting its obligations under the act. Policy makers may also wish to consider if some of the NTIA's shared obligations might be effectively and efficiently transferred to its partners, allowing the NTIA to focus on communications policies that are considered by many to be key to future economic growth and development.

Chapter 13 - The Satellite Television Extension and Localism Act of 2010 (STELA), P.L. 111-175, modified the copyright and carriage rules for satellite and cable retransmission of broadcast television signals. The legislation was needed to reauthorize (through December 31, 2014) certain expiring provisions in the Copyright Act and the Communications Act and to update the language in those acts to reflect the transition from analog to digital transmission of broadcast signals, as well as to address certain public policy issues. Had the expiring provisions not been reauthorized, satellite operators would have lost access to a statutory compulsory copyright license and to statutory relief from retransmission consent requirements. This would have made it difficult, if not impossible, for them to retransmit certain distant broadcast signals to their subscribers, including signals providing otherwise unavailable broadcast network programming.

The Copyright Act and Communications Act distinguish between the retransmission of local signals—the broadcast signals of stations located in the same local market as the subscriber—and distant signals. Statutory provisions block or restrict the retransmission of many distant broadcast signals in order to foster local programming. These provisions typically take the form of defining which households are "served" or "unserved" by local broadcasters, with unserved households eligible to receive distant signals. But there are many grandfather clauses and other exceptions built into the rules that allow households to receive otherwise proscribed distant signals. STELA generally retained, and in some cases expanded upon, these grandfathered and exceptional cases.

STELA provided broadcasters two new incentives to use their digital technology to broadcast multiple video streams (to "multicast"). It clarified that royalty fees are payable to copyright owners of the materials on non-primary digital voice streams as well as primary streams, thus encouraging broadcasters (who often hold some of those copyrights) to expand their multicasting. STELA specifically gave broadcasters the incentive to undertake such multicasting to offer otherwise unprovided network programming in so-called "short markets"—markets that do not have network affiliates for all four major networks. It did this by defining households that can receive the programming of a particular network from the non-primary multicast video stream of a local broadcaster as being served, rather than unserved, with respect to that network, thus prohibiting satellite operators from retransmitting to those households distant signals that carry that network's programming. The local broadcaster can then seek retransmission consent payments from satellite operators. Several other provisions in STELA also were intended to reduce the number of short markets or increase flow of distant network signals into short markets. Today, satellite operators are allowed, but not required, to offer subscribers the signals of the broadcast stations in their local market. Until enactment of STELA, the satellite operators chose not to offer this "local-into-local" service in many small markets, preferring to use their satellite capacity to provide additional high definition and other programming to larger, more lucrative markets. The costs associated with providing local-into-local service in small markets may exceed the revenues. STELA provided DISH Network, which had been subject to a permanent court injunction that in effect prohibited it from retransmitting to its subscribers the signals of distant broadcast stations, the opportunity to have that injunction waived if it provided local-into- local service in all 210 local markets in the United States, which it began doing on June 3, 2010. STELA did not address the issue of "orphan counties"—counties located in one state that are assigned to a local market, as defined by the Nielsen Media Research designated market areas, for which the principal city and most or all of the local broadcast stations are in another state.

In: Transformations in Telecommunications and Media   ISBN: 978-1-62948-413-6
Editor: Irwin Cavazos   © 2013 Nova Science Publishers, Inc.

Chapter 1

TELECOMMUNICATIONS AND MEDIA CONVERGENCE: SELECTED ISSUES FOR CONSIDERATION[*]

Angele A. Gilroy

SUMMARY

The passage of the 1996 Telecommunications Act (P.L. 104-104) resulted in a major revision of the Communications Act of 1934 (47 U.S.C. 151 et seq.) to address the emergence of competition in what were previously considered to be monopolistic markets. Since its passage, however, the advancement of broadband technology to supply data, voice, and video; the growing convergence of the telecommunications and media sectors; and the growth in demand for usable radio-frequency spectrum has led to a consensus that the laws that govern these sectors have become inadequate to address this rapidly changing environment and have, according to a growing number of policymakers, made it necessary to consider revising the current regulatory framework.

This report provides an overview of selected topics that, while far from a definitive list, provide a broad overview of issues that are central to the telecommunications/media convergence debate. The issues covered in this report include broadband deployment, broadband regulation and access, broadcast media ownership rules, funding for the Corporation for Public Broadcasting, emergency communications, legal issues regarding facilities siting, Federal Communications Commission oversight and reform, Internet governance and the domain name system, reauthorization of statutory copyright and communications provisions in the Satellite Television Extension and Localism Act, spectrum policy and wireless broadband deployment, and Universal Service Fund reform.

Rather than addressing the specific legislative, regulatory, and industry activities, this report provides an overview of these major issues.

---

[*] This is an edited, reformatted and augmented version of Congressional Research Service, Publication No. R43178, dated August 14, 2013.

INTRODUCTION

The rapid pace of technological advances, including the shift from voice to data, from wireline to wireless, and from copper to fiber is redefining the parameters of the telecommunications and media markets. As these changes dramatically transform the marketplace, there is a growing consensus that existing laws and regulations be reexamined to address this transformation.

In general terms, the regulatory debate focuses on a number of issues including the extent to which existing regulations should be applied to traditional providers as they enter new markets where they do not hold market power, the extent to which existing regulations should be imposed on new entrants as they compete with traditional providers in the same markets, and the appropriate regulatory framework to be imposed on new and/or converging technologies that are not easily classified under the present framework.

If, and to what extent, the role of the Federal Communications Commission and state regulatory bodies should be modified as networks transition from a circuit-switched to an Internet Protocol network[1] and how to ensure that the core values (e.g., consumer protection, public safety, disability access, and competition) are preserved in this new environment are also being addressed.

How traditional policy goals, such as the advancement of universal service mandates in the provision of telecommunications services and the media market's long-standing policy objectives of localism, diversity of voices, and competition, should be applied, and/or revised, as these markets transform is also under consideration.

The deployment, adoption, and regulatory treatment of broadband technologies continues to hold a major focus in the policy debate. Some policymakers feel it is necessary to take steps to ensure access to the Internet for content, services, and applications providers, as well as consumers, while others feel that such actions will slow deployment of and access to the Internet, and limit innovation. The transition of the Universal Service Fund to the Connect America Fund to support broadband deployment and adoption has generated concerns regarding consequences for small rate-of-return carriers. The impact of broadband deployment on the media sector as consumers change their viewing patterns and adopt new delivery technologies is also central to the debate.

The allocation and regulation of radio-frequency spectrum has also become a crucial component in the policy debate. The ability of new wireless technologies to deliver a variety of communications services and the increasing demand for mobility has placed increased pressure on usable spectrum as consumer demand fuels commercial demand for spectrum. The public sector also requires spectrum for a variety of government and emergency uses. Policymakers are increasingly being called upon to balance the needs of both the public and the private sector.

BROADBAND DEPLOYMENT[2]

Broadband—whether delivered via fiber, cable modem, copper wire, satellite, or wirelessly—is increasingly the technology underlying telecommunications services such as voice, video, and data. Since the initial deployment of broadband in the late 1990s, Congress has viewed broadband infrastructure deployment as a means towards improving regional economic development, and in the long term, to create jobs. According to the Federal Communications Commission's (FCC's) National Broadband Plan, the lack of adequate broadband availability is most pressing in rural America, where the costs of serving large geographical areas, coupled with low population densities, often reduce economic incentives for telecommunications providers to invest in and maintain broadband infrastructure and service. The National Broadband Plan also identified broadband adoption as a problem, whereby one in three Americans have broadband available but choose not to subscribe. Populations continuing to lag behind in broadband adoption include people with low incomes, seniors, minorities, the less-educated, non-family households, and the non-employed.

The 113th Congress may address a range of broadband-related issues. These include the transition of the telephone-era Universal Service Fund to the broadband-focused Connect America Fund, reauthorization of broadband loan programs in the 2013 farm bill, and the development of new wireless spectrum policies. Additionally, the 113th Congress may choose to examine existing regulatory structure and consider possible revision of the 1996 Telecommunications Act and its underlying statute, the Communications Act of 1934. Both the convergence of telecommunications providers and markets and the transition to an Internet protocol (IP) based network have, according to a growing number of policymakers, made it necessary to consider revising the current regulatory framework. How a possible revision might create additional incentives for investment in, deployment of, and subscribership to broadband infrastructure is likely to be just one of many issues under consideration.

To the extent that Congress may consider various options for further enhancing broadband deployment, a key issue is how to develop and implement federal policies intended to increase the nation's broadband availability and adoption, while at the same time minimizing any deleterious effects that government intervention in the marketplace may have on competition and private sector investment.

BROADBAND ACCESS AND "NET NEUTRALITY"[3]

As policymakers continue to debate telecommunications reform, a major point of contention is whether action is needed to ensure unfettered access to the Internet. The move to place restrictions on the owners of the networks that compose and provide access to the Internet, to ensure equal access and non-discriminatory treatment, is referred to as "net neutrality." While there is no single accepted definition of "net neutrality," most agree that any such definition should include the general principles that owners of the networks that compose and provide access to the Internet should not control how consumers lawfully use that network, and should not be able to discriminate against content provider access to that network.

A major focus in the debate is concern over whether it is necessary for policymakers to take steps to ensure access to the Internet for content, services, and applications providers, as well as consumers, and if so, what these steps should be. Some policymakers contend that more specific regulatory guidelines may be necessary to protect the marketplace from potential abuses which could threaten the net neutrality concept. Others contend that existing laws and policies are sufficient to deal with potential anti-competitive behavior and that additional regulations would have negative effects on the expansion and future development of the Internet.

What, if any, action should be taken to ensure "net neutrality" has become a major focal point in the debate over broadband regulation. As the marketplace for broadband continues to evolve, some contend that no new regulations are needed, and if enacted will slow deployment of and access to the Internet, as well as limit innovation. Others, however, contend that the consolidation of broadband providers, coupled with their diversification into content, has the potential to lead to discriminatory behaviors which conflict with net neutrality principles. The two potential behaviors most often cited are the network providers' ability to control access to and the pricing of broadband facilities, and the incentive to favor network-owned content, thereby placing unaffiliated content providers at a competitive disadvantage.

The December 21, 2010, adoption, and November 20, 2011, implementation, by the Federal Communications Commission (FCC) of its Open Internet Order has focused attention on the issue. A consensus on the net neutrality issue has remained elusive and support for the FCC's Open Internet Order has been mixed. While some Members of Congress support the action, and in some cases would have supported an even stronger approach, others feel that the FCC has overstepped its authority and that the regulation of the Internet is not only unnecessary, but harmful. Internet regulation and the FCC's authority to implement such regulations, which is currently facing court challenge, is an issue of growing importance in the wide ranging debate over broadband regulation.

BROADCAST MEDIA OWNERSHIP RULES[4]

The Federal Communications Commission's (FCC's) broadcast media ownership rules, which place restrictions on the number of media outlets that a single entity can *own or control* in a local market or nationally, are intended to foster the three long-standing goals of U.S. media policy— competition, localism, and diversity of voices. The FCC is statutorily required to review these rules every four years to determine whether they continue to serve the public interest or should be modified or eliminated. One part of these rules, the FCC's attribution rules, identify criteria for determining when an entity holds sufficient ownership or control of a broadcast station that such ownership or control should be attributed to the entity for the purposes of applying the media ownership rules.

In December 2011, the FCC proposed a number of rule changes, which it has not yet adopted. It proposed eliminating its Radio/Television Cross-Ownership rule because it is no longer needed to foster the goals of diversity of voices and localism. It also proposed modifying its Newspaper/Broadcast Cross-Ownership rule to allow certain types of combinations in the 20 largest markets. It proposed a technical change in its Local Television Ownership Rule, but otherwise would continue to prohibit ownership of two stations in a local market unless one is not among the four highest-ranked stations in the market and, after the combination, there would still be eight independently owned and operating commercial full-power television stations. The FCC proposed that its Local Radio Ownership and Dual Network rules be retained as is. The FCC also sought public comment on how to define the criteria for an entity to be eligible for programs intended to promote the diversity of media ownership, and, in particular, to promote ownership by women and minorities.

In recent years, many television stations have entered into sharing arrangements with other stations in their local markets to jointly sell advertising and/or produce local news programming, typically with one station managing that shared operation and perhaps providing most or all of the staffing and other resources. The FCC sought public comment on how, for the purposes of the media ownership rules, to attribute control of a broadcast television station that has entered into such a sharing arrangement. Currently, the only sharing agreement-related attribution rule for television stations covers local marketing agreements in which one station purchases blocks of time from another station in the same market and sells the advertising for the purchased time (that is, the broker station provides both the programming and the advertising) for at least 15% of the brokered station's broadcasting time. The FCC has enforced this as a bright-line rule. As long as (1) the block of time covered by an agreement does not exceed 15% of the brokered station's programming time, and (2) the agreement contains a certification and perhaps other language indicating that the licensee of the brokered station maintains ultimate control over station finances, personnel, and programming, the agreement will not trigger the attribution rule. Other evidence is considered immaterial. As a result, in many cases the FCC has not deemed a station to have control over another station in the same market even if such control is considered to exist, and must be reported, under generally accepted accounting practices. Such agreements create what is known in the industry as "virtual duopolies."

In late 2012, the FCC released (and made available for public comment) a report on broadcast ownership by gender, ethnicity, and race, and invited the public to comment on how its proposed ownership rule changes might affect female and minority ownership. It delayed adoption of new broadcast ownership rules until those public comments could be analyzed. Responding to that report, the Minority Media and Telecommunications Council asked the FCC for an additional delay so it could conduct a study of the likely impact of the FCC's proposed rule changes on female and minority ownership. The National Association of Broadcasters supports that delay and the FCC has agreed to it. The FCC is expected to adopt new rules later in 2013.

CORPORATION FOR PUBLIC BROADCASTING[5]

Since 1967, the Corporation for Public Broadcasting (CPB) has been the funding vehicle to provide federal support to local public television and radio broadcasting entities through the country. The CPB was created to provide a non-profit entity that could disburse federal grants without political interference, and without direct federal control of who receives the funding.

The CPB receives virtually all of its funding through federal appropriations; overall, about 15% of all public television and radio broadcasting funding comes from the federal appropriations that CPB distributes. CPB's appropriation is allocated through a distribution formula established in its authorizing legislation and has historically received two-year advanced appropriations. On March 22, 2013, President Obama signed a Continuing Resolution (CR) of federal funding for FY2013 into law (H.J.Res. 117, P.L. 112-175). It maintains CPB's advanced appropriations for FY2013 at $445 million from October 1, 2012, through October 1, 2013. However, the federal government is also under a sequestration action mandated under the American Taxpayer Relief Act of 2012 (P.L. 112-40). Under this law, the CPB's appropriation is reduced by 5%, or $22.25 million. Therefore, the CPB has an appropriated budget of $422.75 million for FY2013. On March 26, 2013, the President signed into law the Consolidated and Further Continuing Appropriations Act, 2013 (P.L. 113-6), which provides continuing federal funding, under sequestration limits, through the end of FY2013.

Congressional policymakers are increasingly interested in the federal role in supporting CPB due to concerns over the federal debt, the role of the federal government funding for public radio and television, and whether public broadcasting provides a balanced and nuanced approach to covering news of national interest. It is also important to note that many congressional policymakers defend the federal role of funding public broadcasting. They contend that it provides news and information to large segments of the population that seek to understand complex policy issues in depth, and, in particular for children's television broadcasting, that it has a significant and positive impact on early learning and education for children.

On June 20, 2012, the Corporation for Public Broadcasting released a report, *Alternative Sources of Funding for Public Broadcasting Stations*. The report was undertaken in response to the conference report accompanying the Military Construction and Veterans Affairs and Related Appropriations Act of 2012 (incorporated into the Consolidated Appropriations Act, FY2012, H.R. 2055, P.L. 112-74). The CPB engaged the consulting firm of Booz & Company to explore possible alternatives to the federal appropriation to CPB. Among its findings, the report stated that ending federal funding for public broadcasting would severely diminish, if not destroy, public broadcasting service in the United States.

EMERGENCY COMMUNICATIONS[6]

The three pillars of emergency communications are wireless networks for first responders and other emergency personnel; 9-1-1 calls and dispatching; and emergency alerts such as the Emergency Alert System (EAS), delivered over television and radio, and Wireless Emergency Alerts (WEA) on mobile devices. Increasingly, emergency communications rely on using the same network architecture and protocols as the Internet (IP-enabled network) to provide interoperability within and among networks.

Previous Congresses have passed key laws to improve emergency communications. The 113th Congress is likely to continue legislative initiatives and to conduct oversight of programs that are underway in response to earlier legislation. In Title VI (Spectrum Act) of the Middle Class Tax Relief and Job Creation Act of 2012 (P.L. 112-96), Congress has addressed some of the needs of first responders and 9-1-1 call centers.

To provide seamless communications among first responders and emergency personnel at the scene of a major disaster, for example, Congress included provisions in the Spectrum Act for planning, building, and managing a new, nationwide broadband network for public safety communications, and assigned additional spectrum to accommodate the new network.

The Spectrum Act also has included provisions to improve 9-1-1 services and technology. It reestablished the federal 9-1-1 Implementation Coordination Office (ICO) to advance planning for next-generation systems and to administer a grant program. Previously, recognizing the importance of providing effective 9-1-1 service, Congress passed three major bills supporting improvements in the handling of 9-1-1 calls. The Wireless Communications and Public Safety Act of 1999 (P.L. 106-81) established 9-1-1 as the number to call for emergencies and gave the Federal Communications Commission (FCC) authority to regulate many aspects of the service. The NET 9-1-1 Improvement Act of 2008 (P.L. 110-283) required the preparation of a National Plan for migrating to an IP-enabled emergency network. Responsibility for the plan was assigned to ICO, originally created to meet requirements of an earlier law, the ENHANCE 911 Act of 2004 (P.L. 108-494).

EAS messages, a crucial part of emergency alerts, are being incorporated into the Integrated Public Alert and Warning System (IPAWS), which is being built to serve as a communications backbone to receive and relay alerts to designated geographical areas. In addition to broadcast, satellite, cable, and radio communications through EAS, IPAWS can deliver messages to any IPenabled network including, for example, electronic highway signs.

In cooperation with wireless carriers, IPAWS is supporting WEA to mobile devices. The development of WEA, originally known as the Commercial Mobile Alert System (CMAS), was mandated by Congress as part of the Warning, Alert, and Response Network (WARN) Act, Title VI of P.L. 109-37.

FACILITIES SITING—LEGAL ISSUES[7]

An integral part of the mission of Congress and the Federal Communications Commission (FCC) to encourage broadband deployment is the effort to increase both wireless and wireline broadband capacity. To that end, both Congress and the FCC have taken steps to streamline the process by which companies providing wired and wireless broadband services may place their equipment on already existing poles or structures designed to host such equipment (known as collocation).

For its part, Congress included Section 704 of the Telecommunications Act of 1996 (47 U.S.C. §224), which governs federal, state, and local regulation of the siting of "personal wireless service facilities." Under Section 704, state and local governments are prohibited from unreasonably discriminating among "providers of functionally equivalent services," nor can they adopt policies that have the effect of prohibiting wireless services. These prohibitions grant states and localities flexibility in deciding where towers should be placed within their communities, while ensuring that these governing bodies cannot prevent the provision of personal wireless services in an area.

Notwithstanding the flexibility for the siting of brand new facilities provided by Section 704, Section 6409 of the Middle Class Tax Relief and Job Creation Act of 2012 (P.L. 112-96) (47 U.S.C. §1455) amended the Communications Act to require state and local governments to grant requests for modifications of existing wireless towers or base stations if the request would not substantially change the physical dimensions of the tower or base station. Presumably, this provision is intended to increase the speed of wireless infrastructure deployment. Section 6409 is not without its ambiguities, however. No definition is provided in the statute for the terms "tower" or "base station." Furthermore, no definition is provided for what it might mean to "substantially change the physical dimensions" of a tower. These ambiguities may cause difficulty in applying the new provision to future collocation requests. They may be resolved either by federal courts during litigation or by the FCC in a declaratory rulemaking to define the terms.

While Congress attempted to streamline the process of collocation on existing wireless towers, the FCC has engaged in a similar effort to streamline the process of collocation of equipment on existing poles owned by utility companies. Section 224 of the Communications Act (47 U.S.C. §224) grants the Commission the authority to regulate the rates, terms, and conditions for pole attachments, which are defined by the statute as "any attachment by a cable television system or provider of a telecommunications service to a pole, duct, conduit, or right-of-way owned or controlled by a utility." In 2011, the FCC issued an order revising its interpretation of Section 224 to allow incumbent local exchange carriers (ILECs) for the first time to share some of the benefits of Section 224; reformulate (i.e., lower) the rates utilities could charge telecommunications carriers, bringing those rates closer to the rates charged to cable providers; and reformulate the timing of the calculation of refunds when attachers are overcharged. (*In the Matter of the Implementation of Section 224 of the Act, Report and Order and Order on Reconsideration*, 26 FCC Rcd. 5240 (April 7, 2011).)

Utility companies challenged the FCC's authority to make these changes, claiming that ILECs were excluded from the definition of telecommunications service providers under Section 224 and could not be eligible for pole attachment rights under Section 224 as a result. The Court of Appeals for the D.C. Circuit disagreed, upholding the FCC's interpretation of the statute. (*American Electric Power Service Corp. v. Federal Communications Commission*, 2013 U.S. App. LEXIS 3924 (D.C. Circuit, 2013).) The court found that while Section 224 did exclude ILECs from the definition of telecommunications carriers, that exclusion only applied to Section 224(e), which permits the FCC to regulate charges for pole attachments to telecommunications carriers when the parties fail to resolve a dispute regarding those charges. The FCC was permitted to interpret Section 224(a), which allows the Commission to regulate pole attachments for providers of telecommunications services more generally, to apply to ILECs.

The utility companies also challenged the FCC's decision to adopt telecom rates that were substantially equivalent to cable rates for pole attachments and the amendments to the calculation of the so-called "refund period." The court accorded deference to the FCC's interpretation of Section 224 in both of those instances, as well, and denied the utility companies' petition to review the FCC's order amending its regulations under Section 224 in full.

FEDERAL COMMUNICATIONS COMMISSION— OVERSIGHT AND REFORM[8]

The Federal Communications Commission (FCC) is an independent federal agency with its five members appointed by the President, subject to confirmation by the Senate. It was established by the Communications Act of 1934 (1934 Act) and is charged with regulating interstate and international communications by radio, television, wire, satellite, and cable. The mission of the FCC is to ensure that the American people have available—at reasonable cost and without discrimination—rapid, efficient, nation- and world-wide communication services, whether by radio, television, wire, satellite, or cable.

Although the FCC has restructured over the past few years to better reflect the industry, it is still required to adhere to the statutory requirements of its governing legislation, the Communications Act of 1934. The 1934 Act requires the FCC to regulate the various industry sectors differently. Some congressional policymakers have been critical of the FCC and the manner in which it regulates various sectors of the telecommunications industry—telephone, cable television, radio and television broadcasting, and some aspects of the Internet. These policymakers have called for varying degrees and types of reform to the FCC to better reflect the current state of the telecommunications industry. Most proposals fall into two categories: (1) procedural changes made within the FCC or through congressional action that would affect the agency's operations or (2) substantive policy changes requiring congressional action that would affect how the agency regulates different services and industry sectors.

The FCC's budget is derived from regulatory fees collected by the agency rather than through a direct appropriation. The fees, often referred to as "Section (9) fees," are collected from license holders and certain other entities (e.g., cable television systems) and deposited into an FCC account. The law gives the FCC authority to review the regulatory fees and to adjust the fees to reflect changes in its appropriation from year to year. It may also add, delete, or reclassify services under certain circumstances.

INTERNET GOVERNANCE AND THE DOMAIN NAME SYSTEM[9]

The Internet is comprised of international and decentralized networks largely owned and operated by private sector entities. As the Internet becomes more pervasive in all aspects of society, the question of how it should be governed becomes more pressing. Currently, an important aspect of the Internet is governed by a private sector, international organization called the Internet Corporation for Assigned Names and Numbers (ICANN), which manages the domain name system and Internet addressing. ICANN makes its decisions using a multistakeholder model of governance, in which a collaborative policy development process is open to all Internet stakeholders.

National governments have increasingly recognized the importance of ICANN policy decisions, especially in cases where Internet policy intersects with national laws addressing such issues as intellectual property, privacy, law enforcement, Internet freedom, and cybersecurity. Some governments are advocating an increased level of intergovernmental influence over the way the Internet is governed, while other governments (such as the United States and the European Union) oppose intergovernmental jurisdiction over the Internet. This debate surfaced during consideration of the revised International Telecommunication Regulations (ITR) treaty held by the International Telecommunication Union (a United Nations agency) during the December 2012 World Conference on International Telecommunications (WCIT) in Dubai. Ultimately, the United States (and 54 other nations) chose not to sign the final treaty, citing an unacceptable expansion of ITR jurisdiction over the Internet.

As part of its input into the WCIT debate, the 112th Congress unanimously passed S.Con.Res. 50, which expressed the sense of Congress that the Administration should promote a global Internet free from intergovernmental control, and should preserve and advance the successful multistakeholder model of Internet governance. A key issue for the 113th Congress is whether and how the U.S. government should continue to maximize its influence over ICANN's multistakeholder Internet governance process, while at the same time effectively resisting proposals for an increased role by international governmental institutions such as the United Nations. An ongoing concern is, to what extent will future intergovernmental telecommunications conferences constitute an opportunity for some nations to increase intergovernmental control over the Internet, and how effectively will the Administration work to counteract that threat?

REAUTHORIZATION OF STATUTORY COPYRIGHT AND COMMUNICATIONS PROVISIONS IN THE SATELLITE TELEVISION EXTENSION AND LOCALISM ACT[10]

The Satellite Television Extension and Localism Act of 2010 (STELA, P.L. 111-175) reauthorized, through December 31, 2014, several provisions in the Copyright Act and in the Communications Act relating to the retransmission of broadcast television signals by satellite television operators. Several of these provisions make it easier for satellite operators to import the programming of distant network television stations to households that cannot receive that programming from a local television station, by creating a low cost compulsory license for the copyrighted works contained in the network programming and by allowing the satellite operator to retransmit the programming without first obtaining the consent of the distant television station. In most other situations, satellite operators (and cable operators) must obtain the prior consent of broadcasters when retransmitting their signals. Other sunsetting provisions, intended to minimize the blackout of broadcast programming, require broadcast stations, satellite operators, and cable operators to negotiate this retransmission consent in good faith.

Congress will have to decide whether it wishes to retain these provisions or modify or eliminate them. This is likely to generate policy debates about statutory copyright licenses and retransmission consent.

There is a separate compulsory license for the retransmission of broadcast signals by cable operators, but that license is created in a statutory provision that is not subject to sunset. Copyright holders generally oppose compulsory licenses because they believe negotiated copyright rates would be higher. They therefore oppose extension of the compulsory satellite license and also would eliminate the compulsory cable license. There would be competitive implications if the satellite copyright license were eliminated, but not the cable copyright license. Currently, there is no statutory compulsory copyright license provision for the retransmission of broadcast signals by online video distributors and therefore online distributors seeking to provide their subscribers with broadcast signals must negotiate copyright fees directly with the copyright holders. This places online video distributors at a competitive disadvantage, and in the debate about the compulsory satellite license they are likely to seek parity with cable and satellite operators.

Cable and satellite operators would prefer not to have to pay retransmission consent fees for the right to retransmit broadcast signals and thus they seek to eliminate the retransmission consent requirement or, at the least, to require that retransmission consent negotiations that hit an impasse be subject to mandatory arbitration and that the contested broadcast signals continue to be retransmitted during the arbitration process. The broadcasters oppose those proposals.

To foster the long-standing goal of localism, in most situations cable and satellite operators are restricted from retransmitting to their subscribers the signals of distant television stations if the same network programming is provided by a local station. But some counties around the country are assigned to local markets for which all the television stations are actually located across the state border, so that they do not receive news, sports, and other programming pertaining to their state. There has been a long, unsettled debate about how to better serve these "orphan counties" that is likely to arise again as Congress addresses reauthorization of these statutory provisions.

SPECTRUM POLICY AND WIRELESS BROADBAND DEPLOYMENT[11]

Wireless broadband, with its rich array of services and content, requires new spectrum capacity to accommodate growth. Spectrum capacity is necessary to deliver mobile broadband to consumers and businesses and also to support the communications needs of industries that use fixed wireless broadband to transmit large quantities of information quickly and reliably.

Electromagnetic spectrum, commonly referred to as radio frequency spectrum or wireless spectrum, refers to the properties in air that transmit electric signals and, with applied technology, can deliver voice, text, and video communications. Access to radio frequency spectrum is controlled by assigning rights to specific license holders or to certain classes of users. The assignment of spectrum rights does not convey ownership. Radio frequency spectrum is managed by the Federal Communications Commission (FCC) for commercial and other non-federal uses and by the National Telecommunications and Information Administration (NTIA) for federal government use.

Although radio frequency spectrum (air) is abundant, usable spectrum is currently limited by the constraints of applied technology. Spectrum policy therefore includes making decisions about how radio frequencies will be allocated, who will have access to them, and how technology may enhance service and increase capacity and accessibility.

Spectrum policy issues that may be addressed in the 113th Congress include assuring that new capacity is made available for wireless broadband. Congress may also choose to explore emerging technologies that promise to enhance broadband capacity and spur innovation.

The Middle Class Tax Relief and Job Creation Act of 2012 (P.L. 112-96), signed February 22, 2012, contained provisions in Title VI to increase the availability of spectrum for commercial use. The provisions in Title VI—also known as the Spectrum Act—included expediting auctions of licenses for spectrum designated for mobile broadband; authorizing incentive auctions, which would permit television broadcasters to receive compensation for steps they might take to release some of their airwaves for mobile broadband; requiring that specified federal holdings be auctioned or reassigned for commercial use; providing for the availability of spectrum for unlicensed use; and assigning additional spectrum to support the construction of a new, interoperable broadband radio network for first responders and others. The act also included provisions to apply future spectrum license auction revenues toward deficit reduction and to establish a planning and governance structure to deploy public safety broadband networks, using some spectrum license auction proceeds for that purpose.

UNIVERSAL SERVICE FUND REFORM

Since its creation in 1934 the Federal Communications Commission (FCC) has been tasked with "mak[ing] available, so far as possible, to all the people of the United States ... a rapid, efficient, Nation-wide, and world-wide wire and radio communications service with adequate facilities at reasonable charges." (Communications Act of 1934, as amended, Title I §1 [47 U.S.C. 151].) This mandate led to the development of what has come to be known as the universal service concept.

The universal service concept, as originally designed, called for the establishment of policies to ensure that telecommunications services are available to all Americans, including those in rural, insular, and high cost areas, by ensuring that rates remain affordable. The Telecommunications Act of 1996 (P.L. 104-104; 47 U.S.C., 1996 act) codified the long-standing commitment by U.S. policymakers to ensure universal service in the provision of telecommunications services (§254), and the FCC established, in 1997, a federal Universal Service Fund (USF) to meet the objectives and principles contained in the act. The USF was designed to provide subsidies for voice telecommunications services for eligible high-cost telecommunications carriers (High Cost Program) and economically needy individuals (Low Income Program); access for telecommunications services and broadband access for schools and libraries (Schools and Libraries Program); and access to telecommunications, advanced telecommunications, and information services for public and non-profit rural health care providers (Rural Health Care Program).

One of the major policy debates surrounding universal service in the last decade was whether access to advanced telecommunications services (i.e., broadband) should be incorporated into universal service objectives. With the growing importance and acceptance of broadband and Internet access, gaps in access to such services, particularly in rural areas, generated concern. A growing number of policymakers felt that the USF should play a role in helping to alleviate this availability gap. This debate was put to rest when provisions contained in the American Recovery and Reinvestment Act of 2009 (ARRA) called for the FCC to develop, and submit to Congress, a national broadband plan (NBP) to ensure that every American has "access to broadband capability." (American Recovery and Reinvestment Act of 2009, P.L. 111-5, §6001 (k)(2)(D).) This plan, Connecting America: The National Broadband Plan, submitted to Congress on March 16, 2010, called for the USF to play a major role in achieving this goal. However, with the exception of funding for schools and libraries and rural health care providers, the USF was not designed to directly support broadband.

The FCC, in an October 2011 decision, adopted an order that calls for the USF to be transformed, in stages, over a multi-year period, from a mechanism to support voice telephone service to one that supports the deployment, adoption, and utilization of both fixed and mobile broadband. More specifically, the High Cost Program is to be phased out and a new fund, the Connect America Fund (CAF), which includes the targeted Mobility Fund and new Remote Areas Fund, is to be created to replace it; and the Low Income, Schools and Libraries, and Rural Health Care programs are to be modified and given wider responsibilities.

This transition is a vast undertaking and has caused considerable debate as policymakers balance the myriad goals and objectives to modernize the USF. As the USF undergoes this major and unprecedented transition it is anticipated that Congress will continue to assess the impact of these reforms and the FCC's progress in their implementation.

End Notes

[1] The circuit-switched network refers to the legacy telecommunications network based on copper wires and switching which was built to handle one-to-one voice communication and is referred to as the public switched telephone network. The Internet Protocol, or IP network, refers to the high-speed next-generation digital network which is capable of handling multimedia as well as voice and more.

[2] Lennard G. Kruger, Specialist in Science and Technology Policy, and Angele A. Gilroy, Specialist in Telecommunications Policy, Resources, Science, and Industry Division.

[3] Angele A. Gilroy, Specialist in Telecommunications Policy, Resources, Science, and Industry Division.

[4] Charles B. Goldfarb, Specialist in Telecommunications Policy, Resources, Science, and Industry Division.

[5] Glenn J. McLoughlin, Section Research Manager, Resources, Science, and Industry Division and Mark Gurevitz, Information Research Specialist, Knowledge Services Group.

[6] Linda K. Moore, Specialist in Telecommunications Policy, Resources, Science, and Industry Division.

[7] Kathleen Ann Ruane, Legislative Attorney, American Law Division.

[8] Patricia Moloney Figliola, Specialist in Internet and Telecommunications Policy, Resources, Science, and Industry Division.

[9] Lennard G. Kruger, Specialist in Science and Technology Policy, Resources, Science, and Industry Division.

[10] Charles B. Goldfarb, Specialist in Telecommunications Policy, Resources, Science, and Industry Division.

[11] Linda K. Moore, Specialist in Telecommunications Policy, Resources, Science, and Industry Division.

ISBN: 978-1-62948-413-6

Chapter 2

BROADBAND INTERNET ACCESS AND THE DIGITAL DIVIDE: FEDERAL ASSISTANCE PROGRAMS*

Lennard G. Kruger and Angele A. Gilroy

SUMMARY

The "digital divide" is a term that has been used to characterize a gap between "information haves and have-nots," or in other words, between those Americans who use or have access to telecommunications and information technologies and those who do not. One important subset of the digital divide debate concerns high-speed Internet access and advanced telecommunications services, also known as broadband. Broadband is provided by a series of technologies (e.g., cable, telephone wire, fiber, satellite, wireless) that give users the ability to send and receive data at volumes and speeds far greater than traditional "dial-up" Internet access over telephone lines.

Broadband technologies are currently being deployed primarily by the private sector throughout the United States. While the numbers of new broadband subscribers continue to grow, studies and data suggest that the rate of broadband deployment in urban/suburban and high income areas is outpacing deployment in rural and low-income areas. Some policymakers, believing that disparities in broadband access across American society could have adverse economic and social consequences on those left behind, assert that the federal government should play a more active role to avoid a "digital divide" in broadband access.

With the conclusion of the grant and loan awards established by the American Recovery and Reinvestment Act of 2009 (P.L. 111-5), there remain two ongoing federal vehicles which direct federal money to fund broadband infrastructure: the broadband and telecommunications programs at the Rural Utilities Service (RUS) of the U.S. Department of Agriculture and the Universal Service Fund (USF) programs under the Federal Communications Commission (FCC). Although the USF's High Cost Program does not explicitly fund broadband infrastructure, subsidies are used, in many cases, to

---

* This is an edited, reformatted and augmented version of a Congressional Research Service publication, CRS Report for Congress RL30719, prepared for Members and Committees of Congress, from www.crs.gov, dated July 17, 2013.

upgrade existing telephone networks so that they are capable of delivering high-speed services. Additionally, subsidies provided by USF's Schools and Libraries Program and Rural Health Care Program are used for a variety of telecommunications services, including broadband access. Currently the USF is undergoing a major transition to the Connect America Fund, which is targeted to the deployment, adoption, and utilization of both fixed and mobile broadband.

To the extent that the 113th Congress may consider various options for further encouraging broadband deployment and adoption, a key issue is how to strike a balance between providing federal assistance for unserved and underserved areas where the private sector may not be providing acceptable levels of broadband service, while at the same time minimizing any deleterious effects that government intervention in the marketplace may have on competition and private sector investment.

INTRODUCTION

The "digital divide" is a term used to describe a perceived gap between "information haves and have-nots," or in other words, between those Americans who use or have access to telecommunications and information technologies and those who do not.[1] Whether or not individuals or communities fall into the "information haves" category depends on a number of factors, ranging from the presence of computers in the home, to training and education, to the availability of affordable Internet access.

Broadband technologies are currently being deployed primarily by the private sector throughout the United States. While the numbers of new broadband subscribers continue to grow, studies and data suggest that the rate of broadband deployment in urban/suburban and high income areas is outpacing deployment in rural and low-income areas.

STATUS OF BROADBAND DEPLOYMENT IN THE UNITED STATES

Prior to the late 1990s, American homes accessed the Internet at maximum speeds of 56 kilobits per second by dialing up an Internet Service Provider (such as AOL) over the same copper telephone line used for traditional voice service. A relatively small number of businesses and institutions used broadband or high speed connections through the installation of special "dedicated lines" typically provided by their local telephone company. Starting in the late 1990s, cable television companies began offering cable modem broadband service to homes and businesses. This was accompanied by telephone companies beginning to offer DSL service (broadband over existing copper telephone wireline). Growth has been steep, rising from 2.8 million high speed lines reported as of December 1999, to 243 million connections as of June 30, 2012.[2] Of the 243 million high speed connections reported by the FCC, 197 million serve residential users.[3]

Table 1 depicts the relative deployment of different types of broadband technologies. A distinction is often made between "current generation" and "next generation" broadband (commonly referred to as next generation networks or NGN). "Current generation" typically refers to currently deployed cable, DSL, and many wireless systems, while "next generation" refers to dramatically faster download and upload speeds offered by fiber technologies and also by successive generations of cable, DSL, and wireless technologies.[4] In general, the greater the download and upload speeds offered by a broadband connection, the more sophisticated (and potentially valuable) the application that is enabled.

Based on the latest FCC broadband connection data, *Table 2* shows the percentages of households with broadband connections by state, both for download connections over 200 kbps and for connections of at least 3 Mbps (which approximates the FCC's National Broadband Availability target). According to the FCC, high speed connections over 200 kbps are reported in 67% of households nationwide, while connections of at least 3 Mbps (download) and 768 kbps (upload) are reported in 38% of households nationwide.

Table 1. Percentage of Broadband Technologies by Types of Connection

	Connections over 200 kbps in at least one direction	Residential connections over 200 kbps in at least one direction	Connections at least 3 Mbps downstream and 768 kbps upstream	Residential connections at least 3 Mbps downstream and 768 kbps upstream
cable modem	20.4%	23.9%	37.7%	45.3%
DSL	12.8%	14.1%	12.8%	14.7%
Mobile wireless	63.0%	58.2%	42.9%	32.5%
Fiber	2.6%	3.0%	6.0%	7.2%
All other	1.2%	0.8%	0.6%	0.4%

Source: FCC, Internet Access Services: Status as of June 30, 2012, pp. 23-26.

Table 2. Percentage of Households With Broadband Connections by State (as of June 30, 2012)

	Connections over 200 kbps	Connections at least 3 mbps downstream and 768 kbps upstream
Alabama	57%	30%
Alaska	71%	34%
Arizona	69%	54%
Arkansas	54%	25%
California	76%	44%
Colorado	76%	61%
Connecticut	79%	58%
Delaware	78%	69%
District of Columbia	72%	65%
Florida	74%	51%
Georgia	65%	42%
Hawaii	*	*
Idaho	61%	30%
Illinois	67%	44%
Indiana	64%	41%
Iowa	66%	25%
Kansas	67%	33%
Kentucky	62%	38%
Louisiana	60%	30%

Table 2. (Continued)

	Connections over 200 kbps	Connections at least 3 mbps downstream and 768 kbps upstream
Maine	75%	26%
Maryland	76%	70%
Massachusetts	81%	75%
Michigan	65%	46%
Minnesota	69%	48%
Mississippi	50%	21%
Missouri	62%	29%
Montana	64%	38%
Nebraska	68%	45%
Nevada	66%	50%
New Hampshire	81%	65%
New Jersey	82%	76%
New Mexico	59%	37%
New York	76%	54%
North Carolina	69%	17%
North Dakota	70%	41%
Ohio	68%	25%
Oklahoma	58%	28%
Oregon	69%	55%
Pennsylvania	71%	58%
Rhode Island	76%	*
South Carolina	64%	29%
South Dakota	67%	47%
Tennessee	58%	39%
Texas	65%	33%
Utah	73%	56%
Vermont	77%	61%
Virginia	70%	60%
Washington	72%	61%
West Virginia	60%	49%
Wisconsin	67%	30%
Wyoming	66%	44%
National subscribership ratio	69%	45%

Source: FCC, Internet Access Services: Status as of June 30, 2012, pp. 34-35.
Notes: Asterisk (*) indicates data withheld by the FCC to maintain firm confidentiality. Subscribership
ratio is the number of reported residential high speed lines (broadband connections) divided by the
number of households in each state.

Meanwhile, the National Broadband Map, which is composed of state broadband data
and compiled by NTIA, provides data on where broadband is and is not available. The latest
update of these data indicate that 98.2% of the U.S. population have available advertised
speeds of at least 3 Mbps (download) and 768 kbps (upload), 96.2% have download speeds 6
Mbps or greater, and 94.4% have download speeds 10 Mbps or greater.[5] The FCC's *Eighth
Broadband Progress Report*, released on August 21, 2012, used National Broadband Map
data to estimate that 19 million Americans living in 7 million households lack access to fixed

broadband at speeds of 4 Mbps (download)/1 Mbps (upload) or greater.[6] *Table 3* shows a state-by-state breakdown of the percentage of population without access to fixed broadband at the FCC's benchmark speed of 4 Mbps/1Mbps.

**Table 3. Americans Without Access to Fixed Broadband by State
(access to speeds of at least 4 Mbps download/1 Mbps upload)**

	% of population without access	% of population without access, nonrural areas	% of population without access, rural areas
United States	6.0%	1.8%	23.7%
Alabama	11.4%	1.6%	25.5%
Alaska	19.6%	4.4%	48.9%
Arizona	4.7%	1.2%	35.8%
Arkansas	13.6%	1.8%	28.8%
California	3.3%	1.6%	35.2%
Colorado	4.3%	1.0%	25.3%
Connecticut	0.7%	0.5%	2.6%
Delaware	3.1%	1.1%	13.0%
District of Columbia	0.0%	0.0%	N/A
Florida	3.1%	2.0%	14.3%
Georgia	3.4%	1.3%	9.9%
Hawaii	1.5%	0.1%	17.7%
Idaho	13.1%	1.3%	41.4%
Illinois	3.3%	0.4%	25.6%
Indiana	4.3%	1.3%	12.4%
Iowa	7.1%	0.7%	18.7%
Kansas	7.7%	1.0%	27.0%
Kentucky	10.5%	1.5%	23.0%
Louisiana	8.8%	1.3%	29.6%
Maine	4.7%	1.2%	7.0%
Maryland	3.2%	0.9%	19.2%
Massachusetts	1.0%	0.5%	6.4%
Michigan	6.3%	0.8%	22.4%
Minnesota	8.0%	0.8%	27.7%
Mississippi	12.1%	1.2%	22.8%
Missouri	7.5%	0.6%	24.2%
Montana	26.7%	4.0%	55.4%
Nebraska	10.1%	1.9%	33.0%
Nevada	2.3%	0.6%	30.2%
New Hampshire	7.5%	2.5%	15.2%
New Jersey	0.7%	0.4%	5.6%
New Mexico	14.2%	4.8%	46.7%
New York	1.3%	0.0%	10.4%
North Carolina	6.4%	2.1%	15.0%
North Dakota	15.9%	2.5%	36.2%
Ohio	3.4%	0.5%	14.0%
Oklahoma	16.2%	2.9%	42.5%
Oregon	3.4%	0.2%	17.3%
Pennsylvania	1.7%	0.3%	6.8%

Table 3. (Continued)

	% of population without access	% of population without access, nonrural areas	% of population without access, rural areas
Rhode Island	0.2%	0.0%	2.3%
South Carolina	11.7%	4.9%	25.1%
South Dakota	21.1%	3.2%	44.6%
Tennessee	6.8%	0.9%	18.6%
Texas	5.9%	2.0%	27.6%
Utah	1.8%	0.3%	16.7%
Vermont	9.4%	0.2%	15.2%
Virginia	10.9%	2.2%	37.6%
Washington	3.2%	0.5%	17.4%
West Virginia	45.9%	31.4%	59.8%
Wisconsin	6.9%	0.1%	23.0%
Wyoming	13.2%	1.1%	35.4%
U.S. Territories	54.0%	41.5%	85.2%
American Samoa	78.6%	30.9%	92.0%
Northern Marianas	100.0%	100.0%	100.0%
Guam	54.3%	0.1%	76.1%
Puerto Rico	51.6%	40.3%	84.8%
U.S. Virgin Islands	100.0%	100.0%	100.0%

Source: FCC, Eighth Broadband Progress Report, Appendix C.

In contrast to broadband *availability*, which refers to whether or not broadband service is offered, broadband *adoption* refers to the extent to which American households actually subscribe to and use broadband. The U.S. Department of Commerce, based on October 2012 survey data, found that 72.4% of U.S. households have adopted broadband.[7] Similarly, the FCC's *Eighth Broadband Progress Report* found that 64% of American households with broadband available to them adopt broadband service offering speeds faster than 768 kbps/200 kbps, while 40% adopt speeds faster than the FCC benchmark of 4 Mbps/1Mbps. The FCC found that the "broadband adoption rates for American households are lower, on average, in the counties with the lowest median household income, in areas outside of urban areas, on Tribal lands, and in U.S. Territories."[8]

According to the Department of Commerce report *Exploring the Digital Nation: America's Emerging Online Experience* (based on July 2011 U.S. Census Bureau survey data), the three main reasons cited for not having broadband at home are that it is perceived as not needed, too expensive, and lack of a home computer.[9] The Department of Commerce report, the FCC's National Broadband Plan, and a survey conducted by the Pew Internet and American Life Project[10] also found disparities in broadband adoption among demographic groups. Populations continuing to lag behind in broadband adoption include people with low incomes, seniors, minorities, the less-educated, non-family households, and the non-employed.

BROADBAND IN RURAL AREAS[11]

While the number of new broadband subscribers continues to grow, the rate of broadband deployment in urban areas appears to be outpacing deployment in rural areas.

While there are many examples of rural communities with state of the art telecommunications facilities,[12] recent surveys and studies have indicated that, in general, rural areas tend to lag behind urban and suburban areas in broadband deployment. For example:

- According to the FCC's *Eighth Broadband Progress Report*, released in August 2012, of the 19 million Americans who live where fixed broadband is unavailable, 14.5 million live in rural areas.[13]
- According to June 2012 data from the National Broadband Map, 99.6% of the population in urban areas have access to available broadband download speeds of at least 6 Mbps, as opposed to 81.8% of the population in rural areas.[14]
- The Department of Commerce report, *Exploring the Digital Nation: America's Emerging Online Experience*, found that while the digital divide between urban and rural areas has lessened since 2007, it still persists with 72% of urban households adopting broadband service in 2011, compared to 58% of rural households.[15]

The comparatively lower population density of rural areas is likely the major reason why broadband is less deployed than in more highly populated suburban and urban areas. Particularly for wireline broadband technologies—such as cable modem and DSL—the greater the geographical distances among customers, the larger the cost to serve those customers. Thus, there is often less incentive for companies to invest in broadband in rural areas than, for example, in an urban area where there is more demand (more customers with perhaps higher incomes) and less cost to wire the market area.[16]

Some policymakers believe that disparities in broadband access across American society could have adverse consequences on those left behind, and that advanced telecommunications applications critical for businesses and consumers to engage in e-commerce are increasingly dependent on high speed broadband connections to the Internet. Thus, some say, communities and individuals without access to broadband could be at risk to the extent that connectivity becomes a critical factor in determining future economic development and prosperity. A February 2006 study done by the Massachusetts Institute of Technology for the Economic Development Administration of the Department of Commerce marked the first attempt to quantitatively measure the impact of broadband on economic growth. The study found that "between 1998 and 2002, communities in which mass-market broadband was available by December 1999 experienced more rapid growth in employment, the number of businesses overall, and businesses in IT-intensive sectors, relative to comparable communities without broadband at that time."[17]

A June 2007 report from the Brookings Institution found that for every one percentage point increase in broadband penetration in a state, employment is projected to increase by 0.2% to 0.3% per year. For the entire U.S. private non-farm economy, the study projected an increase of about 300,000 jobs.[18]

Subsequently, a July 2009 study commissioned by the Internet Innovation Alliance found net consumer benefits of home broadband on the order of $32 billion per year, up from an estimated $20 billion in consumer benefits from home broadband in 2005.[19]

Some also argue that broadband is an important contributor to U.S. future economic strength with respect to the rest of the world. Data from the Organization for Economic Cooperation and Development (OECD) found the U.S. ranking 15th among OECD nations in broadband access per 100 inhabitants as of December 2011.[20] By contrast, in 2001 an OECD study found the U.S. ranking fourth in broadband subscribership per 100 inhabitants (after Korea, Sweden, and Canada).[21] While many argue that declining U.S. performance in international broadband rankings is a cause for concern,[22] others maintain that the OECD data undercount U.S. broadband deployment,[23] and that cross-country broadband deployment comparisons are not necessarily meaningful and are inherently problematic.[24] Finally, an issue related to international broadband rankings is the extent to which broadband speeds and prices differ between the United States and the rest of the world.[25]

ARE BROADBAND DEPLOYMENT DATA ADEQUATE?

Obtaining an accurate snapshot of the status of broadband deployment is problematic. Anecdotes abound of rural and low-income areas which do not have adequate Internet access, as well as those which are receiving access to high-speed, state-of-the-art connections. Rapidly evolving technologies, the constant flux of the telecommunications industry, the uncertainty of consumer wants and needs, and the sheer diversity and size of the nation's economy and geography make the status of broadband deployment very difficult to characterize. The FCC periodically collects broadband deployment data from the private sector via "FCC Form 477"—a standardized information gathering survey. Statistics derived from the Form 477 survey are published every six months. Additionally, data from Form 477 are used in the FCC's (to date) eight broadband deployment reports.

The FCC is working to refine the data used in future reports in order to provide an increasingly accurate portrayal. In its March 17, 2004, Notice of Inquiry for the *Fourth Report*, the FCC sought comments on specific proposals to improve the FCC Form 477 data gathering program.[26] On November 9, 2004, the FCC voted to expand its data collection program by requiring reports from all facilities based carriers regardless of size in order to better track rural and underserved markets, by requiring broadband providers to provide more information on the speed and nature of their service, and by establishing broadband-over-power line as a separate category in order to track its development and deployment. The FCC Form 477 data gathering program was extended for five years beyond its then March 2005 expiration date.[27]

On April 16, 2007, the FCC announced a Notice of Proposed Rulemaking which sought comment on a number of broadband data collection issues, including how to develop a more accurate picture of broadband deployment; gathering information on price, other factors determining consumer uptake of broadband, and international comparisons; how to improve data on wireless broadband; how to collect information on subscribership to voice over Internet Protocol service (VoIP); and whether to modify collection of speed tier information.[28]

On March 19, 2008, the FCC adopted an order that substantially expands its broadband data collection capability. Specifically, the order expands the number of broadband reporting speed tiers to capture more information about upload and download speeds offered in the marketplace, requires broadband providers to report numbers of broadband subscribers by census tract, and improves the accuracy of information collected on mobile wireless broadband deployment.

Additionally, in a Further Notice of Proposed Rulemaking, the FCC sought comment on broadband service pricing and availability.[29] The July 2009 data release (providing data as of June 30, 2008) was the final data set gathered under the old FCC Form 477. The February 2010 data report (December 31, 2008, data) reflected the new Form 477 data collection requirements. Meanwhile, during the 110th Congress, state initiatives to collect broadband deployment data in order to promote broadband in underserved areas were viewed as a possible model for governmental efforts to encourage broadband.

The Broadband Data Improvement Act was enacted by the 110th Congress and became P.L. 110-385 on October 10, 2008. The law requires the FCC to collect demographic information on unserved areas, data comparing broadband service with 75 communities in at least 25 nations abroad, and data on consumer use of broadband. The act also directs the Census Bureau to collect broadband data, the Government Accountability Office to study broadband data metrics and standards, and the Department of Commerce to provide grants supporting state broadband initiatives.

P.L. 111-5, the American Recovery and Reinvestment Act, provided NTIA with an appropriation of $350 million to implement the Broadband Data Improvement Act and to develop and maintain a national broadband inventory map. The National Broadband Map was first released on February 17, 2011 (http://www.broadbandmap.gov), and is updated every six months.[30]

Finally, the FCC's National Broadband Plan addressed the broadband data issue, recommending that the FCC and the U.S. Bureau of Labor Statistics (BLS) should collect more detailed and accurate data on actual availability, penetration, prices, churn, and bundles offered by broadband service providers to consumers and businesses, and should publish analyses of these data.

BROADBAND AND THE FEDERAL ROLE

The Telecommunications Act of 1996 (P.L. 104-104) addressed the issue of whether the federal government should intervene to prevent a "digital divide" in broadband access. Section 706 requires the FCC to determine whether "advanced telecommunications capability [i.e., broadband or high-speed access] is being deployed to all Americans in a reasonable and timely fashion."

Since 1999, the FCC has adopted and released eight reports pursuant to Section 706. The first five reports formally concluded that the deployment of advanced telecommunications capability to all Americans is reasonable and timely. Unlike the first five 706 reports, the sixth, seventh, and eighth reports concluded that broadband is not being deployed to all Americans in a reasonable and timely fashion. According to the *Eighth Broadband Progress Report*:

Our analysis shows that the nation's broadband deployment gap remains significant and is particularly pronounced for Americans living in rural areas and on Tribal lands. We find that as of June 30, 2011, approximately 19 million Americans did not have access to fixed broadband. Significantly, approximately 76 percent of these Americans reside in rural areas. Our analysis further shows that Americans residing on Tribal lands disproportionately lack access to fixed broadband. And the available international broadband data, though not perfectly comparable to U.S. data, suggest that the availability and deployment of broadband in the United States may lag behind a number of other developed countries in certain respects, although we also compare favorably to some developed countries in other respects. Moreover, as many as 80 percent of E-rate recipients say that their broadband connections do not fully meet their needs, and 78 percent of recipients say that they need additional bandwidth. These data combined with our findings concerning availability above provide further indication that broadband is not yet being reasonably and timely deployed to all Americans.[31]

FCC Commissioners Robert McDowell and Ajit Pai issued dissenting statements, maintaining that there is insufficient justification for the 706 report conclusion that broadband is not being deployed in a reasonable and timely fashion. For example, the dissents argued that the report did not sufficiently account for the dramatic growth in the availability and deployment of mobile broadband, and that gaps in broadband adoption should not be used to determine whether or not broadband is being sufficiently deployed.[32]

The National Broadband Plan

As mandated by the ARRA, on March 16, 2010, the FCC publically released its report, *Connecting America: The National Broadband Plan*.[33] The National Broadband Plan (NBP) seeks to "create a high-performance America," which the FCC defines as "a more productive, creative, efficient America in which affordable broadband is available everywhere and everyone has the means and skills to use valuable broadband applications."[34] In order to achieve this mission, the NBP recommends that the country set six goals for 2020:

- Goal No. 1: At least 100 million U.S. homes should have affordable access to actual download speeds of at least 100 megabits per second and actual upload speeds of at least 50 megabits per second.
- Goal No. 2: The United States should lead the world in mobile innovation, with the fastest and most extensive wireless networks of any nation.
- Goal No. 3: Every American should have affordable access to robust broadband service, and the means and skills to subscribe if they so choose.
- Goal No. 4: Every American community should have affordable access to at least 1 gigabit per second broadband service to anchor institutions such as schools, hospitals, and government buildings.
- Goal No. 5: To ensure the safety of the American people, every first responder should have access to a nationwide, wireless, interoperable broadband public safety network.

- Goal No. 6: To ensure that America leads in the clean energy economy, every American should be able to use broadband to track and manage their real-time energy consumption.

The National Broadband Plan is categorized into three parts:

- *Part I (Innovation and Investment)*, which "discusses recommendations to maximize innovation, investment and consumer welfare, primarily through competition. It then recommends more efficient allocation and management of assets government controls or influences."[35] The recommendations address a number of issues, including spectrum policy, improved broadband data collection, broadband performance standards and disclosure, special access rates, interconnection, privacy and cybersecurity, child online safety, poles and rightsof-way, research and experimentation (R&E) tax credits, and R&D funding.
- *Part II (Inclusion)*, which "makes recommendations to promote inclusion—to ensure that all Americans have access to the opportunities broadband can provide."[36] Issues include reforming the Universal Service Fund, intercarrier compensation, federal assistance for broadband in Tribal lands, expanding existing broadband grant and loan programs at the Rural Utilities Service, enabling greater broadband connectivity in anchor institutions, and improved broadband adoption and utilization especially among disadvantaged and vulnerable populations.
- *Part III (National Purposes)*, which "makes recommendations to maximize the use of broadband to address national priorities. This includes reforming laws, policies and incentives to maximize the benefits of broadband in areas where government plays a significant role."[37] National purposes include health care, education, energy and the environment, government performance, civic engagement, and public safety. Issues include telehealth and health IT, online learning and modernizing educational broadband infrastructure, digital literacy and job training, smart grid and smart buildings, federal support for broadband in small businesses, telework within the federal government, cybersecurity and protection of critical broadband infrastructure, copyright of public digital media, interoperable public safety communications, next generation 911 networks, and emergency alert systems.

The release of the National Broadband Plan is seen by many as a precursor towards the development of a national broadband policy—whether comprehensive or piecemeal—that will likely be shaped and developed by Congress, the FCC, and the Administration.[38] Congress will likely play a major role in implementing the National Broadband Plan, both by considering legislation to implement NBP recommendations, and by overseeing broadband activities conducted by the FCC and executive branch agencies.

CURRENT FEDERAL BROADBAND PROGRAMS

With the conclusion of grant and loan awards established by the American Recovery and Reinvestment Act of 2009 (P.L. 111-5),[39] there remain two ongoing federal vehicles which direct federal money to fund broadband infrastructure: the broadband and telecommunications programs at the Rural Utilities Service (RUS) of the U.S. Department of Agriculture and the Universal Service Fund (USF) programs under the Federal Communications Commission (FCC). Although the USF's High Cost Program does not explicitly fund broadband infrastructure, subsidies are used, in many cases, to upgrade existing telephone networks so that they are capable of delivering high-speed services. Additionally, subsidies provided by USF's Schools and Libraries Program and Rural Health Care Program are used for a variety of telecommunications services, including broadband access. Currently the USF is undergoing a major transition to the Connect America Fund, which is targeted to the deployment, adoption, and use of both fixed and mobile broadband.

Table 4 (at the end of this report) shows selected federal domestic assistance programs throughout the federal government that currently can be associated with broadband and telecommunications development. The table categorizes the programs in three ways: programs exclusively devoted to the deployment of broadband infrastructure; programs which have traditionally focused on deployment of telecommunications infrastructure generally (which typically can and does include broadband); and applications-specific programs which fund some aspect of broadband access or adoption as a means towards supporting a particular application, such as distance learning or telemedicine.

Rural Utilities Service Programs

RUS implements two programs specifically targeted at providing assistance for broadband infrastructure deployment in rural areas: the Rural Broadband Access Loan and Loan Guarantee Program and Community Connect Broadband Grants.[40] The 110th Congress reauthorized and reformed the Rural Broadband Access Loan and Loan Guarantee program as part of the 2008 farm bill (P.L. 110-234). The 112th Congress considered reauthorization of the program as part of the 2012 farm bill.[41]

RUS also has a rural telephone loan program (dating back to 1949, now called Telecommunications Infrastructure Loans) that has historically supported infrastructure for telephone voice service, but has now evolved into support for broadband-capable service provided by traditional telephone borrowers. Additionally, the Distance Learning and Telemedicine Grant Program supports broadband-based applications.[42]

The Universal Service Concept and the FCC[43]

Since its creation in 1934 the Federal Communications Commission (FCC) has been tasked with "mak[ing] available, so far as possible, to all the people of the United States ... a rapid, efficient, Nation-wide, and world-wide wire and radio communications service with adequate facilities at reasonable charges."[44] This mandate led to the development of what has come to be known as the universal service concept.

The universal service concept, as originally designed, called for the establishment of policies to ensure that telecommunications services are available to all Americans, including those in rural, insular and high cost areas, by ensuring that rates remain affordable. Over the years this concept fostered the development of various FCC policies and programs to meet this goal. The FCC offers universal service support through a number of direct mechanisms that target both providers of and subscribers to telecommunications and, more recently, broadband services.[45]

Universal Service and the Telecommunications Act of 1996

Passage of the Telecommunications Act of 1996 (P.L. 104-104) codified the long-standing commitment by U.S. policymakers to ensure universal service in the provision of telecommunications services.

The Schools and Libraries, and Rural Health Care Programs

Congress, through the 1996 act, not only codified, but also expanded the concept of universal service to include, among other principles, that elementary and secondary schools and classrooms, libraries, and rural health care providers have access to telecommunications services for specific purposes at discounted rates. (See §§254(b)(6) and 254(h)of the 1996 Telecommunications Act, 47 U.S.C. 254.)

1. The Schools and Libraries Program. Under universal service provisions contained in the 1996 act, elementary and secondary schools and classrooms and libraries are designated as beneficiaries of universal service discounts. Universal service principles detailed in Section 254(b)(6) state that "Elementary and secondary schools and classrooms ... and libraries should have access to advanced telecommunications services." The act further requires in Section 254(h)(1)(B) that services within the definition of universal service be provided to elementary and secondary schools and libraries for education purposes at discounts, that is at "rates less than the amounts charged for similar services to other parties."

The FCC established the Schools and Libraries Division within the Universal Service Administrative Company (USAC) to administer the schools and libraries or "E (education)-rate" program to comply with these provisions. Under this program, eligible schools and libraries receive discounts ranging from 20% to 90% for telecommunications services depending on the poverty level of the school's (or school district's) population and its location in a high cost telecommunications area. The FCC established a funding ceiling, or cap, of $2.25 billion, adjusted for inflation prospectively beginning with funding year 2010. Three categories of services are eligible for discounts: internal connections (e.g., wiring, routers and servers); Internet access; and telecommunications and dedicated services, with the third category receiving funding priority. According to data released by program administrators, approximately $31 billion in funding has been committed over the first 14 years of the program with funding released to all states, the District of Columbia and all territories. Funding commitments for funding Year 2012 (July 1, 2012-June 30, 2013), the 15th year of the program, totaled $ 2.6 billion as of July 5, 2013.[46]

2. The Rural Health Care Programs. Section 254(h) of the 1996 act requires that public and nonprofit rural health care providers have access to telecommunications services necessary for the provision of health care services at rates comparable to those paid for similar services in urban areas. Subsection 254(h)(1) further specifies that "to the extent technically feasible and economically reasonable" health care providers should have access to advanced telecommunications and information services. The FCC established the Rural Health Care Division (RHCD) within the USAC to administer the universal support program to comply with these provisions. Under FCC established rules only public or non-profit health care providers are eligible to receive funding. Eligible health care providers, with the

exception of those requesting only access to the Internet, must also be located in a rural area. The funding ceiling, or cap, for this support was established at $400 million annually. The funding level for Year One of the program (January 1998-June 30, 1999) was set at $100 million. Due to less than anticipated demand, the FCC established a $12 million funding level for the second year (July 1, 1999 to June 30, 2000) of the program but has since returned to a $400 million yearly cap. As of March 31, 2012, a total of $514.3 million has been committed since the program's inception in 1998. The primary use of the funding is to provide reduced rates for telecommunications and information services necessary for the provision of health care.[47] In addition, the FCC established, in 2007, the "Rural Health Care Pilot Program" to help public and non-profit health care providers build state and region-wide broadband networks dedicated to the provision of health care services. There are 50 projects in the program with $387.9 million in authorized funds. As of February 29, 2010, $232.6 million of the funds have been committed to the 50 FCC designated projects. The FCC in a December 12, 2012, order, created a new program, the Healthcare Connect Fund, which will expand health care provider access to broadband, particularly in rural areas, and replace the Rural Health Care Pilot Program with a permanent program. Non-rural participation is limited to consortia of which at least 50% must be located in a rural area. The upfront funding cap for the Healthcare Connect Fund is $150 million annually and participants will be required to contribute 35% of the costs. The FCC also established, as part of the Healthcare Connect Fund, a new pilot program, to expand broadband connections to skilled nursing facilities. Funding for this pilot program, which will begin in 2014, will be up to $50 million total over three years. The total funding cap for all of the above mentioned USF rural health care programs will remain at $400 million annually.[48]

Universal Service and Broadband

One of the policy debates surrounding universal service in the last decade was whether access to advanced telecommunications services (i.e., broadband) should be incorporated into universal service objectives. The term universal service, when applied to telecommunications, refers to the ability to make available a basket of telecommunications services to the public, across the nation, at a reasonable price. As directed in the 1996 Telecommunications Act (§254[c]) a federal-state Joint Board was tasked with defining the services which should be included in the basket of services to be eligible for federal universal service support; in effect using and defining the term "universal service" for the first time. The Joint Board's recommendation, which was subsequently adopted by the FCC in May 1997, included the following in its universal service package: voice grade access to and some usage of the public switched network; single line service; dual tone signaling; access to directory assistance; emergency service such as 911; operator services; and access and interexchange (long distance) service. Some policy makers expressed concern that the FCC-adopted definition is too limited and does not take into consideration the importance and growing acceptance of advanced services such as broadband and Internet access. They point to a number of provisions contained in the Universal Service section of the 1996 act to support their claim. Universal service principles contained in Section 254(b)(2) state that "Access to advanced telecommunications services should be provided to all regions of the Nation." The subsequent principle (b)(3) calls for consumers in all regions of the nation including "low-income" and those in "rural, insular, and high cost areas" to have access to telecommunications and information services including "advanced services" at a comparable level and a comparable rate charged for similar services in urban areas. Such provisions, they state, dictate that the FCC expand its universal service definition. The 1996 act does take into consideration the changing nature of the telecommunications sector and allows for the universal service definition to be modified if future conditions warrant. Section 254(c)of the act states that "universal service is an evolving level of telecommunications services" and the FCC is tasked with "periodically" reevaluating this definition "taking into account advances in telecommunications and information technologies and services." Furthermore, the Joint Board is given specific authority to recommend "from time to time" to the FCC modification in the definition of the services to be included for federal universal service support. The Joint Board, on November 19, 2007, concluded such an inquiry and recommended that the FCC change the mix of services eligible for universal service support. The Joint Board recommended, among other things, that "the universal availability of broadband Internet services" be included in the nation's communications goals and hence be supported by federal universal service funds.[49] This debate was put to rest when provisions contained in the American Recovery and Reinvestment Act of 2009 (ARRA) called for the FCC to develop, and submit to Congress, a national broadband plan to ensure that every American has "access to broadband capability."[50] The FCC in its national broadband plan, *Connecting America: the National Broadband Plan,* recommended that access to and adoption of broadband be a national goal. Furthermore the national broadband plan proposed that the Universal Service Fund be restructured to become a vehicle to help reach this goal. The FCC, in an October 2011 decision, adopted an Order that calls for the USF to be transformed, in stages, over a multi-year period, from a mechanism to support voice telephone service to one that supports the deployment, adoption, and utilization of both fixed and mobile broadband.

This transformation includes the phase out of the USF's legacy High Cost Program and the creation of a new fund, the Connect America Fund, to replace it.[51]

LEGISLATION IN THE 110TH CONGRESS

In the 110th Congress, legislation was enacted to provide financial assistance for broadband deployment. Of particular note is the reauthorization of the Rural Utilities Service (RUS) broadband loan program, which was enacted as part of the 2008 farm bill (P.L. 110-234). In addition to reauthorizing and reforming the RUS broadband loan program, P.L. 110-234 contains provisions establishing a National Center for Rural Telecommunications Assessment and requiring the FCC and RUS to formulate a comprehensive rural broadband strategy. The Broadband Data Improvement Act (P.L. 110-385) was enacted by the 110th Congress and required the FCC to collect demographic information on unserved areas, data comparing broadband service with 75 communities in at least 25 nations abroad, and data on consumer use of broadband. The act also directed the Census Bureau to collect broadband data, the Government Accountability Office to study broadband data metrics and standards, and the Department of Commerce to provide grants supporting state broadband initiatives.

Meanwhile, the America COMPETES Act (H.R. 2272) was enacted (P.L. 110-69) and contained a provision authorizing the National Science Foundation (NSF) to provide grants for basic research in advanced information and communications technologies. Areas of research included affordable broadband access, including wireless technologies. P.L. 110-69 also directs NSF to develop a plan that describes the current status of broadband access for scientific research purposes.

LEGISLATION IN THE 111TH CONGRESS

In the 111th Congress, legislation was enacted that sought to provide financial assistance for broadband deployment. Of particular note, provisions in the American Recovery and Reinvestment Act of 2009 (P.L. 111-5) provided grants and loans to support broadband access and adoption in unserved and underserved areas.

P.L. 111-5: The American Recovery and Reinvestment Act of 2009

On February 17, 2009, President Obama signed P.L. 111-5, the American Recovery and Reinvestment Act (ARRA). Broadband provisions of the ARRA provided a total of *$7.2 billion*, for broadband grants, loans, and loan/grant combinations.

The total consisted of $4.7 billion to NTIA/DOC for a newly established Broadband Technology Opportunities Program (grants) and $2.5 billion to the RUS/USDA Broadband Initiatives Program (grants, loans, and grant/loan combinations).[52]

Regarding the $2.5 billion to RUS/USDA broadband programs, the ARRA specified that at least 75% of the area to be served by a project receiving funds shall be in a rural area without sufficient access to high speed broadband service to facilitate economic development, as determined by the Secretary of Agriculture.

Priority was given to projects that provide service to the most rural residents that do not have access to broadband services. Priority was also given to borrowers and former borrowers of rural telephone loans.

Of the $4.7 billion appropriated to NTIA:

- $4.35 billion was directed to a competitive broadband grant program, of which not less than $200 million shall be available for competitive grants for expanding public computer center capacity (including at community colleges and public libraries); not less than $250 million to encourage sustainable adoption of broadband service; and $10 million transferred to the Department of Commerce Office of Inspector General for audits and oversight; and
- $350 million was directed for funding the Broadband Data Improvement Act (P.L. 110-385) and for the purpose of developing and maintaining a broadband inventory map, which shall be made accessible to the public no later than two years after enactment. Funds deemed necessary and appropriate by the Secretary of Commerce may be transferred to the FCC for the purposes of developing a national broadband plan, which shall be completed one year after enactment.

Final BTOP and BIP program awards were announced by September 30, 2010.

Other Enacted Broadband Legislation in the 111th Congress

P.L. 111-8 (H.R. 1105). Omnibus Appropriations Act, 2009. Appropriates to RUS/USDA $15.619 million to support a loan level of $400.487 million for the Rural Broadband Access Loan and Loan Guarantee Program, and $13.406 million for the Community Connect Grant Program.

To the FCC, designates not less than $3 million to establish and administer a State Broadband Data and Development matching grants program for state-level broadband demand aggregation activities and creation of geographic inventory maps of broadband service to identify gaps in service and provide a baseline assessment of statewide broadband deployment. Passed House February 25, 2009. Passed Senate March 10, 2009. Signed by President, March 12, 2009.

P.L. 111-32 (H.R. 2346). Supplemental Appropriations Act, 2009. Provides not less than $3 million to the FCC to develop a national broadband plan pursuant to the American Recovery and Reinvestment Act of 2009. Introduced May 12, 2009; referred to Committee on Appropriations. Passed House May 14, 2009; passed Senate May 21, 2009. Signed by President, June 24, 2009.

P.L. 111-80 (H.R. 2997). Agriculture, Rural Development, Food and Drug Administration, and Related Agencies Appropriations Act, 2010. For Rural Utilities Service, U.S. Department of Agriculture, provides $28.96 million to support a loan level of $400 million for the broadband loan program, and $17.97 million for broadband community connect grants. Introduced June 23, 2009; referred to Committee on Appropriations. Reported by Committee on Appropriations June 23, 2009. Passed House July 9, 2009. Passed Senate August 4, 2009. Conference Report (H.Rept. 111-279) printed September 30, 2009. Signed by President October 21, 2009.

LEGISLATION IN THE 112TH CONGRESS

The 112th Congress examined the efficacy of federal broadband assistance programs and how they may fit into the context of a national broadband policy.

P.L. 112-10 (H.R. 1473). Department of Defense and Full-Year Continuing Appropriations Act, 2011. Rescinds existing unobligated past-year funding for the Rural Broadband Access Loan and Loan Guarantee Program and the Community Connect Grants at the Rural Utilities Service. For FY2011, appropriates $22.3 million to the Rural Broadband Access Loan and Loan Guarantee Program for the cost of broadband loans, and $13.4 million to Community Connect Grants. Signed by President, April 15, 2011.

P.L. 112-55 (H.R. 2112). Consolidated and Further Continuing Appropriations Act, 2012. Provides FY2012 appropriations for Rural Utilities Service broadband loan program and broadband community connect grants: $6 million for the broadband loan program (subsidizing a loan level of $212 million) and $10.372 million for Community Connect grants. Introduced June 3, 2011; referred to Committee on Appropriations. Reported by Committee on Appropriations June 3, 2011 (H.Rept. 112-101). Passed House June 16, 2011. Reported by Senate Appropriations Committee September 7, 2011 (S.Rept. 112-73). Signed by President, November 18, 2011.

H.R. 1083 (Owens). Rural Broadband Initiative Act. Establishes an Office of Rural Broadband Initiatives in the Department of Agriculture which would administer the RUS broadband loan and grant programs, and would develop a comprehensive rural broadband strategy. Introduced March 15, 2011; referred to Committee on Agriculture and in addition to the Committee on Energy and Commerce.

H.R. 1343 (Bass). To return unused or reclaimed funds made available for broadband awards in the American Recovery and Reinvestment Act of 2009 to the Treasury of the United States. Introduced April 4, 2011; referred to Committee on Energy and Commerce and to Committee on Agriculture. Reported (amended) by the Committee on Energy and Commerce (H.Rept. 112-228) on September 29, 2011. Passed House October 5, 2011. Referred to Senate Committee on Commerce, Science and Transportation October 6, 2011.

H.R. 1695 (Eshoo). Broadband Conduit Deployment Act of 2011. Directs the Secretary of Transportation to require that broadband conduit be installed as part of certain highway construction projects. Introduced May 3, 2011; referred to Committee on Transportation and Infrastructure.

H.R. 2163 (Matsui). Broadband Affordability Act of 2011. Amends the Communications Act of 1934 to establish a Lifeline Assistance Program for universal broadband adoption. Introduced June 14, 2011; referred to Committee on Energy and Commerce.

H.R. 5973 (Kingston). Agriculture, Rural Development, Food and Drug Administration, and Related Agencies Appropriations Act, 2013. For Rural Utilities Service, U.S. Department of Agriculture, provides $2 million to support a loan level of $21 million for the broadband loan program, and $10 million for broadband community connect grants. Introduced June 20, 2012; referred to Committee on Appropriations. Reported by Committee on Appropriations June 20, 2012.

H.R. 6083 (Lucas). Federal Agriculture Reform and Risk Management Act of 2012. Reauthorizes rural broadband loan program at $25 million per year through FY2017. Introduced July 9, 2012; referred to Committee on Agriculture. Ordered to be reported by committee July 11, 2012.

S. 257 (Landrieu). Small Business Broadband and Emerging Information Technology Enhancement Act of 2011. Seeks to improve certain programs of the Small Business Administration to better assist small business customers in accessing broadband technology. Introduced February 2, 2011; referred to Committee on Small Business and Entrepreneurship.

S. 1659 (Ayotte). To return unused or reclaimed funds made available for broadband awards in the American Recovery and Reinvestment Act of 2009 to the Treasury of the United States. Introduced October 5, 2011; referred to Committee on Commerce, Science and Transportation.

S. 1939 (Klobuchar). Broadband Conduit Deployment Act of 2011. Directs the Secretary of Transportation to require that broadband conduit be installed as part of certain highway construction projects. Introduced December 1, 2011; referred to Committee on Environment and Public Works.

S. 2298 (Brown of Ohio). Connecting Rural America Act. Amends the Rural Electrification Act of 1936 to improve the program of access to broadband telecommunications services in rural areas. Introduced April 18, 2002; referred to Committee on Agriculture, Nutrition, and Forestry.

S. 2375 (Kohl). Agriculture, Rural Development, Food and Drug Administration, and Related Agencies Appropriations Act, 2013. For Rural Utilities Service, U.S. Department of

Agriculture, provides $6 million to support a loan level of $63 million for the broadband loan program, and $10 million for broadband community connect grants. Introduced April 26, 2012; referred to Committee on Appropriations. Reported by Committee on Appropriations April 26, 2012.

S. 3240 (Stabenow). Agriculture Reform, Food, and Jobs Act of 2012. Authorizes broadband loan and grant program at $50 million per year through FY2017. Introduced May 24, 2012; referred to Committee on Agriculture, Nutrition and Forestry. Reported to Senate May 24, 2012. Passed Senate (amended) June 21, 2012.

S. 3439 (Snowe). Federal Wi-Net Act. Directs the Administrator of General Services to install Wi-Fi hotspots and wireless neutral host systems in all federal buildings in order to improve in-building wireless communications coverage and commercial network capacity by offloading wireless traffic onto wireline broadband networks. Introduced July 25, 2012; referred to Committee on Environment and Public Works.

LEGISLATION IN THE 113TH CONGRESS

The following is a listing of broadband legislation directly related to the issue of federal assistance for broadband deployment in unserved areas.[53]

P.L. 113-6 (H.R. 933). Consolidated and Further Continuing Appropriations Act, 2013.Funds the broadband loan program at $4 million (supporting a loan level of approximately $42 million) and the Community Connect grant program at $10.372 million. Signed by President March 26, 2013.

H.R. 2163 (Matsui). Broadband Adoption Act of 2013. Amends the Communications Act of 1934 to reform and modernize the Universal Service Fund Lifeline Assistance Program. Introduced April 26, 2013; referred to Committee on Energy and Commerce.

H.R. 1947 (Lucas). Federal Agriculture Reform and Risk Management Act of 2013. Section 6105 would reauthorize the broadband loan and loan guarantee program through FY2018 at the current level of $25 million per year. Introduced May 13, 2013; reported by Committees on Agriculture and Judiciary.

H.R. 1639 (Gibson). Amends the Rural Electrification Act of 1936 to authorize loan/grant combinations under RUS broadband program. Introduced April 18, 2013; referred to Committee on Agriculture and Committee on Energy and Commerce.

H.R. 2410 (Aderholt). Agriculture, Rural Development, Food and Drug Administration, and Related Agencies Appropriations Act, 2014. Provides $5.5 million to subsidize a loan level of $42.146 million for the broadband loan program, and $10.111 million for the Community Connect grant program. Introduced June 18, 2013; reported by House Committee on Appropriations (H.Rept. 113-116).

S. 954 (Stabenow). Agriculture Reform, Food, and Jobs Act of 2013. Includes the establishment of a new grant program in combination with the existing loan and loan guarantee program authorization, which is extended at $50 million per year through FY2018. Passed by Senate June 10, 2013.

Table 4. Selected Federal Domestic Assistance Programs Related to Broadband and Telecommunications Development

Program	Agency	Description	Funding Amount (est. FY2012 unless otherwise noted)	Web Links
		Broadband Infrastructure Deployment Programs		
Broadband Technology Opportunities Program (BTOP)	National Telecommunications and Information Administration, Dept. of Commerce	Provides competitive grants to public and private sector entities in order to provide broadband access in unserved and underserved areas; provide broadband support and services to strategic institutions; improve broadband access by public safety agencies; and stimulate broadband demand, economic growth, and job creation.	$4 billion (ARRA, P.L. 111-5) (2009)	http://www.ntia.doc.gov/ broadbandgrants/
Broadband Initiatives Program (BIP)	Rural Utilities Service, U.S. Dept. of Agriculture	Provides competitive grants, loans, and loan/grant combinations to public and private sector entities in order to provide broadband access in unserved and underserved rural areas.	$2.5 billion for the cost of loans, grants, and loan/grant combinations (ARRA, P.L. 111-5) (2009)	http://www.rurdev.usda. gov/utp_bip.html
Rural Broadband Access Loan and Loan Guarantee Program	Rural Utilities Service, U.S. Dept. of Agriculture	Provides loan and loan guarantees for facilities and equipment providing broadband service in rural communities	$169 million for cost of money loans	http://www.rurdev.usda. gov/utp_farmbill.html
Community Connect Broadband Grants	Rural Utilities Service, U.S. Dept. of Agriculture	Provides grants to applicants proposing to provide broadband service on a "community-oriented connectivity" basis to rural communities of under 20,000 inhabitants.	$10 million	http://www.rurdev.usda. gov/utp_commconnect.ht ml
		Telecommunications Broadband Infrastructure Deployment Programs		
Telecommunications Infrastructure Loan Program	*Rural Utilities Service, U.S. Dept. of Agriculture*	Provides long-term direct and guaranteed loans to qualified organizations for the purpose of financing the improvement, expansion, construction, acquisition, and operation of telephone lines, facilities, or systems to furnish and improve telecommunications service in rural areas. All facilities financed must be capable of supporting broadband services.	$145 million for hardship loans; $250 million for cost of money loans; and $295 million for FFB Treasury loans	http://www.rurdev.usda. gov/utp_infrastructure.ht ml

Program	Agency	Description	Funding Amount (est. FY2012 unless otherwise noted)	Web Links
Distance Learning and Telemedicine Loans and Grants	*Rural Utilities Service, U.S. Dept. of Agriculture*	Provides seed money to rural community facilities (e.g., schools, libraries, hospitals) for advanced telecommunications systems that can provide health care and educational benefits to rural areas	$21 million	http://www.rurdev.usda.gov/ UTP_DLT.html
Universal Service High Cost Program[a]	*Federal Communications Commission*	Provides funding to eligible telecommunications carriers to help pay for telecommunications services in high-cost, rural, and insular areas so that prices charged to customers are reasonably comparable across all regions of the nation.	$4.5 billion (annually through 2017)	http://www.usac.org/hc/
Universal Service Schools and Libraries Program (i.e., E-rate)	*Federal Communications Commission*	Provides discounts for affordable telecommunications and Internet access services to ensure that schools and libraries have access to affordable telecommunications and information services.	$2.25 billion (annually adjusted for inflation)	http://www.universalservice.org/sl/
Universal Service Rural Health Care Pilot Program[b]	*Federal Communications Commission*	Provides funds to cover 85% of the cost of constructing statewide or regional broadband telehealth networks and of connecting those projects to dedicated nationwide broadband telehealth networks and the public Internet.	$418 million (annually)	http://www.usac.org/rhc-pilot-program/
Appalachian Area Development Program	*Appalachian Regional Commission*	Project grants to support self-sustaining economic development in the region's most distressed counties and areas. Includes funds for a Telecommunications Initiative involving projects that enable communities to capitalize on broadband access.	$56 million	http://www.arc.gov/index.do?nodeId=21
States' Economic Development Assistance Program	*Delta Regional Authority*	Grants for self-sustaining economic development projects of eight states in Mississippi Delta region.	$11 million (2011)	http://grants.dra.gov/
Investments for Public Works and Economic Development Facilities	*Economic Development Administration, Dept. of Commerce*	Provides funding for construction of infrastructure in areas that are not attractive to private investment; most funding is for water and sewer infrastructure but some has been designated for telecommunications and broadband projects.	$112 million	http://www.eda.gov/

Table 4. (Continued)

Program	Agency	Description	Funding Amount (est. FY2012 unless otherwise noted)	Web Links
Library Services and Technology Act Grants to States	*Institute of Museum and Library Services, National Foundation on the Arts and the Humanities*	Provides funds for a wide range of library services including installation of fiber and wireless networks that provide access to library resources and services.	$156 million	http://www.imls.gov/pro grams/ programs.shtm
Native American Library Services	*Institute of Museum and Library Services, National Foundation on the Arts and the Humanities*	Grants to support library services including electronically linking libraries to networks.	$4 million	http://www.imls.gov/app licants/grants/nativeAme rican.shtm
Choice Neighborhood Implementation Grants	*Office of the Assistant Secretary for Public and Indian Housing and Office of Multifamily Housing Programs, Dept. of Housing and Urban Development*	Helps communities transform neighborhoods by revitalizing severely distressed public and/or assisted housing. Grantees may use funds to provide unit-based broadband Internet connectivity.	$110 million	http://www.hud.gov/cn/
Special Education—Technology and Media Services for Individuals with Disabilities	*Office of Special Education and Rehabilitative Services, Dept. of Education*	Supports development and application of technology and education media activities for disabled children and adults	$30 million	http://www.ed.gov/about /offices/list/osers/index.h tml?src=mr/
Telehealth Network Grants	*Health Resources and Services Administration, Department of Health and Human Services*	Grants to develop sustainable telehealth programs and networks in rural and frontier areas, and in medically unserved areas and populations.	$6 million	http://www.hrsa.gov/tele health/

Program	Agency	Description	Funding Amount (est. FY2012 unless otherwise noted)	Web Links
Telehealth Resource Center Grant Program	*Health Resources and Services Administration, Department of Health and Human Services*	Provides grants that support establishment and development of telehealth resource centers to assist health care providers in the development of telehealth services, including decisions regarding the purchase of advanced telecommunications services.	$4 million	http://www.hrsa.gov/tele health/
Licensure Portability Grant Program	*Health Resources and Services Administration, Department of Health and Human Services*	Provides support for state professional licensing boards to develop and implement state policies that will reduce statutory and regulatory barriers to telemedicine.	$0.35 million	http://www.hrsa.gov/tele health/
NLM Extramural Programs	*National Library of Medicine, National Institutes of Health,*	Provides funds to train professional personnel; strengthen library and information services; facilitate access to and delivery of	$62 million	http://www.nlm.nih.gov/ ep/extramural.html
	Department of Health and Human Services	health science information; plan and develop advanced information networks; support certain kinds of biomedical publications; and conduct research in medical informatics and related sciences.		
National Environmental Information Exchange Network Grant Program	*Environmental Protection Agency*	Provides funding to states, territories, and federally recognized Indian Tribes to support the development of an Environmental Information Exchange Network, including broadband infrastructure.	$10 million	http://epa.gov/exchangen etwork/grants/

Source: Compiled by CRS from FY2013 budget documents, the Catalog of Federal Domestic Assistance, and grants.

[a] The High Cost program will be phased out, and replaced in stages by the Connect America Fund. The program provides funding to eligible service providers to support the provision of affordable voice and broadband services, both fixed and mobile.

[b] The Rural Health Care program will be replaced by the Healthcare Connect Fund.

CONCLUSION

To the extent that the 113th Congress may consider various options for encouraging broadband deployment and adoption, a key issue is how to strike a balance between providing federal assistance for unserved and underserved areas where the private sector may not be providing acceptable levels of broadband service, while at the same time minimizing any deleterious effects that government intervention in the marketplace may have on competition and private sector investment.

In addition to loans, loan guarantees, and grants for broadband infrastructure deployment, a wide array of policy instruments are available to policymakers, including universal service reform, tax incentives to encourage private sector deployment, broadband bonds, demand-side incentives (such as assistance to low income families for purchasing computers), regulatory and deregulatory measures, and spectrum policy to spur roll-out of wireless broadband services. In assessing federal incentives for broadband deployment, the 113th Congress may consider the appropriate mix of broadband deployment incentives to create jobs in the short and long term, the extent to which incentives should target next-generation broadband technologies, the extent to which "underserved" areas with existing broadband providers should receive federal assistance, and whether broadband stimulus projects are being efficiently managed and how they may fit into the context of overall goals for a national broadband policy.

End Notes

[1] The term "digital divide" can also refer to international disparities in access to communications and information technology. This report focuses on domestic issues only.

[2] FCC, Internet Access Services: Status as of June 30, 2012, released May 2013, p. 17. Available at http://transition.fcc.gov/Daily_Releases/Daily_Business/2013/db0520/DOC-321076A1. pdf.

[3] Ibid.

[4] Initially, and for many years following, the FCC defined broadband (or more specifically "high-speed lines") as over 200 kilobits per second (kbps) in at least one direction, which was roughly four times the speed of conventional dialup Internet access. In recent years, the 200 kbps threshold was considered too low, and on March 19, 2008, the FCC adopted a report and order (FCC 08-89) establishing new categories of broadband speed tiers for data collection purposes. Specifically, 200 kbps to 768 kbps is considered "first generation," 768 kbps to 1.5 Mbps is "basic broadband tier 1," and increasingly higher speed tiers are broadband tiers 2 through 7 (tier seven is greater than or equal to 100 Mbps in any one direction). Tiers can change as technology advances.

[5] NTIA, U.S. Broadband Availability: June 2010 – June 2012, May 2013, p.5.available at http://www.ntia.doc.gov/files/ ntia/publications/usbb_avail_report_05102013.pdf.

[6] Federal Communications Commission, Eighth Broadband Progress Report, FCC 12-90, released August 21, 2012, p. 29, available at http://transition.fcc.gov/Daily_Releases/Daily_ Business/2012/db0827/FCC-12-90A1.pdf.

[7] U.S. Department of Commerce, National Telecommunications and Information Administration, Blog, "Household Broadband Adoption Climbs to 72.4 Percent," June 6, 2013, available at http://www.ntia.doc.gov/blog/2013 /householdbroadband-adoption-climbs-724-percent.

[8] Eighth Broadband Progress Report, p. 54.

[9] U.S. Department of Commerce, National Telecommunications and Information Administration, Exploring the Digital Nation: America's Emerging Online Experience, June 2013, p. 26, available at http://www.ntia.doc.gov/files/ntia/ publications/exploring_the_digital_nation_-_americas_emerging_online_ experience.pdf.

[10] Smith, Aaron, Pew Internet & American Life Project, Home Broadband 2010, August 11, 2010, available at http://www.pewinternet.org/~/media//Files/Reports/2010/ Home% 20broadband%202010.pdf.

[11] For more information on rural broadband and broadband programs at the Rural Utilities Service, see CRS Report RL33816, Broadband Loan and Grant Programs in the USDA's Rural Utilities Service, by Lennard G. Kruger.

[12] See for example: National Exchange Carrier Association (NECA), Trends 2006: Making Progress With Broadband, 2006, 26 p. Available at http://www.neca.org/media/trends_ brochure_website.pdf.

[13] Federal Communications Commission, Eighth Broadband Progress Report, FCC 12-90, released August 21, 2012, p. 5, available at http://transition.fcc.gov/Daily_Releases/ Daily_Business/2012/db0827/FCC-12-90A1.pdf.

[14] NTIA, National Broadband Map, Broadband Statistics Report: Broadband Availability in Urban vs. Rural Areas, January 2013, p. 7, available at http://www.broadbandmap.gov/ download/ Broadband%20Availability%20 in%20Rural%20vs%20Urban%20Areas.pdf. Also see NTIA, U.S. Broadband Availability: June 2010–June 2012, May 2013, p. 10-11, available at http://www.ntia.doc.gov/files/ntia/publications/ usbb_avail_report_ 05102013. pdf.

[15] U.S. Department of Commerce, National Telecommunications and Information Administration, Exploring the Digital Nation: America's Emerging Online Experience, June 2013, p. 26, available at http://www.ntia.doc.gov/files/ ntia/publications/exploring_ the_digital_nation_-_americas_emerging_online_ experience.pdf.

[16] The terrain of rural areas can also be a hindrance to broadband deployment because it is more expensive to deploy broadband technologies in a mountainous or heavily forested area. An additional added cost factor for remote areas can be the expense of "backhaul" (e.g., the "middle mile") which refers to the installation of a dedicated line which transmits a signal to and from an Internet backbone which is typically located in or near an urban area.

[17] Gillett, Sharon E., Massachusetts Institute of Technology, Measuring Broadband's Economic Impact, report prepared for the Economic Development Administration, U.S. Department of Commerce, February 28, 2006, p. 4.

[18] Crandall, Robert, William Lehr, and Robert Litan, The Effects of Broadband Deployment on Output and Employment: A Cross-sectional Analysis of U.S. Data, June 2007, 20 pp. Available at http://www3.brookings.edu/ views/papers/crandall/200706litan.pdf.

[19] Mark Dutz, Jonathan Orszag, and Robert Willig, The Substantial Consumer Benefits of Broadband Connectivity for U.S. Households, Internet Innovation Alliance, July 2009, p. 4, http://internetinnovation.org/files/special-reports/ CONSUMER_BENEFITS_OF_BROADBAND.pdf.

[20] OECD, OECD Broadband Portal. Available at http://www.oecd.org/sti/ict/broadband.

[21] OECD, Directorate for Science, Technology and Industry, The Development of Broadband Access in OECD Countries, October 29, 2001, 63 pp. For a comparison of government broadband policies, also see OECD, Directorate for Science, Technology and Industry, Broadband Infrastructure Deployment: The Role of Government Assistance, May 22, 2002, 42 pp.

[22] See Turner, Derek S., Free Press, Broadband Reality Check II: The Truth Behind America's Digital Divide, August 2006, pp. 8-11. Available at http://www.freepress.net/files/bbrc2-final.pdf; and Turner, Derek S., Free Press, 'Shooting the Messenger' Myth vs. Reality: U.S. Broadband Policy and International Broadband Rankings, July 2007, 25 pp., available at http://www.freepress.net/files/shooting_the_messenger.pdf.

[23] National Telecommunications and Information Administration, Fact Sheet: United States Maintains Information and Communication Technology (ICT) Leadership and Economic Strength, at http://www.ntia.doc.gov /ntiahome/press/ 2007/ICTleader_042407.html.

[24] See Wallsten, Scott, Progress and Freedom Foundation, Towards Effective U.S. Broadband Policies, May 2007, 19 pp. Available at http://www.pff.org/issues-pubs/pops/pop14.7 usbroadbandpolicy.pdf. Also see Ford, George, Phoenix Center, The Broadband Performance Index: What Really Drives Broadband Adoption Across the OECD?, Phoenix Center Policy Paper Number 33, May 2008, 27 pp.; available at http://www.phoenix-center.org/pcpp/PCPP33Final.pdf.

[25] See price and services and speed data on OECD Broadband Portal, available at http://www.oecd.org/sti/ict/broadband; see also Federal Communications Commission, Third Annual International Broadband Data Report, IB Docket No. 10-171, DA 12-1334, August 21, 2012, available at http://hraunfoss.fcc.gov/edocs_public/attachmatch/DA12-1334A1.pdf.

[26] Federal Communications Commission, Notice of Inquiry, "Concerning the Deployment of Advanced Telecommunications Capability to All Americans in a Reasonable and Timely Fashion, and possible Steps to

Accelerate Such Deployment Pursuant to Section 706 of the Telecommunications Act of 1996," FCC 04-55, March 17, 2004, p. 6.

[27] FCC News Release, FCC Improves Data Collection to Monitor Nationwide Broadband Rollout, November 9, 2004. Available at http://hraunfoss.fcc.gov/edocs_public/ attachmatch/DOC-254115A1.pdf.

[28] Federal Communications Commission, Notice Proposed Rulemaking, "Development of Nationwide Broadband Data to Evaluate Reasonable and Timely Deployment of Advanced Services to All Americans, Improvement of Wireless Broadband Subscribership Data, and Development of Data on Interconnected Voice Over Internet Protocol (VoIP) Subscribership," WC Docket No. 07-38, FCC 07-17, released April 16, 2007, 56 pp.

[29] FCC, News Release, "FCC Expands, Improves Broadband Data Collection," March 19, 2008. Available at http://hraunfoss.fcc.gov/edocs_public/attachmatch/DOC-280909A1.pdf.

[30] For more information on the national broadband mapping program and the State Broadband Data and Development Program, see http://www.ntia.doc.gov/broadband grants/ broadbandmapping.html.

[31] Eighth Broadband Progress Report, p. 59-60.

[32] Ibid., p.171, 177.

[33] Available at http://www.broadband.gov/plan/For more information on the National Broadband Plan, see CRS Report R41324, The National Broadband Plan, by Lennard G. Kruger et al..

[34] Federal Communications Commission, Connecting America: The National Broadband Plan, March 17, 2010, p. 9.

[35] Ibid., p. 11.

[36] Ibid.

[37] Ibid.

[38] See for example, Office of Science and Technology Policy and National Economic Council, The White House, Four Years of Broadband Growth, June 2013, 26 pages, available at http://www.whitehouse.gov /sites/default/files/ broadband_report_final.pdf.

[39] See CRS Report R40436, Broadband Infrastructure Programs in the American Recovery and Reinvestment Act, by Lennard G. Kruger.

[40] For more information on these programs, see CRS Report RL33816, Broadband Loan and Grant Programs in the USDA's Rural Utilities Service, by Lennard G. Kruger.

[41] Ibid.

[42] See CRS Report R42524, Rural Broadband: The Roles of the Rural Utilities Service and the Universal Service Fund, by Angele A. Gilroy and Lennard G. Kruger.

[43] The section on universal service was prepared by Angele Gilroy, Specialist in Telecommunications, Resources, Science and Industry Division. For more information on universal service, see CRS Report RL33979, Universal Service Fund: Background and Options for Reform, by Angele A. Gilroy.

[44] Communications Act of 1934, As Amended, Title I §1 [47 U.S.C. 151].

[45] Many states participate in or have programs that mirror FCC universal service mechanisms to help promote universal service goals within their states.

[46] For additional information on this program, including funding commitments, see the E-rate website: http://www.universalservice.org/sl/.

[47] For additional information on this program, including funding commitments, see the RHCD website: http://www.universalservice.org/rhc/.

[48] For more details on the USF rural health care support mechanism and the newly established Healthcare Connect Fund see In the Matter of Rural Health Care Support Mechanism, WC Docket No. 02-60, Federal Communications Commission, adopted December 12, 2012. Available at http://hraunfoss.fcc.gov/edocs_ public /attachmatch/FCC-12- 150A1.pdf.

[49] The Joint Board recommended that the definition of those services that qualify for universal service support be expanded and that the nation's communications goals include the universal availability of: mobility services (i.e., wireless voice); broadband Internet services; and voice services at affordable and comparable rates for all rural and non-rural areas. For a copy of this recommendation see http://hraunfoss.fcc.gov/edocs_public/ attachmatch/FCC-07J4A1.pdf.

[50] American Recovery and Reinvestment Act of 2009, P.L. 111-5, sec. 6001 (k)(2)(D).

[51] For a detailed discussion of this Order and USF transition see CRS Report R42524, Rural Broadband: The Roles of the Rural Utilities Service and the Universal Service Fund, by Angele A. Gilroy and Lennard G. Kruger.

[52] For information on existing broadband programs at RUS, see CRS Report RL33816, Broadband Loan and Grant Programs in the USDA's Rural Utilities Service, by Lennard G. Kruger.

[53] For information on public safety wireless broadband legislation, see CRS Report R41842, *Funding Emergency Communications: Technology and Policy Considerations*, by Linda K. Moore.

In: Transformations in Telecommunications and Media
Editor: Irwin Cavazos
ISBN: 978-1-62948-413-6
© 2013 Nova Science Publishers, Inc.

Chapter 3

ACCESS TO BROADBAND NETWORKS: THE NET NEUTRALITY DEBATE[*]

Angele A. Gilroy

SUMMARY

As congressional policymakers continue to debate telecommunications reform, a major point of contention is the question of whether action is needed to ensure unfettered access to the Internet. The move to place restrictions on the owners of the networks that compose and provide access to the Internet, to ensure equal access and non-discriminatory treatment, is referred to as "net neutrality." While there is no single accepted definition of "net neutrality," most agree that any such definition should include the general principles that owners of the networks that compose and provide access to the Internet should not control how consumers lawfully use that network, and they should not be able to discriminate against content provider access to that network.

A major focus in the debate is concern over whether it is necessary for policymakers to take steps to ensure access to the Internet for content, services, and applications providers, as well as consumers, and if so, what these steps should be. Some policymakers contend that more specific regulatory guidelines may be necessary to protect the marketplace from potential abuses which could threaten the net neutrality concept. Others contend that existing laws and policies are sufficient to deal with potential anti-competitive behavior and that additional regulations would have negative effects on the expansion and future development of the Internet. The December 21, 2010, adoption, and November 20, 2011, implementation, by the Federal Communications Commission (FCC) of its Open Internet Order has focused attention on the issue. Although most concede that networks have always needed and will continue to need some management, the use of prioritization tools, such as deep packet inspection, as well as the initiation of metered/usagebased billing practices have further fueled the debate.

A consensus on the net neutrality issue has remained elusive and support for the FCC's Open Internet Order has been mixed. While some Members of Congress support the action, and in some cases would have supported an even stronger approach, others

---

[*] This is an edited, reformatted and augmented version of Congressional Research Service, Publication No. R40616, dated July 22, 2013.

feel that the FCC has overstepped its authority and that the regulation of the Internet is not only unnecessary, but harmful. Internet regulation and the FCC's authority to implement such regulations has been a topic of legislation (H.R. 96, H.R. 166, S. 74, H.R. 2434, H.R. 1, H.R. 3630, H.J.Res. 37, and S.J.Res. 6) and hearings in the 112th Congress. The House, on April 8, 2011, passed (240-179) H.J.Res. 37, to state disapproval of and remove the force and effect of the FCC's Open Internet Order. However, an identical resolution of disapproval (S.J.Res. 6) failed to pass the Senate on November 10, 2011, by a 52-46 vote. Attempts to prohibit implementation through the appropriations process, by the withholding of FCC funds for such purposes, have also been unsuccessful. It is anticipated that the issue of Internet access will be of continued interest to policymakers.

The net neutrality issue has also been narrowly addressed within the context of the American Recovery and Reinvestment Act of 2009 (ARRA, P.L. 111-5). Provisions required the National Telecommunications and Information Administration (NTIA), in consultation with the FCC, to establish "nondiscrimination and network interconnection obligations" as a requirement for grant participants in the Broadband Technology Opportunities Program (BTOP). These obligations were released, July 1, 2009, in conjunction with the issuance of a notice of funds availability soliciting applications. Recipients of these awards have been selected and continued congressional oversight is expected.

INTRODUCTION

As congressional policymakers continue to debate telecommunications reform, a major point of contention is the question of whether action is needed to ensure unfettered access to the Internet. The move to place restrictions on the owners of the networks that compose and provide access to the Internet, to ensure equal access and non-discriminatory treatment, is referred to as "net neutrality." There is no single accepted definition of "net neutrality." However, most agree that any such definition should include the general principles that owners of the networks that compose and provide access to the Internet should not control how consumers lawfully use that network, and they should not be able to discriminate against content provider access to that network.

What, if any, action should be taken to ensure "net neutrality" has become a major focal point in the debate over broadband regulation. As the marketplace for broadband continues to evolve, some contend that no new regulations are needed, and if enacted will slow deployment of and access to the Internet, as well as limit innovation. Others, however, contend that the consolidation and diversification of broadband providers into content providers has the potential to lead to discriminatory behaviors which conflict with net neutrality principles. The two potential behaviors most often cited are the network providers' ability to control access to and the pricing of broadband facilities, and the incentive to favor network-owned content, thereby placing unaffiliated content providers at a competitive disadvantage.

FEDERAL COMMUNICATIONS COMMISSION ACTIVITY

The Information Services Designation and Title I

In 2005 two major actions dramatically changed the regulatory landscape as it applied to broadband services, further fueling the net neutrality debate. In both cases these actions led to the classification of broadband Internet access services as Title I information services, thereby subjecting them to a less rigorous regulatory framework than those services classified as telecommunications services. In the first action, the U.S. Supreme Court, in a June 2005 decision (*National Cable & Telecommunications Association v. Brand X Internet Services*), upheld the Federal Communications Commission's (FCC's) 2002 ruling that the provision of cable modem service (i.e., cable television broadband Internet) is an interstate information service and is therefore subject to the less stringent regulatory regime under Title I of the Communications Act of 1934.[1] In a second action, the FCC, in an August 5, 2005, decision, extended the same regulatory relief to telephone company Internet access services (i.e., wireline broadband Internet access, or DSL), thereby also defining such services as information services subject to Title I regulation.[2] As a result neither telephone companies nor cable companies, when providing broadband services, are required to adhere to the more stringent regulatory regime for telecommunications services found under Title II (common carrier) of the 1934 act.[3] However, classification as an information service does not free the service from regulation. The FCC continues to have regulatory authority over information services under its Title I, ancillary jurisdiction.[4] Similarly classification under Title II does not mean that an entity will be subject to the full range of regulatory requirements as the FCC is given the authority, under Section 10 of the Communications Act of 1934, to forbear from regulation.

The 2005 Internet Policy Statement

Simultaneous to the issuing of its August 2005 information services classification order, the FCC also adopted a policy statement (Internet Policy Statement) outlining four principles to "encourage broadband deployment and preserve and promote the open and interconnected nature of [the] public Internet." The four principles are (1) consumers are entitled to access the lawful Internet content of their choice; (2) consumers are entitled to run applications and services of their choice (subject to the needs of law enforcement); (3) consumers are entitled to connect their choice of legal devices that do not harm the network; and (4) consumers are entitled to competition among network providers, application and service providers, and content providers. Then-FCC Chairman Martin did not call for their codification. However, he stated that they would be incorporated into the policymaking activities of the commission.[5] For example, one of the agreed upon conditions for the October 2005 approval of both the Verizon/MCI and the SBC/AT&T mergers was an agreement made by the involved parties to commit, for two years, "to conduct business in a way that comports with the commission's (2005) Internet policy statement."[6] In a further action AT&T included in its concessions to gain FCC approval of its merger to BellSouth to adhering, for two years, to significant net neutrality requirements. Under terms of the merger agreement, which was approved on

December 29, 2006, AT&T agreed to not only uphold, for 30 months, the FCC's Internet policy statement principles, but also committed, for two years (expired December 2008), to stringent requirements to "maintain a neutral network and neutral routing in its wireline broadband Internet access service."[7]

FCC Chairman Genachowski announced, in a September 21, 2009, speech,[8] a proposal to consider the expansion and codification of the 2005 Internet Policy Statement and suggested that this be accomplished through a notice of proposed rulemaking (NPR) process. Shortly thereafter an NPR on preserving the open Internet and broadband industry practices was adopted by the FCC in its October 22, 2009, meeting. (See "The FCC Open Internet Order," below.)

The FCC August 2008 Comcast Decision

In perhaps one of its most significant actions relating to its Internet Policy Statement to date, the FCC, on August 1, 2008, ruled that Comcast Corp., a provider of Internet access over cable lines, violated the FCC's policy statement when it selectively blocked peer-to-peer connections in an attempt to manage its traffic.[9] This practice, the FCC concluded, "unduly interfered with Internet users' rights to access the lawful Internet content and to use the applications of their choice." Although no monetary penalties were imposed, Comcast was required to stop these practices by the end of 2008. Comcast complied with the order, and developed a new system to manage network congestion. Comcast no longer manages congestion by focusing on specific applications (such as peer-to-peer), nor by focusing on online activities, or protocols, but identifies individual users within congested neighborhoods that are using large amounts of bandwidth in real time and slows them down, by placing them in a lower priority category, for short periods.[10] This new system complies with the FCC Internet principles in that it is application agnostic; that is, it does not discriminate against or favor one application over another but manages congestion based on the amount of a user's real-time bandwidth usage. As a result of a April 6, 2010, court ruling the FCC's order was vacated. Comcast, however, has stated that it will continue to comply with the Internet principles issued in the FCC's August 2005 Internet policy statement.[11] (See "Comcast v. FCC," below.)

Comcast v. FCC

Despite compliance, however, Comcast filed an appeal[12] in the U.S. Court of Appeals for the District of Columbia, claiming that the FCC did not have the authority to enforce its Internet policy statement, therefore making the order invalid. The FCC argued that while it did not have express statutory authority over such practices, it derived such authority based on its ancillary authority contained in Title I of the 1934 Communications Act.[13] The court, in an April 6, 2010, decision, ruled (3-0) that the FCC did not have the authority to regulate an Internet service provider's (in this case Comcast's) network management practices and vacated the FCC's order.[14] The court ruled that the exercise of ancillary authority must be linked to statutory authority and that the FCC did not in its arguments prove that connection; it cannot exercise ancillary authority based on policy alone. More specifically, the Court ruled that the FCC "failed to tie its assertion of ancillary authority over Comcast's Internet service to any ["statutorily mandated responsibility"]."[15] Based on that conclusion the court granted the petition for review and vacated the order.

The impact of this decision on the FCC's ability to regulate broadband services and implement its broadband policy goals remains unclear. Regardless of the path that is taken FCC Chairman Genachowski has stated that the court decision "does not change our broadband policy goals, or the ultimate authority of the FCC to act to achieve those goals." He further stated that "[T]he court did not question the FCC's goals; it merely invalidated one, technical, legal mechanism for broadband policy chosen by prior Commissions."[16] Consistent with this statement, the FCC in a December 21, 2010, action adopted the Open Internet Order to establish rules to maintain network neutrality (see "The FCC Open Internet Order").

The FCC Open Internet Order

The FCC adopted, on December 21, 2010, an Open Internet Order establishing rules to govern the network management practices of broadband Internet access providers.[17] The order, which was passed by a 3-2 vote,[18] intends to maintain network neutrality by establishing three rules covering transparency,[19] no blocking, and no unreasonable discrimination. More specifically:

- fixed and mobile broadband Internet service providers are required to publically disclose accurate information regarding network management practices, performance, and commercial terms to consumers and as well as content, application, service, and device providers;
- fixed and mobile broadband Internet service providers are both subject, to varying degrees, to no blocking requirements. Fixed providers are prohibited from blocking lawful content, applications, services, or non-harmful devices, subject to reasonable network management. Mobile providers are prohibited from blocking consumers from accessing lawful websites, subject to reasonable network management, nor can they block applications that compete with the provider's voice or video telephony services, subject to reasonable network management;
- fixed broadband Internet service providers are subject to a "no unreasonable discrimination rule" that states that they shall not unreasonably discriminate in transmitting lawful network traffic over a consumer's broadband Internet access service. Reasonable network management shall not constitute unreasonable discrimination.[20]

Additional provisions in the order include those which provide for ongoing monitoring of the mobile broadband sector and create an Open Internet Advisory Committee[21] to track and evaluate the effects of the rules and provide recommendations to the FCC regarding open Internet policies and practices; while not banning paid prioritization, state it is unlikely to satisfy the "no unreasonable discrimination" rule; raise concerns about specialized services and while not "adopting policies specific to such services at this time," will closely monitor such services; call for review, and possible adjustment, of all rules in the order no later than two years from their effective date; and detail a formal and informal complaint process. The order, however, does not prohibit tiered or usage-based pricing (see "Metered/Usage-Based Billing," below). According to the order, the framework "... does not prevent broadband providers from asking subscribers who use the network less to pay less, and subscribers who use the network more to pay more" since prohibiting such practices "... would force lighter end users of the network to subsidize heavier end users" and "... would also foreclose practices that may appropriately align incentives to encourage efficient use of networks."[22]

The authority to adopt the order abandons the "third way approach" previously endorsed by Chairman Genachowski and other Democratic commissioners,[23] and treats broadband Internet access service as an information service under Title I. The order relies on a number of provisions contain in the 1934 Communications Act, as amended, to support FCC authority. According to the order the authority to implement these rules lies in Section 706 of the 1996 Telecommunications Act, which directs the FCC to "encourage the deployment on a reasonable and timely basis" of "advanced telecommunications capability" to all Americans and to take action if it finds that such capability is not being deployed in a reasonable and timely fashion;[24] Title II of the Communications Act and its role in protecting competition and consumers of telecommunications services; Title III, which gives the FCC the authority to license spectrum, subject to terms that serve the public interest, used to provide fixed and mobile wireless services; and Title VI, which gives the FCC the duty to protect competition in video services.

The order went into effect November 20, 2011, which was 60 days after its publication in the *Federal Register.*[25] Since the Order's publication multiple appeals were filed and subsequently consolidated for review in the U.S. Court of Appeals, D.C. Circuit.[26] Verizon Communications is the remaining challenger seeking review[27] claiming, among issues, that it is a violation of free speech and that the FCC has exceeded its authority in establishing the rules.[28] Oral arguments are scheduled for September 9, 2013.

The American Recovery and Reinvestment Act of 2009

The FCC has also been called upon to address net neutrality principles within the context of the implementation of the American Recovery and Reinvestment Act of 2009 (ARRA, P.L. 111-5). Provisions require the National Telecommunications and Information Administration (NTIA), in consultation with the FCC, to establish "nondiscrimination and network interconnection obligations" as a requirement for grant participants in the Broadband Technology Opportunities Program (BTOP). These obligations were issued July 1, 2009, in conjunction with the release of the notice of funds availability (NOFA) soliciting applications for the program.[29] The NOFA requires that recipients of both ARRA programs (the Rural Utilities Service Broadband Initiative Program (BIP) as well as the mandated BTOP program) adhere to these requirements,[30] and expands requirements beyond those contained in the FCC's 2005 Internet Policy Statement. More specifically award recipients are required to adhere to the FCC's 2005 Internet Policy Statement; not favor any lawful Internet applications and content over others; display network management policies on their web pages and provide notice to customers of changes to these policies; connect to the public Internet directly or indirectly (that is, the project cannot be an entirely private closed network); and "offer interconnection, where technically feasible without exceeding current or reasonably anticipated capacity limitations, on reasonable rates and terms to be negotiated with requesting parties." Recipients of these awards have been selected, projects are being deployed, and congressional oversight is ongoing.

The FCC's National Broadband Plan

The ARRA also required the FCC to submit a report, containing a national broadband plan, to both the House and Senate Commerce Committees. The report, *Connecting America: The National Broadband Plan* (NBP), was released on March 16, 2010.[31] The NBP did not contain specific recommendations regarding the debate over access to broadband networks, but Chapter 4 did discuss the value of an open Internet. The NBP referred to the FCC's then-ongoing notice of proposed rulemaking on Preserving the Open Internet (see "The FCC Open Internet Order," above) and stated that "broadband's ability to derive the many benefits discussed in this plan depend[s] on its continued openness."[32]

One other issue relevant to the open access/net neutrality debate focuses on the regulatory classification of broadband services. Chapter 17 of the NBP provides a discussion of the legal framework for the plan's implementation. While the discussion does not reach any conclusions regarding the appropriate framework, it does outline the debate over whether broadband services should retain its Title I classification as an information service, or should be classified as a telecommunications service under Title II.[33] (See "The Information Services Designation and Title I," above.) While the NBP does not reach a conclusion regarding classification, some feel it does open up the door for discussion[34] by concluding that "the FCC will consider these and related questions as it moves forward to implement the plan."[35] Since the NBP's release, however, the FCC, in its Open Internet Order, adopted in December 2010, concluded that such services would remain under Title I classification. (See "The FCC Open Internet Order," above.)

Additional Activity

In a June 17, 2010, action the FCC adopted a notice of inquiry (NOI), which is still pending, to examine the framework for broadband Internet service. The NOI (General Docket No.10-127) seeks comment on issues such as broadband Internet classification, and the proper role of the states with respect to broadband Internet service.[36] Separately, in an April 2007 action, the FCC released a notice of inquiry (WC Docket No. 07-52), on broadband industry practices seeking comment on a wide range of issues including whether the August 2005 Internet policy statement should be amended to incorporate a new principle of nondiscrimination and if so, what form it should take.[37] On January 14, 2008, the FCC issued three public notices seeking comment on issues related to network management (including the now-completed Comcast ruling, discussed above) and held two (February 25 and April 17, 2008) public hearings specific to broadband network management practices.

Certain restrictions on the operation and management of Comcast's Internet facilities were agreed to as a condition of the January 18, 2011, approval by the Department of Justice (DOJ) and the FCC, of the merger between Comcast Corp. and NBC Universal Inc.[38] For example, Section V.G of the DOJ Final Judgment enumerates restrictions that Comcast has agreed to abide by regarding its Internet facilities. Open access requirements, consistent with the FCC's Open Internet Order, were agreed to as part of the settlement. More specifically, Comcast is prohibited from unreasonably discriminating in the transmission of an OVD's (online video distributors) lawful network traffic to a Comcast broadband customer.[39] Additional restrictions include those which: prohibit Comcast from excluding its own services from any caps, tiers, metering, or other usage based plans and requires that OVD traffic be counted in the same way as Comcast's traffic to ensure that billing plans are not used to disadvantage an OVD; prohibits Comcast from offering specialized services that are comprised substantially or entirely of its own or its affiliates services; and if offering specialized services must offer similar specialized services on a nondiscriminatory basis. The DOJ Final judgment and the FCC Order stay in force for seven years (January 2018).

INDUSTRY INITIATIVES

Industry stakeholders have also taken the initiative to address broadband policy issues by establishing voluntary discussion groups and frameworks to further the debate. For example, a voluntary working group comprised of Internet service providers, content, applications, hardware makers, and community representatives announced the establishment of a technical advisory group of engineers to address technical issues surrounding the net neutrality debate. The major mission of this working group, called the Broadband Internet Technical Advisory Group (BITAG), is to develop consensus on voluntary industry guidelines to address industry technical standards relating to broadband network management practices or other related issues that can affect users' Internet experience. The BITAG mission could also include "(1) educating policymakers on technical issues; (2) attempting to address specific technical matters in an effort to minimize related policy disputes; and (3) serving as a sounding board for new ideas and network management practices."[40] BITAG, an independent non-profit organization, announced on December 16, 2010, the appointment of an interim board of directors and the commencement of a Technical Working Group to address substantive issues.[41]

Two major stakeholders, Verizon and Google, announced on August 9, 2010, a proposal containing a suggested "open Internet framework for the consideration of policymakers and the public."[42] Some of the key elements of the proposal, which was offered in the form of a suggested "legislative framework," include

- broadband Internet access service providers would be prohibited from preventing their users from sending and receiving lawful content of their choice, running lawful applications and using lawful services of their choice, and connecting their choice of legal devices;
- broadband Internet access providers would be prohibited from engaging in undue discrimination against any lawful Internet content, application, or service that causes meaningful harm to competition or users;
- providers of broadband Internet access service would be subject to disclosure and transparency requirements so that consumers and others could make informed choices;

- broadband Internet access service providers are permitted to engage in reasonable network management;
- a provider who is complying with these principles could offer any other additional or differentiated services that could include traffic prioritization;
- the FCC would enforce consumer protection and nondiscrimination requirements on a case-by-case basis and could impose a forfeiture of up to $2 million for knowing violations;
- the FCC would have exclusive authority over broadband Internet access service but would have no authority over Internet software applications, content, or services;
- broadband Internet access service and traffic or services using Internet protocol would be considered exclusively interstate in nature;
- broadband Internet access would be eligible for Federal universal service support to spur deployment in unserved areas and adoption by low-income populations; and
- wireless networks would only be subject to the transparency principle at this time.

Industry stakeholders have also participated in talks conducted by the FCC and designated congressional committees of jurisdiction. The FCC talks, which consisted of a series of meetings with various industry stakeholders to discuss communications issues with a particular focus on the broadband reclassification issue, concluded in the summer of 2010, without reaching a consensus. Congressional sessions held in the 111th Congress, by the Senate Commerce and the House Energy and Commerce Committees and their Communications Subcommittees, covered the topics of broadband regulation/consumer protection and FCC authority; spectrum policy; and broadband deployment and adoption; no further action was taken.

NETWORK MANAGEMENT

As consumers expand their use of the Internet and new multimedia and voice services become more commonplace, control over network quality and pricing is an issue. The ability of data bits to travel the network in a nondiscriminatory manner (subject to reasonable management practices), as well as the pricing structure established by broadband service providers for consumer access to that data, have become significant issues in the debate.

Prioritization

In the past, Internet traffic has been delivered on a "best efforts" basis. The quality of service needed for the delivery of the most popular uses, such as e-mail or surfing the web, is not as dependent on guaranteed quality. However, as Internet use expands to include video, online gaming, and voice service, the need for uninterrupted streams of data becomes important. As the demand for such services continues to expand, network broadband operators are moving to prioritize network traffic to ensure the quality of these services. Prioritization may benefit consumers by ensuring faster delivery and quality of service and may be necessary to ensure the proper functioning of expanded service options. However, the move on the part of network operators to establish prioritized networks, although embraced by some, has led to a number of policy concerns.

There is concern that the ability of network providers to prioritize traffic may give them too much power over the operation of, and access to, the Internet. If a multi-tiered Internet develops where content providers pay for different service levels, the potential to limit competition exists if smaller, less financially secure content providers are unable to afford to pay for a higher level of access. Also, if network providers have control over who is given priority access, the ability to discriminate among who gets such access is also present. If such a scenario were to develop, the potential benefits to consumers of a prioritized network would be lessened by a decrease in consumer choice and/or increased costs, if the fees charged for premium access are passed on to the consumer. The potential for these abuses, however, is significantly decreased in a marketplace where multiple, competing broadband providers exist. If a network broadband provider blocks access to content or charges unreasonable fees, in a competitive market, content providers and consumers could obtain their access from other network providers. As consumers and content providers migrate to these competitors, market share and profits of the offending network provider will decrease, leading to corrective action or failure. However, this scenario assumes that every market will have a number of equally competitive broadband options from which to choose, and all competitors will have equal access to, if not identical, at least comparable content.

Deep Packet Inspection

The use of one management tool, deep packet inspection (DPI), illustrates the complexity of the net neutrality debate. DPI refers to a network management technique that enables network operators to inspect, in real time, both the header and the data field of the packets.[43] As a result DPI can allow network operators to not only identify the origin and destination points of the data packet, but also enables the network operator to determine the application used and content of that packet. The information that DPI provides enables the network operator to differentiate, or discriminate, among the packets travelling over its network. The ability to discriminate among packets enables the network operator to treat packets differently. This ability itself is not necessarily viewed in a negative light. Network managers use DPI to assist them in performing various functions that are necessary for network management and that contribute to a positive user experience. For example, DPI technology is used in filters and firewalls to detect and prevent spam, viruses, worms, and malware. DPI is also used to gain information to help plan network capacity and diagnostics, as well as to respond to law enforcement requests.[44] However, the ability to discriminate based on the information gained via DPI also has the potential to be misused.[45] It is the potential negative impact that DPI use can have on consumers and suppliers that raises concern for policymakers. For example, the information gained could be used to discriminate against a competing service causing harm to both the competitor and consumer choice. This could be accomplished by routing a network operator's own, or other preferred content, along a faster priority path, or selectively slowing down competitor's traffic. DPI also has the potential to extract personal information about the data that it inspects, generating concerns about consumer privacy.[46]

Therefore it is not the management tool itself that is under scrutiny, but how it is applied. The DPI technology, in itself, is not what is of concern. It is the behavior that potentially may occur as a result of the information that DPI provides. How to develop a policy that permits some types of discrimination (i.e., "good" discrimination) that may be beneficial to network operation and improve the user experience, while protecting against what would be considered "harmful" or anticompetitive discrimination becomes the crux of the policy debate.

Metered/Usage-Based Billing

The move by some network broadband operators towards the use of metered or usage-based billing has caused considerable controversy. Under such a plan, users subscribe to a set monthly bandwidth cap, for an established fee, and are charged additional fees or could be denied service, if that usage level is exceeded. The use of such billing practices, on both a trial and permanent basis, is becoming more commonplace. Comcast announced the adoption of usage caps for all of its residential customers effective October 1, 2008.[47] Comcast amended its Acceptable Use Policy to establish a specific monthly data usage threshold of 250 GB/month per account for all Xfinity Internet residential customers. Usage above that cap would be considered "excessive" and Comcast will notify and ask the subscriber to moderate their usage.[48] However in May 2012 Comcast announced that it is replacing its 250 GB/per month usage threshold with new flexible usage trials in selected test markets and suspending enforcement of the 250 GB/per month cap in all remaining markets.[49] AT&T adopted usage caps, effective May 2, 2011, for its DSL and U-Verse residential subscribers. DSL subscribers will be subject to a 150 GB/per month usage cap and U-Verse subscribers will be subject to a 250 GB/per month data usage cap. Subscribers who exceed the cap three times across the life of the account, not per month, must pay $10 per every 50 GB above the subscribed cap.[50] Similarly CenturyLink announced effective February 2012, the decision to place download limits (or caps) on its residential high-speed Internet plans. Usage caps will vary based on the subscriber's plan with a monthly maximum of 150 GB for the 1.5 Mbps plan and a monthly cap of 250 GB for plans greater than 1.5 Mbps. CenturyLink will contact those who exceed their usage caps, allow you time to reduce your usage, or allow you to upgrade to a higher data service. There are no overage fees or charges for exceeded usage but CenturyLink reserves the right to disconnect service after the third month of excessive usage in a rolling 12-month period.[51]

Some Internet service providers have also initiated usage trials. For example, in 2008, Time Warner Cable established a usage trial in Beaumont, TX, that offered a range of service tiers. The move by Time Warner Cable to expand these trials to four additional locations[52] caused considerable controversy and was deferred.[53] Since then, however, Time Warner has initiated a voluntary usage-based trial in its southern Texas markets. This trial, which was announced in February 2012, addresses customers who are "light users" by offering an optional usage-based plan, called "Essentials." Subscribers can use up to 5 GB per month for a $5 discount from the customer's current bill; if they exceed the cap they will be charged $1 for every additional gigabyte, with an overage fee cap of $25 per month. A usage meter, capable of tracking consumption hour by hour, is provided. This is a voluntary trial and an unlimited option, at a flat monthly rate, will continue to be offered and customers will be permitted to switch back and forth between options.[54] Smaller, more regional providers have stated that usage-based pricing models are growing in popularity and will be necessary in the future as the demand for high bandwidth applications increases.[55] For example, one provider, Knology of Kansas, uses such a pricing model. Knology of Kansas offers three service levels at 3, 50, or 250 GB per month, with a $1 per Gigabyte overcharge which is levied only after a second over usage.[56]

Reaction to the imposition of data usage caps has been mixed. Supporters of such billing models state that a small percentage of users consume a disproportionately high percentage of bandwidth and that some form of usage-based pricing may benefit the majority of subscribers, particularly those who are light users.[57] Furthermore, they state that offering a range of service tiers at varying prices offers consumers more choice and control over their usage and subsequent costs. The major growth in bandwidth usage, they also claim, places financial pressure on existing networks for both maintenance and expansion, and establishing a pricing system which charges high bandwidth users is more equitable.

Opponents to such billing plans claim that such practices will stifle innovation in high bandwidth applications and are likely to discourage the experimentation with and adoption of new applications and services. Some concerns have also been expressed that a move to metered/usagebased pricing will help to protect the market share for video services, offered in packaged bundles by network broadband service providers, that compete with new applications and if such caps must exist, should be applied to all online video sources. The move to usage-based pricing, they state, will unfairly disadvantage competing online video services and stifle a nascent market since video applications are more bandwidth-intensive. Opponents have also questioned the accuracy of meters, and specific usage limits and overage fees established in specific trials, stating that the former seem to be "arbitrarily low" and the latter "arbitrarily high."[58] Furthermore they state that since network congestion only occurs in specific locations and is temporary, monthly data caps are not a good measure of congestion causation. Citing the generally falling costs of network equipment and the stability of profit margins, they also question the claims of network broadband operators that increased revenues streams are needed to supply the necessary capital to invest in new infrastructure to meet the growing demand for high bandwidth applications.[59]

THE POLICY DEBATE

Questions over the FCC's authority to regulate broadband services under its Title I ancillary authority, and what is perceived by some as inadequacies in the Open Internet Order, have caused some policymakers to support more specific regulatory guidelines to protect the marketplace from potential abuses; a consensus on what these should specifically entail, however, has yet to form. Others feel that that the FCC has overstepped its authority and that the regulation of the Internet is not only unnecessary but harmful. They claim that existing laws regarding competitive behavior are sufficient to deal with potential anti-competitive behavior.

The issue of net neutrality, and whether legislation is needed to ensure access to broadband networks and services, has become a major focal point in the debate over telecommunications reform.[60] Those opposed to the enactment of legislation to impose specific Internet network access or "net neutrality" mandates claim that such action goes against the long-standing policy to keep the Internet as free as possible from regulation. They have claimed that the imposition of such requirements is not only unnecessary, but would have negative consequences for the deployment and advancement of broadband facilities. For example, further expansion of networks by existing providers and the entrance of new network providers would be discouraged, they claim, as investors would be less willing to finance networks that may be operating under mandatory build-out and/or access requirements.

Application innovation could also be discouraged, they contend, if, for example, network providers are restricted in the way they manage their networks or are limited in their ability to offer new service packages or formats. Such legislation is not needed, they claim, as major Internet access providers have stated publicly that they are committed to upholding the FCC's four policy principles.[61]

Opponents also state that advocates of regulation cannot point to any widespread behavior that justifies the need to establish such regulations and note that competition between telephone and cable system providers, as well as the growing presence of new technologies (e.g., satellite, wireless, and power lines), will serve to counteract any potential anti-competitive behavior.

Furthermore, opponent's claim, even if such a violation should occur, the FCC already has the needed authority to pursue violators. They note, for example, that the FCC has successfully used its existing authority in a March 3, 2005, action against Madison River Communications. In this case, the FCC intervened and resolved, through a consent decree, an alleged case of port blocking by Madison River Communications, a local exchange (telephone) company.[62] The full force of antitrust law is also available, they claim, in cases of discriminatory behavior.

Proponents of net neutrality legislation, however, feel that absent some regulation, Internet access providers will become gatekeepers and use their market power to the disadvantage of Internet users and competing content and application providers. They also cite concerns that the Internet could develop into a two-tiered system favoring large, established businesses or those with ties to broadband network providers. While market forces should be a deterrent to such anti-competitive behavior, they point out that the market for residential broadband delivery has traditionally been dominated by two providers, the telephone and cable television companies.[63]

The need to formulate a national policy to clarify expectations and ensure the "openness" of the Internet is important to protect the benefits and promote the further expansion of broadband, they claim. The adoption of a single, coherent, regulatory framework to prevent discrimination, supporters claim, would be a positive step for further development of the Internet, by providing the marketplace stability needed to encourage investment and innovation which will foster the growth of new services and applications. Furthermore, they state that there have been cases where ISPs have abused their market power[64] and relying on current laws and case-by-case anti-trust-like enforcement, they claim, is too cumbersome, slow, and expensive, particularly for small start-up enterprises.[65]

CONGRESSIONAL ACTIVITY

113th Congress

The issue of net neutrality is expected to be of continued interest. A decision, anticipated in late- 2013 or early-2014, by the US Court of Appeals, D.C. Circuit, challenging the FCC's authority to impose its Open Internet Order is likely to focus additional attention on the issue (*Verizon Communications Inc. v. Federal Communications Commission*, D.C. Cir., No.11-1355). Depending on the outcome interested parties may choose to introduce legislation to either expand or restrict the FCC's authority to regulate broadband (see "The FCC Open Internet Order," above).

112th Congress

A consensus on the net neutrality issue has remained elusive and support for the FCC's Open Internet Order has been mixed. (See "The FCC Open Internet Order," above.) While some Members of Congress support the action and in some cases would have supported an even stronger approach, others feel that the FCC has overstepped its authority and that the regulation of the Internet is not only unnecessary, but harmful. Internet regulation and the FCC's authority to implement such regulations was a topic of legislation (H.R. 96, H.R. 166, S. 74, H.R. 2434, H.R. 1, H.R. 3630, H.J.Res. 37, S.J.Res. 6) and hearings (Senate Commerce Committee, House Communications Subcommittee, and House Intellectual Property, Competition, and the Internet Subcommittee) in the 112th Congress.

Legislation to limit FCC regulation was introduced. H.R. 96, the "Internet Freedom Act," introduced, on January 5, 2011, by Representative Blackburn and 59 additional original cosponsors, prohibits, with exceptions, the FCC from proposing, promulgating, or issuing any regulations regarding the Internet or IP-enabled services, effective the date of the bill's enactment. Exceptions are made for regulations that the FCC determines are necessary to prevent damage to national security, to ensure the public safety, or to assist or facilitate actions taken by a federal or state law enforcement agency. The bill also contains a finding that the Internet and IP-enabled services are services affecting interstate commerce and are not subject to State or municipality jurisdiction. Another measure, H.R. 166, the "Internet Investment, Innovation, and Competition Preservation Act," introduced on January 5, 2011, by Representative Stearns, requires the FCC to prove the existence of a "market failure" before regulating information services or Internet access services. The FCC must also conclude that the "market failure" is causing "specific, identified harm to consumers" and that regulations are necessary to ameliorate that harm. The bill also contains provisions that require any FCC regulation to be the "least restrictive," determine that the benefits exceed the cost, permit network management, not prohibit managed services, be reviewed every two years, and be subject to sunset. Any such regulation is required to be enforced on a nondiscriminatory basis between and among broadband network, service, application, and content providers. A more narrowly focused limitation was contained within H.R. 3630, the "Middle Class Tax Relief and Job Creation Act of 2011" as passed (234-193) by the House on December 13, 2011. Section 4105 of Title IV (spectrum provisions) of the bill prohibits the FCC from imposing network access/management requirements on licensees. More specifically, the provision prohibited the promulgation of auction service rules that restrict a licensee's ability to manage network traffic or prioritize the traffic on its network, or that would require providing network access on a wholesale basis. However, the provision was removed from the bill prior to final passage (P.L. 112-96).

Legislation to strengthen the FCC's ability to regulate open access by amending Title II of the 1934 Communications Act was also. S. 74, the "Internet Freedom, Broadband Promotion, and Consumer Protection Act of 2011," introduced, January 25, 2011, by Senator Cantwell, provides for strengthened open access protections. More specifically the bill contains among its provisions those that codify the four FCC principles issued in 2005 as well as those to require Internet service providers to be nondiscriminatory regarding access and transparent in their network management practices. The bill also requires Internet service providers to provide service to end users upon "reasonable request" and offer stand-alone broadband access at "reasonable rates, terms, and conditions" and prohibits Internet service providers from requiring paid prioritization. The bill's requirements apply to both wireline and wireless platforms; however, the FCC is allowed to take into consideration difference in network technologies when applying requirements. The FCC is tasked with establishing the necessary rules and injured parties can be awarded damages by the FCC or a federal district court.

Other measures, which proved unsuccessful, were considered to prevent, or at least delay, implementation of the FCC's Open Internet Order. Attempts were made, through the appropriations process, to add language that would prevent the FCC from using its funds to implement the Open Internet Order. Language attached to the FY2011 appropriation measure, H.R. 1, to prevent the use of FCC FY2011 funds for implementation of the order was passed by the House. The Continuing Appropriations Act, 2011 (H.R. 1) passed (235-189) by the House on February 19, 2011, contained an amendment, introduced by Representative Walden and passed by the House (244-181), to prohibit the FCC from using any funds made available by the act to implement the FCC's Open Internet Order adopted on December 21, 2010. No such provision, however, was included in the final FY2011 appropriations bill, H.R. 1473, passed by Congress and signed by the President (P.L. 112-10). Similarly language included in the FY2012 Financial Services and General Government Appropriations bill (H.R. 2434), which includes funding for the FCC, contained a provision that barred the FCC from using any funds to implement its Open Internet Order adopted December 21, 2010. This measure passed the House Appropriations Committee on June 23, 2011 (H.Rept. 112-136)[66] but no such provision was included in the final FY2012 consolidated appropriations bill, H.R. 2055, which was signed by President Obama (P.L. 112-74) on December 23, 2011.

Another approach, using the Congressional Review Act to overturn the order,[67] was also under consideration. Identical resolutions of disapproval were introduced, on February 16, 2011, in both the House (H.J.Res. 37) and Senate (S.J.Res. 6). These measures state that Congress disapproves of the rule submitted by the FCC's report and order relating to the matter of preserving the open Internet and broadband industry practices adopted by the FCC on December 21, 2010, and further states that "such rule would have no force or effect." A hearing on H.J.Res. 37 was held by the House Energy and Commerce Communications and Technology Subcommittee on March 9, 2011, and the Subcommittee passed the measure (15-8), on a party-line vote, immediately following the hearing. On March 25, 2011, the House Energy and Commerce Committee passed (30-23) H.J.Res. 37. On April 8, 2011, the full

House considered and passed (240-179) H.J.Res. 37. However an identical resolution of disapproval (S.J.Res. 6) failed to pass the Senate on November 10, 2011, by a 52-46 vote.

Legislation addressing the issue of data usage caps was also introduced. The "Data Cap Integrity Act of 2012" (H.R. 3703), introduced on December 20, 2012, by Senator Wyden, addresses the usage of data caps by Internet service providers (ISPs) and their implementation. Included among the bill's provisions are those that require that an ISP that imposes data caps must be certified by the FCC as to accuracy of data cap measurement; that the cap "functions to reasonably limit network congestion without unnecessarily restricting Internet use"; and that the cap does not discriminate (that is, for purposes of measuring does not provide "preferential treatment of data that is based on the source or content of the data"). The bill also requires ISPs that apply data caps to provide data tools, or identify commercially available data measurement tools, to consumers for monitoring and management. Civil penalties for violations will be used to reimburse those violated and unobligated funds in excess of $5 million (annually) will be transferred from the newly created "Data Cap Integrity Fund," to the U.S. Treasury for deficit reduction.

111th Congress

Although the 111th Congress saw considerable activity addressing the net neutrality debate, no final action was taken. One stand-alone measure (H.R. 3458) that comprehensively addressed the net neutrality debate was introduced in the 111th Congress. H.R. 3458, the "Internet Freedom Preservation Act of 2009," introduced by Representative Edward Markey, and also supported by then-House Energy and Commerce Committee Chairman Waxman, sought to establish a national policy of nondiscrimination and openness with respect to Internet access offered to the public. The bill also required the offering of unbundled, or stand-alone, Internet access service as well as transparency for the consuming public with respect to speed, nature, and limitations on service offerings and the public disclosure of network management practices. The FCC was tasked with promulgating the rules relating to the enforcement and implementation of the legislation. Then-House Communications, Technology, and the Internet Subcommittee Chairman Boucher stated that he continued to work with broadband providers and content providers to seek common ground on network management practices, and chose to pursue that approach.[68] Furthermore, the Senate Commerce and House Energy and Commerce Committees and Communications Subcommittees held a series of staff-led sessions with industry stakeholders to discuss a range of communications policies including broadband regulation and FCC authority.[69]

Two bills (S. 1836, H.R. 3924) were introduced in response to the adoption, by the FCC, of a NPR on preserving the open Internet. S. 1836, introduced on October 22, 2009, by Senator McCain, prohibited, with some exceptions, the FCC from proposing, promulgating, or issuing any further regulations regarding the Internet or IP-enabled services. Exceptions included those relating to national security, public safety, federal or state law enforcement, and Universal Service Fund solvency.[70] Additional provisions reaffirmed that existing regulations, including those relating to CALEA, remain in force and stated as a general principle, that the Internet and all IPenable services are services affecting interstate commerce and are not subject to State or municipal locality jurisdiction. H.R. 3924, introduced by Representative Blackburn on October 26, 2009, was identical to S. 1836, except for title and the omission of the reference to the Universal Service Fund. H.Con.Res. 311, introduced by Representative Gene Green and 49 other House Members on July 30, 2010, affirmed that it is the responsibility of Congress to determine the regulatory authority of the FCC with respect to broadband Internet services and called upon the FCC to suspend any further action on its proceedings until such time as Congress delegates such authority to the FCC.

Another measure (H.R. 5257) introduced by Representative Stearns, addressed the possible reclassification of broadband service and would required, among other provisions, that the FCC prove the existence of a "market failure" before regulating information services or Internet access services. Furthermore the bill required, among other provisions, that the FCC conclude that the market failure is causing "specific, identified harm to consumers" and if devising regulations must adopt those that are the "least restrictive," permit network management, and are subject to sunset. Still another measure (S. 3624), introduced by Senator DeMint, contained provisions that required the FCC to prove consumers are being substantially harmed by a lack of marketplace choice before imposing new regulations and must weigh the potential cost of action against any benefits to consumers or competition. The FCC was given the authority to hear complaints for violations and award damages to injured parties. The bill also required that any rules the FCC adopted would sunset in five years unless it could make the same finding again.

The net neutrality issue was also narrowly addressed within the context of the American Recovery and Reinvestment Act of 2009 (ARRA, P.L. 111-5). The ARRA contains provisions that require the National Telecommunications and Information Administration (NTIA), in consultation with the FCC, to establish "nondiscrimination and network interconnection obligations" as a requirement for grant participants in the Broadband Technology Opportunities Program (BTOP). The law further directs that the FCC's four broadband policy principles, issued in August 2005, are the minimum obligations to be imposed.[71] These obligations were issued July 1, 2009, in conjunction with the release of the notice of funds availability (NOFA) soliciting applications for the program. (See "The American Recovery and Reinvestment Act of 2009," above, for details.) The FCC's *National Broadband Plan* (NBP), which was required to be written in compliance with provisions contained in the ARRA, while making no recommendations, did contain discussions regarding the open Internet and the classification of information services. (See "The FCC's National Broadband Plan," above.)

Concern over the move by some broadband network providers to expand their implementation of metered or consumption-based billing prompted the introduction of legislation (H.R. 2902) to provide for oversight of volume usage service plans. H.R. 2902, the "Broadband Internet Fairness Act," introduced by former Representative Massa, required, among its provisions, that any broadband Internet service provider, serving 2 million or more subscribers, submit any volume usage based service plan, which the provider is proposing or offering, to the Federal Trade Commission (FTC) for approval. The FTC, in consultation with the FCC, was required to review such plans "to ensure that such plans are fairly based on cost." Such plans were subject to agency review and public hearings. Plans determined by the FTC to impose "rates, terms, and conditions that are unjust, unreasonable, or unreasonably discriminatory" were to be declared unlawful. Violators were subject to injunctive relief requiring the suspension, termination, or revision of such plans and were subject to a fine of not more than $1 million.

End Notes

[1] 47 U.S.C. 151 et seq. For a full discussion of the Brand X decision see CRS Report RL32985, *Defining Cable Broadband Internet Access Service: Background and Analysis of the Supreme Court's Brand X Decision*, by Angie A. Welborn and Charles B. Goldfarb.

[2] See http://hraunfoss.fcc.gov/edocs_public/attachmatch/DOC-260433A2.pdf for a copy of former FCC Chairman Martin's statement. For a summary of the final rule see Appropriate Framework for Broadband Access to the Internet Over Wireline Facilities. *Federal Register*, Vol. 70, No. 199, October 17, 2005, p. 60222.

[3] For example, Title II regulations impose rigorous anti-discrimination, interconnection and access requirements. For a further discussion of Title I versus Title II regulatory authority see CRS Report RL32985, *Defining Cable Broadband Internet Access Service: Background and Analysis of the Supreme Court's Brand X Decision*.

[4] Title I of the 1934 Communications Act gives the FCC such authority if assertion of jurisdiction is "reasonably ancillary to the effective performance of [its] various responsibilities." The FCC in its order cites consumer protection, network reliability, or national security obligations as examples of cases where such authority would apply (see paragraph 36 of the final rule summarized in the *Federal Register* cite in footnote 2, above).

[5] See http://www.fcc.gov/headlines2005.html. August 5, 2005. *FCC Adopts Policy Statement on Broadband Internet Access.*

[6] See http://hraunfoss.FCC.gov/edocs_public/attachmatch/DOC-261936A1.pdf. It should be noted that applicants offered certain voluntary commitments, of which this was one.

[7] See http://hraunfoss.fcc.gov/edocs_public/attachmatch/DOC-269275A1.pdf.

[8] "*Preserving a Free and Open Internet: A Platform for Innovation, Opportunity, and Prosperity*," prepared remarks of FCC Chairman Julius Genachowski, at the Brookings Institution, September 21, 2009. Available at http://hraunfoss.fcc.gov/edocs_public/attachmatch/DOC-293568A1.pdf.

[9] See http://hraunfoss.fcc.gov/edocs_public/attachmatch/FCC-08-183A1.pdf.

[10] Comcast, *Frequently Asked Questions and Network Management.* Available at http://help.comcast.net/content/faq / Frequently-Asked-Questions-about-Network-Management.

[11] *Comcast Statement on U.S. Court of Appeals Decision on Comcast v. FCC.* Available at http://www.comcast.com / About/PressRelease/PressReleaseDetail.ashx?PRID=984.

[12] Comcast Corporation v. FCC, No. 08-129 (D.C. Cir. September 4, 2008).

[13] For a legal discussion of the FCC's regulatory authority in light of the Comcast decision see CRS Report R40234, *The FCC's Authority to Regulate Net Neutrality After Comcast v. FCC*, by Kathleen Ann Ruane.

[14] Comcast Corporation v. FCC decided April 6, 2010. Available at http://pacer.cadc.uscourts.gov/common /opinions/ 201004/08-1291-1238302.pdf.

[15] Comcast v. FCC decision, issued April 6, 2010, part V, p. 36.

[16] FCC Announces Broadband Action Agenda, released April 8, 2010. Available at http://hraunfoss.fcc.gov/ edocs_public/attachmatch/DOC-297402A1.pdf.

[17] *In the Matter of Preserving the Open Internet, Broadband Industry Practices*. GN Docket No. 09-191; WC Docket No. 07-52, released December 23, 2010. Available at http://www.fcc.gov/Daily_Releases/Daily_ Business/2010/ db1223/FCC-10-201A1.pdf.

[18] The vote fell along party lines with Chairman Genachowski approving, Commissioner Clyburn approving in part and concurring in part, former Commissioner Copps concurring, and Commissioner McDowell and former Commissioner Baker dissenting.

[19] The FCC, on June 30, 2011, released a public notice offering initial guidance regarding compliance with the transparency rule. *FCC Enforcement Bureau and Office Of General Counsel Issue Advisory Guidance For Compliance With Open Internet Transparency Rule.* Available at http://hraunfoss.fcc.gov/edocs_public /attachmatch/DA-11- 1148A1.pdf.

[20] A network management practice is considered reasonable if "it is appropriate and tailored to achieving a legitimate network management purpose, taking in to account the particular network architecture and technology of the broadband Internet access service." Cited examples include ensuring network security and integrity; providing parental controls; or reducing or mitigating the effects of congestion on the network.

[21] The FCC announced the creation of an Open Internet Advisory Committee April 21, 2011, *Federal Register*, Vol. 76, No. 77, April 21, 2011, p. 22395. The Committee, which includes members from a broad range of industry, academic and community representatives held its first meeting in July 2012. For additional information see http://www.fcc.gov/ encyclopedia/open-internet-advisory-committee.

[22] *In the Matter of Preserving the Open Internet, Broadband Industry Practices*, paragraph 72.

[23] This approach consists of pursuing a bifurcated, or separate, regulatory approach by applying the specific provisions of Title II to the transmission component of broadband access service and subjecting the information component to, at most, whatever ancillary jurisdiction may exist under Title I. See *The Third Way: A Narrowly Tailored Broadband Framework*, FCC Chairman Julius Genachowski, May 6, 2010. Available at http://hraunfoss.fcc.gov/edocs_public/ attachmatch/DOC-297944A1.pdf. Also see *A Third-Way Legal Framework for Addressing the Comcast Dilemma*, Austin Schlick, FCC General Counsel, May 6, 2010. Available at http://hraunfoss.fcc.gov/edocs_public/attachmatch/ DOC-297945A1.pdf.

[24] The FCC made such a finding, that is that "broadband is not being deployed to all Americans in a reasonable and timely fashion" in its *Sixth Broadband Deployment Report*, adopted on July 16, 2010. Available at http://www.fcc.gov/ Daily_Releases/Daily_Business/2010/db0720/FCC-10-129A1.pdf.

[25] Preserving the Open Internet; Final Rule. *Federal Register*, Vol.76, No. 185, September 23, 2011, pp. 59192-59235.

[26] Order Granting Mot. Cons., DC/1:11-ca-01356, (J.P.M.L., October 6, 2011).

[27] Earlier appeals by both companies were filed but dismissed by the court. See *Verizon v. FCC* D.C. Cir. 11-1014, 1/20/2011; and *MetroPCS Communications et. al. v. FCC* D.C. Cir.11-1016, 1/24/2011. The U.S. Court of Appeals, on April 4, 2011, rejected both filings as premature, stating that the Order is a rulemaking and therefore must first be published in the *Federal Register* before it can be subject to judicial review *Verizon v. FCC*, Order Granting Mot. Dismiss, Case No.11-1014 (D.C. Cir. April 4, 2011).

[28] *Verizon Communications Inc. v. FCC*, D.C. Cir. 11-1355, 10/18/2011.

[29] For additional details on the NOFA see Department of Agriculture, Rural Utilities Service, and Department of Commerce, National Telecommunications and Information Administration, "Broadband Initiatives Program; Broadband Technology Opportunities Program; Notice," 74 *Federal Register* 33104 -33134, July 9, 2009.

[30] As of October 1, 2010, all BTOP and BIP award announcements were complete. For a review of ARRA programs and a listing of awards granted see CRS Report R40436, *Broadband Infrastructure Programs in the American Recovery and Reinvestment Act*, by Lennard G. Kruger.

[31] *Connecting America: The National Broadband Plan*. Available at http://hraunfoss.fcc.gov/edocs_public /attachmatch/ DOC-296935A1.pdf.

[32] *Connecting America: The National Broadband Plan*, Chapter 4, Broadband Competition and Innovation Policy, Section 4.4, Competition for Value Across the Ecosystem.

[33] It should be noted that the FCC is given the authority, under §10 of the 1934 Communications Act, to forbear from regulation, therefore, if such a reclassification should occur, all requirements of a Title II classification would not necessarily be imposed.

[34] See, for example, Statement of FCC Commissioner Robert McDowell, before the Committee on Energy and Commerce, Subcommittee on Communications, Technology, and the Internet, hearing on Oversight of the Federal Communications Commission: The National Broadband Plan, March 25, 2010. available at http://hraunfoss.fcc.gov/ edocs_public/attachmatch/DOC-297139A1.pdf.

[35] *Connecting America: The National Broadband Plan*, Chapter 17, Implementation and Benchmarks, Section 17.3, The Legal Framework for the FCC's Implementation of the Plan. The FCC released a "2010 Broadband Action Agenda" on April 8, 2010, containing a timeframe for FCC proceedings to help implement the plan. A summary table of proposed 2010 agenda items is available at http://www.broadband.gov/plan/chart-of-key-broadband-action-agendaitems.pdf.

[36] *In the Matter of Framework for Broadband Internet Service*, General Docket No. 10-127. Available at http://hraunfoss.fcc.gov/edocs_public/attachmatch/FCC-10-114A1.pdf.

[37] *Broadband Industry Practices*, WC Docket No. 07-52, Notice of Inquiry, 22 FCC Record 7894 (2007).

[38] *United States, et. al. v. Comcast Corp., et. al.*; Proposed Final Judgment and Competitive Impact Statement; Notice. Federal Register, Vol. 76, No. 20, January 31, 2011, pp.5440-5464. In the Matter of Applications of Comcast Corporation, General Electric Company and NBC Universal, Inc. For Consent to Assign Licenses and Transfer Control of Licenses, MB Docket No. 10-56. available at http://www.fcc.gov/Fcc-11-4.pdf.

[39] "Reasonable network management shall not constitute unreasonable discrimination."

[40] *Initial Plans for Broadband Internet Technical Advisory Group Announced*. PRNewswire, June 9, 2010. Available at http://www.prnewswire.com/news-releases/initial-plans-for-broadband-internet-technical-advisory-group-announced95950.

[41] BITAG's Interim Board of Directors Announced; First Board Meeting Scheduled for Next Week. Available at http://log.bitag.org/2010/12/bitags-interim-board-of-directors.html.

[42] *Verizon-Google Legislative Framework Proposal*. Available at http://www.scribd.com/doc/35599242/verizongoogle-legislative-framework-proposal.

[43] The header contains the processing information which includes the source and destination addresses, and the data field includes the message content and the identity of the source application.

[44] For a further discussion of the positive uses, by network operators, of DPI technologies see testimony of Kyle McSlarrow, President and CEO National Cable and Telecommunications Association, hearings on "Communications Networks and Consumer Privacy: Recent Developments," House Committee on Energy and Commerce, Subcommittee on Communications, Technology, and the Internet, April 23, 2009. Available at http://energycommercehouse.gov/ Press_111/20090423/testimony_mcslarrow.pdf.

[45] For a further discussion of the potential abuses associated with DPI technology see testimony of Ben Scott, Policy Director, Free Press, hearings on "Communications Networks and Consumer Privacy: Recent Developments," House Committee on Energy and Commerce, Subcommittee on Communications, Technology, and the Internet, April 23, 2009. Available at http://energycommercehouse.gov/Press_111/20090423/testimony_scott.pdf.

[46] For example, concern that information can be gathered, without permission, based on consumer use of the Internet to develop user profiles to provide targeted online advertising, also known as "behavioral advertising," has raised privacy issues. For an examination of this issue see testimony from hearings "Communications Networks and Consumer Privacy: Recent Developments," held April 23, 2009, by the House Energy and Commerce Subcommittee on Communications, Technology, and the Internet. Available at http://energycommerce.house.gov/.

[47] Network Management Policy. *Announcement Regarding An Amendment to Our Acceptable Use Policy*. Available at http://xfinity.comcast.net/terms/network/amendment/.

[48] If the subscriber does not modify their use and/or the subscriber exceeds the cap again within six months service will be subject to termination and eligibility for either residential or commercial Internet service will be suspended for 12 months. According to Comcast the median data usage by their Internet residential customers is approximately 4-6 GB per month and less than 1% of Comcast customers use an "excessive" amount of data. A customer would have to do any one of the following, Comcast states, to reach the monthly 250 GB limit: send 50 million e-mails (at 0.05 KB/email); download 62,500 songs (at 4 MB/song); download 125 standard-definition movies (at 2 GB/movie); or upload 25,000 hi-resolution digital photos (at 10 MB/photo). For additional information on Comcast's excessive use policy see "Frequently asked Questions about Excessive Use," available at http://customer.comcast.com.

[49] http://customer.comcast.com/help-and-support/internet/common-questions-excessive-use/.

[50] For additional information on AT&T usage policy for residential broadband services see *What Are AT&T's Tiered Pricing Plans, and What Do They Mean to Me?* Available at http://www.att.com/esupport/internet/usage.JSP#fbid= AOrWfWmMh8z.

[51] *CenturyLink Excessive Use Policy FAQ*, available at http://qwest.centurylink.com/internethelp/pdf/EUP.pdf.

[52] Time Warner Cable announced, on April 9, 2009, plans to implement usage-based billing trials in Rochester, New York and Greensboro, North Carolina, in August 2009, and Austin and San Antonio, Texas, in October, 2009. See *Statement from Landel Hobbs, Chief Operating Officer, Time Warner Cable Re: Consumption Based Billing Trials*, April 9, 2009. Available at http://www.timewarnercable.com/corporate/announcements /cbb.html.

[53] Citing "misunderstanding about our trials," Time Warner Cable announced plans to defer implementation of usage-based billing trials in Rochester, New York, Greensboro, North Carolina, and Austin and San Antonio, Texas, to enable "consultation with our customers and other interested parties." See *Time Warner Cable Charts a New Course on Consumption Based Billing Measurement Tools to be Made Available,* April 16, 2009. Available at http://www.timewarnercable.com/Corporate/announcements/cbb.html.

[54] *Launching An Optional Usage-Based Broadband Pricing Plan In Southern Texas*, available at http://www.twcableuntangled.com/2012/02/launching-an-optional-usage-based-pricing-plan-in-southern-texas-2/ 27/2012.

[55] For example see *ACA: Metered Bandwidth Pricing Is Coming,* available at http://www.broadcastingcable.com/article/print/210247-ACA_Metered_Bandwidth_Pricing_Is_Coming.php.

[56] Additional bandwidth can be purchased in advance at 10 GB for $10 per month or 50 GB for $25 per month. For additional information on Knology of Kansas bandwidth management see http://www.Kansas.Knology.Com/bandwidth/.

[57] For example, Time Warner states that the top 25% of its users consume 100 times more bandwidth than the bottom 25% and 30% of its high speed Internet service (i.e., Road Runner) customers use less than 1 GB (Gigabyte) per month. See *Consumption Based Billing FAQs*. Available at http://www.timewarnercable.com/corporate/announcements/ cbb_faq.html.

[58] See Free Press letter to House Energy and Commerce Committee, April 22, 2009. Available at http://www.Freepress.net/files/FP_metering_letter.pdf.

[59] *As Costs Fall, Companies Push to Raise Internet Price*, New York Times, April 20, 2009. Available at http://www.nytimes.com/2009/04/20/business/20isp.html?_r=1.

[60] For a more lengthy discussion regarding proponents' and opponents' views see, for example, testimony from Senate Commerce Committee hearings on Net Neutrality, February 7, 2006. Available at http://commerce.senate.gov/public/ index.cfm?FuseAction=Hearings.Hearing&Hearing_ID=1708.

[61] See testimony of Kyle McSlarrow, President and CEO of the National Cable and Telecommunications Association, and Walter McCormick, President and CEO of the United States Telecom Association, hearing on Net Neutrality before the Senate Commerce Committee, February 7, 2006, cited above.

[62] The FCC entered into a consent decree with Madison River Communications to settle charges that the company had deliberately blocked the ports on its network that were used by Vonage Corp. to provide voice over Internet protocol (VoIP) service. Under terms of the decree Madison River agreed to pay a $15,000 fine and not block ports used for VoIP applications. See http://hraunfoss.fcc.gov/edocs_public/attachmatch/DA-05-543A2.pdf. for a copy of the consent decree.

[63] Some, however, point to the growth in mobile wireless subscribers with data plans for full Internet access as a growing third provider. For FCC market share data for high-speed connections see *Internet Access Services: Status as of December 31, 2011*, Federal Communications Commission, Industry Analysis and Technology Division, Wireline Competition Bureau, released February 2013. Available at http://transition.fcc.gov/ Daily_Releases/Daily_Business/ 2013/db0207/DOC-318810A1.pdf.

[64] For example, see the mentioned Comcast and the Madison River cases, discussed above.

[65] For example, see testimony of Vint Cerf, VP Google, Earl Comstock, President and CEO of CompTel, and Jeffrey Citron, Chairman and CEO Vonage, hearing on Net Neutrality, before the Senate Commerce Committee, February 7, 2006, cited above.

[66] The Senate Appropriations subcommittee-passed (September 14, 2011) appropriations measure, S. 1573, did contain a provision to prohibit the FCC from using funds to implement the Open Internet Order, but it did not remain in the full committee passed (September 15, 2011) version (S.Rept. 112-79).

[67] Under the Congressional Review Act (CRA; 5 U.S.C. paras.801-808) Congress is given 60 in-session-days, from publication in the *Federal Register* or submission to Congress, whichever is later, to review and potentially overturn federal agency major rulemakings. For a further discussion of the CRA see CRS Report R40997, *Congressional Review Act: Rules Not Submitted to GAO and Congress*, by Curtis W. Copeland.

[68] *Boucher Opts For Talks, Not Legislation, On Net Neutrality,* National Journal, Congress Daily, February 26, 2009. *Boucher, Stakeholders Working On Network Management Issues,* Telecommunications Reports, March 15, 2009, p. 19.

[69] *Bicameral Bipartisan Telecommunications Update Statement.* U.S. Senate Committee on Commerce, Science and Transportation. Press Release June 18, 2010. Available at http://democrats.energycommerce.house.gov /index.php?q= news/bicameral-bipartisan-telecommunications-update-statement, June 2010.

[70] For a discussion and analysis of issues regarding the Universal Service Fund see CRS Report RL33979, *Universal Service Fund: Background and Options for Reform*, by Angele A. Gilroy.

[71] For a further more detailed discussion of the broadband infrastructure programs contained in P.L. 111-5 see CRS Report R40436, *Broadband Infrastructure Programs in the American Recovery and Reinvestment Act*, by Lennard G. Kruger.

In: Transformations in Telecommunications and Media
Editor: Irwin Cavazos

ISBN: 978-1-62948-413-6
© 2013 Nova Science Publishers, Inc.

Chapter 4

SPECTRUM POLICY IN THE AGE OF BROADBAND: ISSUES FOR CONGRESS[*]

Linda K. Moore

SUMMARY

The convergence of wireless telecommunications technology with the Internet Protocol (IP) is fostering new generations of mobile technologies. This transformation has created new demands for advanced communications infrastructure and radio frequency spectrum capacity that can support high-speed, content-rich uses. Furthermore, a number of services, in addition to consumer and business communications, rely at least in part on wireless links to broadband (highspeed/high-capacity) infrastructure such as the Internet and IP-enabled networks. Policies to provide additional spectrum for mobile broadband services are generally viewed as drivers that would stimulate technological innovation and economic growth.

The Middle Class Tax Relief and Job Creation Act of 2012 (P.L. 112-96, signed February 22, 2012) contained provisions in Title VI that expedite the availability of spectrum for commercial use. The provisions in Title VI —also known as the Public Safety and Spectrum Act, or the Spectrum Act—included expediting auctions of licenses for spectrum designated for mobile broadband; authorizing incentive auctions, which would permit television broadcasters to receive compensation for steps they might take to release some of their airwaves for mobile broadband; requiring that specified federal holdings be auctioned or reassigned for commercial use; and providing for the availability of spectrum for unlicensed use. The act also included provisions to apply future spectrum license auction revenues toward deficit reduction; to establish a planning and governance structure to deploy public safety broadband networks, using some auction proceeds for that purpose; and to assign additional spectrum resources for public safety communications.

Increasing the amount of spectrum available to support new mobile technologies is one step toward meeting future demand for mobile services. This report discusses some of the commercial and federal spectrum policy changes required by the act. It also

---

[*] This is an edited, reformatted and augmented version of Congressional Research Service, Publication No. R40674, dated May 28, 2013.

summarizes new policy directions for spectrum management under consideration in the 113th Congress, such as the encouragement of new technologies that use spectrum more efficiently.

SPECTRUM POLICY

The purpose of spectrum policy, law, and regulation is to manage a natural resource[1] for the maximum possible benefit of the public. Electromagnetic spectrum, commonly referred to as radio frequency spectrum or wireless spectrum, refers to the properties in air that transmit electric signals and, with applied technology, can deliver voice, text, and video communications. Access to radio frequency spectrum is controlled by assigning rights to specific license holders or to certain classes of users. The assignment of spectrum rights does not convey ownership. Radio frequency spectrum is managed by the Federal Communications Commission (FCC) for commercial and other non-federal uses and by the National Telecommunications and Information Administration (NTIA) for federal government use.

Wireless broadband,[2] with its rich array of services and content, requires new spectrum capacity to accommodate growth. Spectrum capacity is necessary to deliver mobile broadband to consumers and businesses and also to support the communications needs of industries that use fixed wireless broadband to transmit large quantities of information quickly and reliably.

Although radio frequency spectrum (air) is abundant, usable spectrum is currently limited by the constraints of applied technology. Spectrum policy therefore requires making decisions about how radio frequencies will be allocated and who will have access to them. Current spectrum policy is based on managing channels of radio frequencies to avoid interference.[3] The FCC, over many years, has developed and refined a system of exclusive licenses for users of specific frequencies. Auctions are a market-driven solution to assigning licenses to use specific frequencies and are a recent innovation in spectrum management and policy. Previously, the FCC granted licenses using a process known as "comparative hearings" (also known as "beauty contests"), and has used lotteries to distribute spectrum licenses. The FCC also allocates spectrum for designated purposes, such as WiFi, without assigning a license to a specific owner (unlicensed spectrum).

As wireless technology moves from channel management to broadband network management, spectrum policy going forward may entail encouraging innovation in spectrum- and network-sharing technologies . An increasing number of policy makers and wireless industry leaders are urging that laws and regulations be revised to reflect significant changes in wireless technology that are creating a new mobile network communications environment.

SPECTRUM POLICY PROVISIONS IN THE MIDDLE CLASS TAX RELIEF AND JOB CREATION ACT OF 2012

Title VI of the Middle Class Tax Relief and Job Creation Act of 2012 (P.L. 112-96, signed into law on February 22, 2012) contains provisions that include reallocation of spectrum, new assignments of spectrum rights, and changes in procedures for repurposing spectrum used by the federal government. Many of the provisions in Title VI, frequently referred to as the Public Safety and Spectrum Act, or Spectrum Act, focus on spectrum assignment within the existing regulatory framework, in which licenses for designated radio frequencies are awarded through competitive bidding systems (auctions).

Major provisions in the Spectrum Act that are summarized in this report cover

- Deficit reduction;
- Directed auctions;
- Incentive auctions for television broadcasters;
- Reallocation of spectrum from federal to commercial use; and
- Unlicensed spectrum.

Other provisions in the act, not covered in this report, include simplifying the approval of zoning requests for modification of cell towers at the state and local level and putting in place measures to facilitate antenna placement on federal property. The act also has provided for the establishment of a new authority to plan and develop a nationwide public safety broadband network and has included other measures in support of improved emergency communications.[4]

Deficit Reduction

The Spectrum Act has addressed the interlaced issues of spectrum access and deficit reduction. The issues are connected because, when radio frequency spectrum licenses are auctioned for commercial purposes by the FCC, the net proceeds are deposited in the U.S. Treasury.[5] The act has extended the FCC's auction authority until the end of FY2022. Because the FCC's authority would have expired at the end of FY2012, revenue from auctions held after FY2012 is considered new revenue.

The legislation that first authorized the FCC to establish "competitive bidding systems"[6] for a limited period was included in the Omnibus Budget Reconciliation Act of 1993 (P.L. 103-66). The Balanced Budget Act of 1997 gave the FCC auction authority until September 30, 2007. This authority was extended to September 30, 2011, by the Deficit Reduction Act of 2005 and to 2012 by the DTV Delay Act (P.L. 111-4). The Deficit Reduction Act of 2005 also specified that $7.363 billion of proceeds from auctions required by the act be applied to deficit reduction.

Distribution of Proceeds from Auctions Required by the Spectrum Act

Most of the proceeds from auctions of licenses in designated spectrum as specified in the act are to be deposited directly into a Public Safety Trust Fund, created by the act, with nearly $28 billion designated for purposes defined in the act,[7] including $20.4 billion for deficit reduction.[8]

Proceeds from the sale of licenses of repurposed federal spectrum identified in the act will be directed first to the Spectrum Relocation Fund, to cover costs of moving federal users, with the balance going to the Public Safety Trust Fund.[9] Proceeds from the sale of advanced wireless service licenses in the other spectrum bands identified by the act will go directly to the Public Safety Trust Fund.[10] Proceeds from the auction of new licenses created by the release of television broadcasting spectrum will go to cover costs specified in the act, with the balance to the Public Safety Trust Fund.[11] Balances remaining in any fund created by the act will revert to the Treasury in 2022.[12]

The Public Safety Trust Fund

The law provides for transfers from a Public Safety Trust Fund that is created by the act to receive revenues from designated auctions of spectrum licenses.[13] The designated amounts are to remain available through FY2022, after which any remaining funds are to revert to the Treasury, to be used for deficit reduction. These sums may be subject to sequestration and consequently reduced.

Auction proceeds are to be distributed in the following priority:

- To the NTIA, to reimburse the Treasury for funds advanced to cover the initial costs of establishing FirstNet: not to exceed $2 billion.
- To the State and Local Implementation Fund for a grant program: $135 million.
- To the Network Construction Fund for costs associated with building the nationwide network and for grants to states that qualify to build their own networks: $7 billion, reduced by the amount advanced to establish FirstNet.
- To NIST for public safety research: $100 million.
- To the Treasury for deficit reduction: $20.4 billion.
- To the NTIA and the National Highway Traffic Safety Administration for a grant program to improve 911 services: $115 million.
- To NIST for public safety research: $200 million.
- To the Treasury for deficit reduction: any remaining amounts from designated auction revenues.

Directed Auctions

The Spectrum Act has required the FCC and the NTIA to identify specific bands for auction from spectrum designated for commercial advanced wireless services and for federal use, and in most cases to commence the auction process within three years. The act has mandated spectrum license auctions for frequencies at 1915-1920 MHz;[14] 1995-2000 MHz; 2155-2180 MHz; an additional 15MHz to be identified by the FCC;[15] and 15MHz of spectrum between 1675 and 1710 MHz, subject to conditions in the act.[16] The Secretary of Commerce was required to submit a report to the President identifying 15 MHz of spectrum between 1675 and 1710 MHz for reallocation from federal to non-federal use.[17] The NTIA had produced a Ten-Year Plan and Timetable that identifies bands of spectrum that might be available for commercial wireless broadband service. As part of its planning efforts, NTIA prepared a "Fast Track Evaluation"[18] of spectrum resources that might be repurposed in the near future. One recommendation was to make available 15 MHz of spectrum from frequencies between 1695 MHz and 1710 MHz. The NTIA has reaffirmed its initial recommendation and submitted a report, as required by the act, recommending that the FCC reallocate the band for commercial use.[19] Some form of sharing between commercial and federal users is deemed likely, however, and recommendations for a regulatory framework for sharing are being developed by the NTIA's Commerce Spectrum Management Advisory Committee (CSMAC).[20]

The act requires that these auctions be completed and licenses issued by February 22, 2015, which would require that the auctions be held by the third quarter of 2014.[21] These licenses would provide an additional 65MHz of spectrum for commercial broadband.

Incentive Auctions

The Spectrum Act has permitted the FCC to conduct incentive auctions, that is, to establish a mechanism whereby spectrum capacity may be relinquished for auction by some license-holders, who would then share in the proceeds.[22] Many commercial wireless licenses can be resold directly by their license-holders for comparable uses; the purpose of incentive auctions is to reward license-holders, such as television broadcasters, who repurpose their spectrum for a different use. Although incentive auctions might be used for other types of license-holders, the act specifically addresses spectrum assignments for over-the-air television broadcasters, possibly limiting the applicability of the law for other license-holders who might wish to relinquish spectrum through an incentive auction.

The act has established procedures for the FCC to follow in reallocating television broadcasting spectrum licenses for commercial auction. Through a reverse auction process, the broadcasters would establish the amount of compensation they are willing to accept for the spectrum they voluntarily release for auction.[23] Additionally, broadcasters that do not voluntarily relinquish spectrum rights but are required to relocate or make other required changes may be compensated for costs incurred.[24] In lieu of cash payment, as compensation for relocation broadcasters may choose to accept regulatory relief that would allow new uses for their spectrum.[25]

Spectrum voluntarily released by TV broadcasters would be repurposed for commercial broadband communications, with licenses sold through what the law refers to as a "forward auction."[26] At least one successful reverse auction is required to set minimum prices for a forward auction. For the results of a forward auction to be valid, auction proceeds must at a minimum cover (1) payments to broadcasters that relinquished spectrum for auction, (2) the costs to the FCC of conducting the auctions, and (3) the estimated costs for relocation of other broadcasters;[27] the latter is not to exceed $1,750 million, deposited in a TV Broadcaster Relocation Fund.[28] If auction revenues do not cover costs as specified in the act, the FCC may not assign new licenses and planned reassignments and reallocations may not occur.[29] If the reverse auction and forward auction conditions are met, the FCC may "make such reassignments of televisions channels" as appropriate in its consideration, subject to certain conditions.[30] Examples of conditions include a general prohibition against reassigning licenses to frequencies from one band to a band below an existing assignment,[31] and obligations to determine that a reassigned channel is not adversely affected by cross-border channel assignment agreements with Canada and Mexico.[32] The auction and channel reassignment process may only occur once.[33]

The Balanced Budget Act of 1997, which mandated the eventual transition to digital television, represented the legislative culmination of over a decade of policy debates and negotiations between the FCC and the television broadcast industry on how to move the industry from analog to digital broadcasting technologies. To facilitate the transition, the FCC provided each qualified broadcaster with 6 MHz of spectrum for digital broadcasting to replace licenses of 6 MHz that were used for analog broadcasting. The analog licenses would be yielded back when the transition to digital television was concluded. The completed transition freed up the 700 MHz band for commercial and public safety communications in 2009.

In its 2010 *National Broadband Plan* (NBP),[34] the FCC revisited the assumptions reflected in the 1997 act and made new proposals based on, among other factors, changes in technology and consumer habits. In particular, because over-the-air digital broadcasting does not necessarily require 6 MHz of spectrum, the NBP proposed that some stations could share a single 6 MHz band without significantly reducing service to over-the-air TV viewers. Among the proposals for how broadcasters might make better use of their TV licenses, the NBP raised the possibility of auctioning unneeded spectrum and sharing the proceeds between the TV license-holder and the U.S. Treasury. The Spectrum Act has provided legislation that would allow this type of incentive auction.[35]

Federal Spectrum Use and Reallocation

The Spectrum Act has addressed how spectrum resources might be repurposed from federal to commercial use through auction or sharing, and how the cost of such reassignment would be defined and compensated, among other provisions. The Commercial Spectrum Enhancement Act of 2004 (P.L. 108-494, Title II) was amended to facilitate the transfer of spectrum rights to commercial purchasers from the agencies relinquishing spectrum. Expenditures incurred by federal agencies for planning may now be included among those costs eligible for reimbursement as part of the transfer of spectrum to the commercial sector. Other reimbursable costs cover a wide range of technical options, including spectrum sharing.[36] Although spectrum sharing to facilitate the transition from federal to commercial use is supported in the act's provisions, the NTIA has been required to give priority to reallocation options that assign spectrum for exclusive, non-federal uses through competitive bidding.[37]

The act has required the establishment of a Technical Panel within the NTIA to review transition plans that each federal agency must prepare in accordance with provisions in the act.[38] The Technical Panel is required to have three members qualified as a radio engineer or technical expert. The Director of the Office of Management and Budget, the Assistant Secretary of Commerce for Communications and Information, and the Chairman of the FCC have been required to appoint one member each.[39] A full discussion and interpretation of provisions of the act as regards the technical panel and related procedural requirements such as dispute resolution have been published by the NTIA as part of the rulemaking process.[40]

Commercial Spectrum Enhancement Act of 2004

The Commercial Spectrum Enhancement Act of 2004 put in place statutory rules for covering the costs to federal agencies of relocating wireless communications facilities to new spectrum assignments. The act created the Spectrum Relocation Fund to provide a means for federal agencies to recover relocation costs directly from auction proceeds when they are required to vacate spectrum slated for auction. In effect, successful commercial bidders cover the costs of relocation. Among key provisions of the act were requirements that the auctions must recoup at least 110% of the costs projected by the NTIA, and that unused funds would revert to the Treasury after eight years. These provisions remain in effect. Specific frequencies were designated for immediate auction[41] by the Commercial Spectrum Enhancement Act but the law was written to apply to any federally used frequencies scheduled for reallocation and possible auction.[42]

NTIA Plans to Make Federal Spectrum Available for Commercial Use

The NTIA, with input from the Policy and Plans Steering Group (PPSG),[43] has produced a 10- year plan and timetable that identifies bands of spectrum that might be available for commercial wireless broadband service. As part of its planning efforts, the NTIA prepared a "Fast Track Evaluation" of spectrum that might be made available in the near future.[44] Specific recommendations were to make available 100 MHz of spectrum within bands from 3550 MHz to 3650 MHz. The fast track evaluation also recommended studying two 20 MHz bands to be identified within 4200-4400 MHz for possible repurposing.

Working through the PPSG, the NTIA studied federal spectrum use by more than 20 agencies with over 3,100 separate frequency assignments in the 1755-1850 MHz band.[45] After evaluating the multiple steps involved in transferring current uses and users to other frequency locations, the NTIA concluded that it would cost $18 billion to clear federal users from all 95 MHz of the band.

Based on this assessment, the NTIA report included recommendations for seeking ways for federal and commercial users to share many of the frequencies, although some frequencies were identified to be cleared for auction to the private sector.

The NTIA assumptions for the estimates of the cost of relocating federal agencies from the 1755- 1850 MHz band were challenged at a hearing of the House Committee on Energy and Commerce, Subcommittee on Communications and Technology,[46] leading to a request to the GAO to examine the process. In particular, the NTIA was criticized during the hearing by some committee members for not separately evaluating the 1755-1780 MHz band, which might be auctioned separately with another spectrum band already available for commercial use. At the hearing, the GAO provided testimony regarding its preliminary findings on spectrum sharing[47] and followed up with a report.[48] Both the hearing and the report indicated that spectrum sharing technology and policies were largely undeveloped. Some of the options to encourage sharing spectrum, as identified by the GAO, include considering spectrum usage fees to provide economic incentive for more efficient use and sharing; identifying more spectrum that could be made available for unlicensed use; encouraging research and development of technologies that can better enable sharing; and improving and expediting regulatory processes related to sharing. Given the challenges for implementing spectrum sharing policies, the GAO found that further study by the NTIA and the FCC was needed.

GAO Cost Estimates for Spectrum Reallocation

In a hearing before the Senate Committee on Armed Services, Subcommittee on Strategic Forces,[49] the GAO presented preliminary findings on Department of Defense (DOD) estimates of reallocation costs from some radio frequencies.[50] The GAO evaluated DOD relocation cost estimates for frequencies at 1755-1850 MHz and reported that the "preliminary cost estimate substantially or partially met GAO's identified best practices." In particular, the GAO noted the variable nature of a number of assumptions for costs and revenues, such as the characteristics of the spectrum to which services would be relocated, the availability of new technology, and market demand for spectrum.

Unlicensed Spectrum

Unlicensed spectrum is not sold to the highest bidder and used for the services provided by the license-holder but is instead accessible to anyone using wireless equipment certified by the FCC for those frequencies. Both commercial and non-commercial entities use unlicensed spectrum to meet a wide variety of monitoring and communications needs. Suppliers of wireless devices must meet requirements for certification to operate on frequency bands designated for unlicensed use. Examples of unlicensed use include garage door openers and WiFi communications. WiFi provides wireless Internet access for personal computers and handheld devices and is also used by businesses to link computer-based communications within a local area. Links are connected to a high-speed landline either at a business location or through hotspots. Hotspots are typically located in homes or convenient public locations.

New technologies, sometimes referred to as Super WiFi, are being developed to expand the usefulness of unlicensed spectrum without causing interference. For example, to use unassigned spectrum, known as white spaces, between broadcasting signals of digital television, geolocation database technology is being put in place to identify unencumbered airwaves. Super WiFi devices are expected to reach the market by 2013.[51]

Similar technologies are being considered to expand the availability of spectrum for unlicensed use at 5 GHz by sharing with existing federal users in those frequencies.[52] Commercial providers, such as for wireless Internet, currently share parts of the spectrum at 5 GHz with federal users. With the objective of improving future WiFi capacity, the Spectrum Act has required new studies and evaluations of frequencies at 5 GHz.[53] These would lay the groundwork to expand commercial use of unlicensed spectrum within the federally managed 5 GHz band. The FCC has been required to commence a proceeding that might open access for some unlicensed devices in the 5350-5470 MHz band.[54] The NTIA was required to prepare an evaluation of spectrum-sharing technologies for the 5350-5470 MHz and 5850-5925 MHz bands.[55]

EMERGING SPECTRUM POLICY ISSUES

The United States currently enjoys a position of world leadership in mobile broadband technology development and deployment.[56] U.S. companies have been at the forefront of mobile wireless and Internet convergence. The introduction of the first iPhone in 2007 is a prominent signpost marking the convergence of these technologies.[57] Innovation in applying the Internet Protocol (IP) to mobile networks has spurred jobs and economic growth. The success of smartphones like the iPhone, in the United States and globally, followed on that of WiFi, which had stoked consumer demand for wireless access to the Internet.[58] New economic and social cultures are being built on the ubiquity and ease of access to the Internet from mobile devices.

This expanding industry requires additional spectrum capacity, a key resource. The Spectrum Act employs three key policy tools for increasing the availability of radio frequency spectrum for wireless broadband: allocating additional spectrum; reassigning spectrum to new users; and opening up spectrum for unlicensed use. Other policy options that may be employed to increase spectrum capacity include requiring that wireless network infrastructure be shared; changing the cost structure of spectrum access; moving to more spectrum-efficient technologies; and sharing spectrum. Facilitating the adoption of new wireless technologies that enable spectrum sharing is emerging as a major policy consideration for spectrum management.

Spectrum Sharing

The Administration of President Obama has identified spectrum sharing as a way to increase spectrum capacity and efficiency. For example, the FCC in its *National Broadband Plan* recommended that a "new, contiguous nationwide band for unlicensed use" be identified by 2020;[59] and that spectrum be provided and other steps taken to "further development and deployment" of new technologies that facilitate sharing.[60]

The President's Council of Advisors on Science and Technology has endorsed increasing spectrum capacity through new technology that increases efficiency and allows for shared use of spectrum resources. In a report, *Realizing the Full Potential of Government Held Spectrum to Spur Economic Growth*, the council has proposed that up to 1000 MHz of additional spectrum capacity could be provided through shared access between the federal government and commercial providers.[61] The report's recommendations include steps to new spectrum policies based on spectrum-sharing.

Although the PCAST report touches on some of the future technical solutions for spectrum sharing, it also proposes that spectrum sharing can begin immediately. It recommends immediate use of the technologies devised to allow sharing between TV broadcast spectrum with super WiFi.

Spectrum-sharing technologies include geolocation databases, smart antenna and cognitive radio—all of which are deployed—and network-centric technologies, such as Dynamic Spectrum Access, that are being tested by research and development facilities such as the Defense Advanced Research Projects Agency.[62] Enabling technologies such as these allow communications to switch instantly among network frequencies that are not in use and therefore available to any radio device equipped with cognitive technology. Future technological breakthroughs in fields such as quantum communications hold the promise of even greater transmission speeds and spectrum efficiencies.[63]

From a policy perspective, actions to speed the arrival of new, spectrally efficient technologies might have significant impact on achieving broadband policy goals over the long term. In particular, support for technologies that enable sharing could pave the way for dramatically different ways of managing the nation's spectrum resources.

CONCLUSION

Major advances in wireless technology have given the United States a competitive edge in communications innovation, fueling industry growth and job creation. Policy makers may wish to consider not only how to maintain this leadership but also how to adapt to changing economic and social expectations. The mobile network communications environment presents new challenges and opportunities in policy areas such as identity theft, privacy protection, street crime (for example, smartphone theft), health services, education, urban management, and electronic payments.[64]

The amount of spectrum needed for fully realized wireless access to broadband is such that meeting the needs of broadband policy goals could be difficult to achieve through the market-driven auction process. To meet current demands of the wireless industry, large amounts of new radio frequencies would need to be identified and released.[65] Without abandoning competitive auctions, spectrum policy could benefit from including additional ways to assign or manage spectrum that might better serve the deployment of wireless broadband and the implementation of a national broadband policy.

Current spectrum policy relies heavily on auctions to assign spectrum rights through licensing. However, the adoption of spectrum-efficient technologies is likely to require a rethinking of spectrum management policies and tools. The assignment and supervision of licenses might give way to policies and procedures for managing pooled resources. Auctioning licenses might be replaced by auctioning access; the static event of selling a license replaced by the dynamic auctioning of spectrum access on a moment-by-moment basis.

Auction winners are deemed to be the companies that can maximize the value of the spectrum to society by maximizing its value as a corporate asset. However, auction-centric spectrum policies appear to have generally focused on assigning licenses to commercial competitors in traditional markets that serve consumers and businesses. Auctioning spectrum licenses may direct assets to end-use customers instead of providing wireless services where the consumer may be the beneficiary but not the customer. Wireless networks are an important component of smart grid communications, for example. Spectrum resources are also needed for railroad safety,[66] for water conservation,[67] for the safe maintenance of critical infrastructure industries,[68] and for many other applications that may not have an immediate commercial value but can provide long-lasting value to society as a whole.

APPENDIX A. COMPETITION AND TECHNOLOGY POLICY

Telephone service was once considered a natural monopoly, and regulated accordingly. The presumption was that redundant telephone infrastructure was inefficient and not in the public interest. State and federal regulators favored granting operating rights to a single company, within a specific facilities territory, to benefit from economies of scale, facilitate interoperability, and maximize other benefits. In return for the monopoly position, the selected provider was expected to fulfill a number of requirements intended to benefit society. Thus, for decades, the regulated monopoly was seen by most policy-makers as (1) ensuring that costly infrastructure was put in place and (2) meeting society's needs, as interpreted by regulations and the law.[69] Past policies to regulate a monopolistic market may have influenced current policies for promoting competition. The FCC's emphasis on efficiency for delivering services to a pre-determined market could be leading wireless competition toward monopoly; new regulatory regimes might be a consequence of this trend, if it continues.

With the introduction of auctions for spectrum licenses in 1994, the United States began to shift away from assigning spectrum licenses based on regulatory decisions and toward competitive market mechanisms. One objective of the Telecommunications Act of 1996 was to open up the communications industry to greater competition among different sectors. One outcome of the growth of competition was the establishment of different regulatory regimes for information networks and for telecommunications.[70] As a consequence of these and other legislative and regulatory changes, the wireless industry has areas of competition (e.g., for spectrum licenses) within a regulatory shell, such as the rules governing the Public Switched Telephone Network (PSTN).[71] As the bulk of wireless communications traffic moves from voice to data, companies will likely modify their business plans in order to remain competitive in the new environment. A shift in infrastructure technology and regulatory environment[72] might open wireless competition to companies with business plans that are not modeled on pre-existing telecommunications industry formulae. Future providers of wireless broadband might include any company with a robust network for carrying data and a business case for serving broadband consumers. Potential new entrants, however, may lack access to radio frequency spectrum, the essential resource for wireless broadband.

In formulating spectrum policy, mainstream viewpoints generally diverge on whether to give priority to market economics or social goals. Regarding access to spectrum, economic policy looks to harness market forces to allocate spectrum efficiently, with spectrum license auctions as the driver. Social policy favors ensuring wireless access to support a variety of social objectives where economic return is not easily quantified, such as improving education, health services, and public safety. Both approaches can stimulate economic growth and job creation.

In evaluating competition within an industry, economists and policy makers examine barriers to entry, among other factors.[73] Barriers might come from high costs for market entry such as investment in infrastructure or there might be legal and regulatory barriers to entry. As part of its evaluation of competition for mobile services, the FCC has identified three factors that could constitute barriers to entry to the commercial mobile communications industry. These barriers affect not only competitiveness but also access to networks and investment in new technology. The factors are "first-mover advantages, large sunk costs, and access to spectrum."[74] All three of these factors are subject to regulations that have been influenced by past or existing policies regarding spectrum allocation and assignment.

First-mover advantages[75] have accrued primarily to the early entrants in the wireless industry. Early in the development of the cell phone industry, the FCC created cellular markets and assigned two spectrum licenses to each market; one license went automatically to the incumbent provider in that market. The second license was made available to a competing service provider (not the market incumbent); the difficulties in choosing the competitors that would receive licenses contributed to the subsequent move to auctions as a means for assigning spectrum rights.[76] These early entrants, and the successor companies that acquired them and their licenses, have maintained their core customer base and benefit from early investments in infrastructure. Many first movers into the wireless market, therefore, acquired their market-leader status through regulatory decisions that provided them with spectrum licenses, not through market competition.

Large sunk costs refer to the high levels of investment needed to enter the wireless market. Not including the price of purchasing spectrum, billions of dollars are required to build new infrastructure. The sunk costs of incumbent wireless service providers set a high bar for new entrants to match if they are to compete effectively in major markets. In the mobile telephone industry, the FCC has observed that most capital expenditures are spent on existing networks: to expand and improve geographic coverage; to increase capacity of existing networks; and to improve network capabilities. Performance requirements for spectrum license-holders, such as the size of a market that must be served or deadlines for completing infrastructure build-outs, are some of the policy decisions that can add to the cost of entry.

Spectrum Auctions and Competition

The FCC was authorized to organize auctions to award spectrum licenses for certain wireless communications services in the Omnibus Budget Reconciliation Act of 1993 (P.L. 103-66). The act amended the Communications Act of 1934 with a number of important provisions affecting the availability of spectrum. The Licensing Improvement section[77] of the act laid out the general requirements for the FCC to establish a competitive bidding methodology and consider, in the process, objectives such as the development and rapid deployment of new technologies.[78] The law prohibited the FCC from making spectrum allocation decisions based "solely or predominately on the expectation of Federal revenues.... ."[79] The Emerging Telecommunications Technologies section[80] directed the NTIA to identify not less than 200 MHz of radio frequencies used by the federal government that could be transferred to the commercial sector through auctions.[81] The FCC was directed to allocate and assign these released frequencies over a period of at least 10 years, and to reserve a significant portion of the frequencies for allocation after the 10-year time span.[82] Similar to the requirements for competitive bidding, the FCC was instructed to ensure the availability of frequencies for new technologies and services, and also the availability of frequencies to stimulate the development of wireless technologies.[83] The FCC was further required to address "the feasibility of reallocating portions of the spectrum from current commercial and other non-federal uses to provide for more efficient use of spectrum" and for "innovation and marketplace developments that may affect the relative efficiencies of different spectrum allocations."[84] Over time, auction rules have been modified in accordance with the changing policy goals of the FCC and Congress but subsequent amendments to the Communications Act of 1934 have not substantively changed the above-noted provisions regarding spectrum allocation.[85]

The rules set by the FCC for using spectrum licenses (service rules) may have been oriented toward the concepts of building and managing networks that were formed in the days of the telephone, favoring traditional telecommunications business plans over those of companies with different business models. Some companies that might be well suited to meet social goals, such as access in rural areas, might have been precluded from bidding at all because of constraints not considered relevant to market-driven allocations. For example, public utilities, municipal cooperatives, commuter railroads, and other public or quasi-public entities face a variety of legal, regulatory, and structural constraints that limit or prohibit their ability to participate in an auction or buy spectrum licenses. Many of these constraints exist at the state level but federal spectrum policy plays a role in perpetuating the status quo.

There are many ways to view competition. Although competitiveness may be evaluated by factors such as barriers to entry or number of market participants, a key measure of whether market competition is working is an assessment of the dynamic of a specific market: its prices, variety, level of service, and other indicators that are considered hallmarks of competitive behavior. The Federal Trade Commission, for example, promotes competition as "the best way to reduce costs, encourage innovation, and expand choices for consumers."[86] Viewpoints about the level of competitiveness in providing wireless services to the U.S. market differ. However, telecommunications business analysts generally describe the U.S. market for wireless services as competitive because consumers benefit in many ways from competition on price, service, coverage, and the availability of new devices.

Both the wireless industry and its regulator have focused on "wireless consumer welfare"[87] in evaluating competition and the effectiveness of spectrum policies for assigning spectrum licenses. Auctions are judged to be an efficient way of assigning spectrum for commercial uses that adhere to traditional business plans.[88]

Spectrum Caps

As part of its preparations for the first spectrum license auctions, the FCC decided to set caps on the amount of spectrum any one company could control in any geographically designated market.[89] The theory behind spectrum capping is that each license has an economic value and a foreclosure value. The economic value is derived from the return on investment in spectrum licenses and network infrastructure. The foreclosure value is the value to a wireless company that already has substantial market share and wants to keep its dominant position by precluding competition. Spectrum caps were chosen as the method to prevent foreclosure bidding. The intent was to ensure multiple competitors in each market and to restrict bidding to only the licenses that could be used in the near term.

Beginning in 2001, spectrum policy placed increased emphasis on promoting spectrum and market efficiency through consolidation. The FCC ruled to end spectrum caps, citing greater spectral efficiency from larger networks as one benefit of the ruling. Spectrum caps were seen as barriers to mergers within the wireless industry, to the growth of existing wireless companies, and to the benefits of scale economies. The spectrum caps were eliminated on January 1, 2003.[90] Auction rules requiring the timely build-out of networks became a key policy tool to deter hoarding. The FCC instituted a policy for evaluating spectrum holdings on a market-by-market, case-by-case basis—a practice referred to as spectrum screening—as a measure of competitiveness.

In 2008, the Rural Telecommunications Group, Inc. (RTG) petitioned the FCC to impose a spectrum cap of 110 MHz for holdings below 2.3 GHz. In October 2008, the FCC sought comments on the RTG petition for rulemaking.[91] RTG argued that competition in the industry was declining as it became more concentrated. It claimed that the larger carriers were warehousing their spectrum holdings in rural areas while rural carriers were struggling to acquire spectrum capacity for mobile broadband and expansion. Rural carriers, RTG reported, were being shut out of opportunities to acquire new spectrum holdings and were being outbid in spectrum auctions.[92] Opponents to the spectrum cap cited data to support their claims that the wireless communications market is competitive. They argued that additional amounts of spectrum are needed to support the growth in mobile broadband and that a spectrum cap could cut off growth and innovation.[93] Implementing spectrum caps as a tool for regulating competition would represent a significant shift in policy for the FCC, were it to take that course.

In comments filed regarding the *National Broadband Plan*, the Department of Justice considered the possibility that "the foreclosure value for incumbents in a given locale could be very high."[94] Although it recognized some form of spectrum caps as an option for assuring new market entrants, it observed that "there are substantial advantages to deploying newly available spectrum in order to enable additional providers to mount stronger challenges to broadband incumbents."[95]

Competition in Rural Markets

Over the years, various legislative and policy initiatives have created a number of requirements to help small and rural carriers acquire spectrum licenses.[96] Some of the FCC's efforts to encourage spectrum license ownership for small, rural, or entrepreneurial businesses are in response to congressional mandates.[97] These and other statutory and regulatory programs may have allowed many small carriers to remain in business even though many others have been absorbed by larger carriers.[98] As wireless traffic, revenue, and profits migrate to broadband, business models that were effective for voice traffic may no longer be viable, especially for companies that have relied on the regulatory environment to protect their markets. This change in operating environment may have disproportionately affected the ability of rural wireless carriers, in particular, to compete effectively.[99] A study of how new technologies might be affecting the competitiveness of small and rural carriers might be useful in reviewing the effectiveness of policies intended to aid them.[100]

The FCC, acting on the statutory authority given to it by Congress, has broad regulatory powers for spectrum management. The FCC was created as part of the Communications Act of 1934[101] as the successor to the Federal Radio Commission, which was formed under the Radio Act of 1927.[102] The first statute covering the regulation of airwaves in the United States was the Radio Act of 1912, which gave the authority to assign usage rights (licenses) to the Secretary of the Department of Commerce and Labor.[103] Licensing was necessary in part because, as radio communications grew, it became crucial that frequencies be reserved for specific uses or users, to minimize interference among wireless transmissions.[104] A key component of spectrum policy is the allocation of bands of frequencies for specific uses and the assignment of licenses within those bands. Allocation refers to the decisions, sometimes reached at the international level, that set aside bands of frequencies for categories of uses or users; assignment refers to the transfer of spectrum rights to specific license-holders.

The rules set by the FCC for using spectrum licenses (service rules) may have been oriented toward the concepts of building and managing networks that were formed in the days of the telephone, favoring traditional telecommunications business plans over those of companies with different business models.

APPENDIX B. SPECTRUM-HUNGRY TECHNOLOGIES

Enabling technologies that are fueling both the demand for mobile broadband services and the need for radio frequency spectrum include Long Term Evolution (LTE); WiMAX; fixed wireless; WiFi; high performance mobile devices such as smartphones and tablets; and cloud computing.

Fixed wireless and WiFi are not new technologies but mobile broadband has given them new roles in meeting consumer demand. Future technologies include network-centric technologies, which include opportunistic solutions such as Dynamic Spectrum Access (DSA).

Long Term Evolution (LTE)

LTE is the projected development of existing 3G networks built on Universal Mobile Telephone System (UMTS) standards.[105] Like all fourth-generation wireless technologies, LTE's core network uses Internet protocols.

The network architecture is intended to facilitate mobile broadband deployment with capabilities that can deliver large amounts of data, quickly and efficiently, to large numbers of simultaneous users. LTE will likely be implemented in stages through modifications to networks using frequencies in bands already allocated for commercial wireless networks.[106] LTE might operate on spectrum bands at 700 MHz, 1.7 GHz, 2.3 GHz, 2.5 GHz, and 3.4 GHz.[107]

WiMAX

WiMAX provides mobile broadband but its earliest applications were for fixed wireless services. WiMAX (Worldwide Interoperability for Microwave Access) refers to both a technology and an industry standard, the work of an industry coalition of network and equipment suppliers.[108]

WiMAX uses multiple frequencies around the world in ranges from 700 MHz to 66 GHz. In the United States, available frequencies include 700 MHz, 1.9 GHz, 2.3 GHz, 2.5 GHz, and 2.7 GHz. The introduction of WiMAX in the United States is being jointly led by Sprint Nextel Corporation and Clearwire Corporation.

Fixed Wireless Services

Fixed wireless services have taken on new importance as a "backhaul" link for 4G. Backhaul is the telecommunications industry term that refers to connections between a core system and a subsidiary node.

An example of backhaul is the link between a network—which could be the Internet or an internetwork that can connect to the Internet—and the cell tower base stations that route traffic from wireless to wired systems. Two backhaul technologies well-suited for mobile

Internet access are fiber optic cable and point-to-point microwave radio relay transmissions.[109] Network expansion plans for WiMAX and LTE include microwave links as a cost-effective substitute for fiber optic wire under certain conditions. Radio frequencies available in the United States for microwave technologies of different types start in the 930 MHz band and range as high as the 90 GHz band.

WiFi

The popularity of WiFi is often cited as a successful innovation that was implemented using unlicensed frequencies.[110] WiFi provides wireless Internet access for personal computers and handheld devices and is also used by businesses to link computer-based communications within a local area.

Links are connected to a high-speed landline either at a business location or through hotspots. Hotspots are typically located in homes or convenient public locations, including airports and café environments such as Starbucks. WiFi uses radio frequencies in the free 2.4 GHz and 5.4/5.7GHz spectrum bands. Many 3G and 4G wireless devices that operate on licensed frequencies can also use the unlicensed frequencies set aside for WiFi.[111]

Cloud Computing

Cloud computing is a catch-all term that is popularly used to describe a range of information technology resources that are separately stored for access through a network, including the Internet. An Internet search on Google, for example, is using cloud computing to access a rich resource of data and information processing. Network connectivity to services is another resource provided by cloud computing. Google Inc., for example, offers word processing, e-mail, and other services through Google Docs. Although off-site data processing and information storage are not new concepts, cloud computing benefits from the significant advances in network technology and capacity that are hallmarks of the broadband era. Cloud computing can provide economies of scale to businesses of all sizes. Small businesses in particular can benefit from forgoing the costs of installing and managing hardware and software by buying what they need from the cloud. Consumers also can benefit because they no longer need to buy personal computers in order to run complex programs or store large amounts of data.

The convergence of 4G wireless technology—with its smartphones and netbooks—and the growing accessibility of cloud computing to businesses and consumers alike will contribute to the predicted explosive growth in demand for wireless bandwidth.

Network-Centric Technologies

The concept of channel management dates to the development of the radio telegraph by Guglielmo Marconi and his contemporaries. In the age of the Internet, however, channel management is an inefficient way to provide spectrum capacity for mobile broadband. Innovation points to network-centric spectrum management as an effective way to provide spectrum capacity to meet the bandwidth needs of fourth-generation wireless devices.[112] Network-centric technologies enable sharing by organizing the transmission of radio signals along the same principle as the Internet. A transmission moves from origination to destination not along a fixed path but by passing from one available node to the next. When radios are networked using network-centric technologies, individual communications nodes continue to operate and can compensate for failed links. The effects of interference are manageable rather than catastrophic. The network is used to overcome radio limitations. With channel management techniques currently in use, if a channel's link fails, the radio is cut off.

Pooling resources, one of the concepts that powers the Internet now, is likely to become the dominant principle for spectrum management in the future. Dynamic Spectrum Access (DSA), Content-Based Networking, and Delay and Disruption Technology Networking, along with cognitive radio, and decision-making software, are examples of technologies that can enable Internet-like management of spectrum resources.

DSA is part of the neXt Generation program, or XG, a technology development project sponsored by the Strategic Technology Office of the Defense Advanced Research Projects Agency (DARPA). The main goals of the program include developing both the enabling technologies and system concepts that dynamically redistribute allocated spectrum.

The Department of Defense (DOD) is working to implement network-centric operations (NCO) through a number of initiatives.[113] Leadership and support to achieve DOD goals in the crucial area of spectrum management is provided by the Defense Spectrum Organization (DSO) created in 2006 within the Defense Information Systems Agency (DISA). The DSO is leading DOD efforts to transform spectrum management in support of future net-centric operations and warfare, and to meet military needs for dynamic, agile, and adaptive access to spectrum. The DSO is guiding DOD spectrum management along a path that envisions moving away from stove-piped systems to network-centric spectrum management.

DARPA has commenced a program to improve radar and communications capabilities by creating technical solutions to enable spectrum sharing. The program, Shared Spectrum Access for Radar and Communications (SSPARC), will focus on frequencies between 2 and 4 GHz. According to DARPA, technologies could be applicable to other frequencies as well. The program seeks to support two types of sharing environments for military radar: one with other military communications networks and one for commercial networks.[114]

End Notes

[1] The Code of Federal Regulations defines natural resources as "land, fish, wildlife, biota, air, water, ground water, drinking water supplies and other such resources belonging to, managed by, held in trust by, appertaining to, or otherwise controlled by the United States.... " (15 CFR 990, Section 990.30).

[2] Broadband refers here to the capacity of the radio frequency channel. A broadband channel can quickly transmit live video, complex graphics, and other data-rich information as well as voice and text messages, whereas a narrowband channel might be limited to handling voice, text, and some graphics.

[3] With technologies that rely on channel management, two signals can interfere with each other even if they are not at the same exact frequency, but are close in frequency. To avoid harmful interference, the signals must have frequencies that are sufficiently different, known as a "minimum separation."

[4] Measures in the act that apply to public safety are covered in CRS Report R42543, *The First Responder Network and Next-Generation Communications for Public Safety: Issues for Congress*, by Linda K. Moore.

[5] 47 USC §308 (j) (8). Net proceeds are the auction revenues minus the FCC's expenses. Congress has twice in the past amended the provision in order to use auction proceeds for other purposes by creating special funds to

hold and disburse auction proceeds. The Commercial Spectrum Enhancement Act, Title II of P.L. 108-494 created the Spectrum Relocation Fund; the Deficit Reduction Act of 2005 created the Public Safety and Digital Television Transition Fund.

[6] 47 USC §308 (j) (3).

[7] H.Rept. 112-96, §6413.

[8] P.L. 112-96, §6413(b) (5). §6413 (a) (2).

[9] P.L. 112-96, §6401 (c) (3) (C).

[10] P.L. 112-96, §6401 (c) (3) (C).

[11] P.L. 112-96, §6401 (c) (4).

[12] P.L. 112-96, §6413 (a) (2).

[13] P.L. 112-96, Section 6413.

[14] Spectrum is segmented into bands of radio frequencies and typically measured in cycles per second, or hertz. Standard abbreviations for measuring frequencies include kHz—kilohertz or thousands of hertz; MHz—megahertz, or millions of hertz; and GHz—gigahertz, or billions of hertz.

[15] P.L. 112-96, §6401 (b).

[16] P.L. 112-96, §6401 (a).

[17] P.L. 112-96, §6401 (a).

[18] NTIA, An Assessment of Near-Term Viability of Accommodating Wireless Broadband Systems in the 1675-1710 MHZ, 1755-1780 MHz, 3500-3650 MHz, and 4200-4220 MHz, 4380-4400 MHZ Bands (President's Spectrum Plan Report), November 15, 2010, at http://www.ntia.doc.gov/report/2010/assessment-near-term-viability-accommodatingwireless-broadband-systems-1675-1710-mhz-17.

[19] Department of Commerce, Identification of 15 Megahertz of Spectrum Between 1675 and 1710 MHz for Reallocation from Federal Use to Non-Federal Use Pursuant to Section 640 (a) of the Middle Class Tax Relief and Job Creation Act of 2012; Report to the President, February 2013, at http://www.ntia.doc.gov/report/2013/report-presidentidentification-15-mhz-spectrum-between-1675-mhz-and-1710-mhz-reallocati.

[20] Information on CSMAC at http://www.ntia.doc.gov/search/node/CSMAC.

[21] See FCC, "Remarks of Commissioner Jessica Rosenworcel, CTIA 2013—The Mobile Marketplace, Las Vegas, Nevada, May 22, 2013, at http://www.fcc.gov/document/commissioner-rosenworcels-speech-ctia-2013.

[22] P.L. 112-96, §6402 "(G) "(i).

[23] P.L. 112-96, §6402 "(G) "(ii) and §6403 (a).

[24] P.L. 112-96, §6403 (b) (4) (A).

[25] P.L. 112-96, §6403 (b) (4) (B).

[26] P.L. 112-96, §6403 (c) (1).

[27] P.L. 112-96, §6403 (c) (2) (B).

[28] P.L. 112-96, §6402 "(G) "(iii) "(I).

[29] P.L. 112-96, §6403 (c) (2) (A).

[30] P.L. 112-96, §6403 (b) (1) (B) (i).

[31] P.L. 112-96, §6403 (b) (3).

[32] P.L. 112-96, §6403 (b) (1) (B).

[33] P.L. 112-96, §6403 (e).

[34] Connecting America: The National Broadband Plan, 2010 at http://www.broadband.gov.

[35] Information about FCC actions to implement incentive auctions at http://www.fcc.gov/incentiveauctions.

[36] P.L. 112-96, §6701 (a) (1) (D) "(3).

[37] P.L. 112-96, §6701 (a) (3) "(j).

[38] P.L. 112-96, §6701 (a) (3) "(h).

[39] P.L. 112-96, §6701 (a) (3) "(h) "(3) "(B).

[40] NTIA, Notice of Proposed Rulemaking, July 17, 2012, and replies, docket no. 110627357-2209-03 at http://www.ntia.doc.gov/federal-register-notice/2012/technical-panel-and-dispute-resolution-board-nprm. Final Rule, January 25, 2013, at http://www.ntia.doc.gov/federal-register-notice/2013/spectrum-relocation-final-rule-technicalpanel-and-dispute-resolution-b.

[41] Following the procedures required by the act, the FCC scheduled an auction for Advanced Wireless Services (AWS), designated Auction 66, which was completed on September 18, 2006. The AWS auction attracted nearly $13.9 billion in completed bids. The cost to move federal agencies to new spectrum locations was set at almost $936 million.

[42] The creation of the Spectrum Relocation Fund is discussed in CRS Report RS21508, Spectrum Management and Special Funds, by Linda K. Moore.

[43] Created in response to Department of Commerce recommendations to improve spectrum efficiency through better management, see http://www.ntia.doc.gov/legacy/reports/specpolini/factsheetspecpolini_06242004.htm.

[44] NTIA, *An Assessment of Near-Term Viability of Accommodating Wireless Broadband Systems in the 1675-1710 MHZ, 1755-1780 MHz, 3500-3650 MHz, and 4200-4220 MHz, 4380-4400 MHZ Bands (President's Spectrum Plan Report)*, November 15, 2010, at http://www.ntia.doc.gov/report/2010/assessment-near-term-viability-accommodatingwireless-broadband-systems-1675-1710-mhz-17.

[45] U.S. Department of Commerce, *An Assessment of the Viability of Accommodating Wireless Broadband in the 1755- 1850 MHz Band,* March 2012, at http://www.ntia.doc.gov/report/2012/assessment-viability-accommodating-wirelessbroadband-1755-1850-mhz-band.

[46] Hearing, House of Representatives, Committee on Energy and Commerce, Subcommittee on Communications and Technology, "Creating Opportunities Through Improved Government Spectrum Efficiency," September 13, 2012.

[47] GAO, *Spectrum Management: Federal Government's Use of Spectrum and Preliminary Information on Spectrum Sharing*, September 13, 2012, GAO-12-1018T at http://www.gao.gov/products/GAO-12-1018T .

[48] GAO, *Spectrum Management: Incentives, Opportunities, and Testing Needed to Enhance Spectrum Sharing*, November 14, 2012, GAO-13-7 at http://gao.gov/products/GAO-13-7.

[49] Hearing, Senate, Committee on Armed Services, Subcommittee on Strategic Forces, "Oversight: Military Space Programs and Views on DoD Usage of the Electromagnetic Spectrum," April 24, 2013.

[50] GAO, *Spectrum Management: Preliminary Findings on Federal Relocation Costs and Auction Revenues*, April 24, 2013, GAO-13-563T at http://www.gao.gov/products/GAO-13-563T. Full report, GAO-13-472 at http://www.gao.gov/ products/GAO-13-472.

[51] "Spectrum Bridge Gains Final FCC Approval, White Spaces Broadband Era to Begin," by Joan Engebretson, telecompetitor.com, December 22, 2011.

[52] These and other frequencies for unlicensed use are discussed in *The Economic Value Generated by Existing and Future Allocations of Unlicensed Spectrum,* Perspective, Ingenious Consulting Network, September 28, 2009; sponsored by Microsoft, Inc.

[53] P.L. 112-96, §6406.

[54] FCC, *Notice of Proposed Rulemaking: 5GHz Unlicensed Spectrum (UNII)*, FCC13-22, released February 20, 2013, http://www.fcc.gov/document/increased-spectrum-available-unlicensed-devices-5-ghz-band.

[55] Department of Commerce, Evaluation of the 5350-5470 MHz and 5850-5925 MHz Bands Pursuant to Section 6406 (b) of the Middle Class Tax Relief and Job Creation Act of 2012, January 2013, http://www.ntia.doc.gov/files/ntia/ publications/ntia_5_ghz_report_01-25-2013.pdf.

[56] See for example spoken and written testimony of witnesses before the House Committee on Science, Space, and Technology, Subcommittee on Technology and Innovation, "Avoiding the Spectrum Crunch: Growing the Wireless Economy Through Innovation," April 18, 2012.

[57] "Spectrum and the Wireless Revolution," Randall Stephenson, *Wall Street Journal*, June 10, 2012.

[58] See "A Brief History of Wi-Fi," *The Economist Technology Quarterly*, June 10, 2004.

[59] *Connecting America*, Recommendation 5.11.

[60] *Connecting America*, Recommendation 5.13.

[61] Recommendations of the President's Council of Advisors on Science and Technology, *Realizing the Full Potential of Government-Held Spectrum to Spur Economic Growth*, released July 20, 2012, http://www.whitehouse.gov/sites/ default/files/microsites/ostp/pcast_spectrum_report_final_july_20_2012.pdf.

[62] Some of the research in network-centric technologies is discussed in Appendix B.

[63] Some of the principles of quantum communications and possibility of a quantum Internet are discussed in "Breakthrough in Quantum Communication," *Science Daily*, April 11, 2012.

[64] Some of these policy issues are discussed in *Socioeconomic Impacts of Wireless Technology*, prepared by BSR for CTIA – The Wireless Association, May 2012. CRS products include "Smartphone Theft and Crime Prevention," Congressional Distribution Memorandum, March 30, 3012, by Linda K. Moore, available on request; CRS Report R41733, *Privacy: An Overview of the Electronic Communications Privacy Act*, by Charles Doyle; CRS Report R41756, *Privacy Protections for Personal Information Online*, by Gina Stevens; CRS Report R42511, *United States v. Jones: GPS Monitoring, Property, and Privacy*, by Richard M. Thompson II; CRS Report R40908, *Advertising Industry in the Digital Age*, by Suzanne M. Kirchhoff; CRS Report R40599, *Identity Theft: Trends and Issues*, by Kristin Finklea; and, CRS Report RL34632, *Text and Multimedia Messaging: Issues for Congress*, by Patricia Moloney Figliola and Gina Stevens.

[65] International Telecommunications Union projects an estimated need for additional spectrum capacity that could reach nearly 1,000 MHz in the United States, as reported in "Summary of Results of ITU-R Report M. 2079,"

p. 13, presented by Cengiz Evci, Chief Frequency Officer, Wireless Business Group, Alcatel-Lucent, August 28, 2007. Available at http://standards.nortel.com/spectrum4IMT/Geneva/R03-WRCAFR07-C-0024.pdf. See also CTIA-The Wireless Association, *Written Ex Parte Communication*, FCC, GN Docket No. 09-51, September 29, 2009, which suggests a goal of at least 800 MHz, based on extrapolations from the ITU research.

[66] The railroad industry uses wireless communications as part of their information networks to monitor activity.

[67] For example, sensors buried at the level of plant roots recognize when watering is needed and communicate this information over wireless networks.

[68] In general, critical infrastructure industries facilitate the production of critical goods and services such as safe drinking water, fuel, telecommunications, financial services, and emergency response. A discussion of key issues appears in CRS Report RL30153, *Critical Infrastructures: Background, Policy, and Implementation*, by John D. Moteff.

[69] The original Communications Act of 1934 codified many regulations for monopolies as practiced at the time.

[70] For a discussion of policy issues, see CRS Report R40234, *The FCC's Authority to Regulate Net Neutrality After Comcast v. FCC*, by Kathleen Ann Ruane, and CRS Report R40616, *Access to Broadband Networks: The Net Neutrality Debate*, by Angele A. Gilroy.

[71] PSTN is a global system; rights of access and usage in the United States are regulated by the FCC.

[72] On December 1, 2009, the FCC published a public notice seeking comments on the "appropriate policy framework to facilitate and respond to the market-led transition in technology and services, from the circuit-switched PSTN system to an IP-based communications world." "Comment Sought on Transition from Circuit-Switched Network to All-IP Network," NBP Public Notice #25, DA 09-2517 at http://hraunfoss.fcc.gov /edocs_public/attachmatch/DA-09- 2517A1.pdf.

[73] For example, U.S. Department of Justice and the Federal Trade Commission, "Horizontal Merger Guidelines," Jointly issued April 2, 1992, revised April 8, 1997.

[74] FCC, "Wireless Telecommunications Bureau Seeks Comment on Commercial Mobile Radio Services Market Competition," Public Notice, February 25, 2008, DA 08-453, WT Docket No. 08-27 at http://hraunfoss.fcc.gov/ edocs_public/attachmatch/DA-08-453A1.pdf. Earlier annual reports have also cited these barriers.

[75] The initial occupant of a market segment may benefit from a number of advantages such as preemption of resources, advantageous relationships with customers and suppliers, and early profits for reinvestment in infrastructure.

[76] The distribution of licenses for cell phone networks from the early days of the technology until the introduction of auctions is described in *Wireless Nation: The Frenzied Launch of the Cellular Revolution in America*, by James B. Murray, Jr., Perseus Press, 2001, 2002.

[77] P.L. 103-66 Title III, Subtitle C, Chapter 1.

[78] 47 U.S.C. §309 (j), especially (1), (3), and (4).

[79] 47 U.S.C. §309 (j) (7) (A).

[80] P.L. 103-66 Title III, Subtitle C, Chapter 2.

[81] 47 U.S.C. §923 (b) (1).

[82] 47 U.S.C. §925 (b) (1).

[83] 47 U.S.C. §925 (b) (2).

[84] 47 U.S.C. §925 (b) (3).

[85] See United States Code Annotated, Title 47, sections as footnoted, WEST Group, 2001 and the 2007 Cumulative Annual Pocket Part.

[86] "Competition in the Technology Marketplace" at http://www.ftc.gov/bc/tech/index.htm.

[87] This phrase is used in the written statement of AT&T Inc. submitted for a hearing before the House of Representatives, Committee on Energy and Commerce, Subcommittee on Communications, Technology, and the Internet, "An Examination of Competition in the Wireless Industry," May 7, 2009. In written testimony submitted by Verizon Wireless for the same hearing, comments stated that wireless providers need suitable and sufficient spectrum because of "consumers' reliance on broadband services."

[88] The GAO has reported this viewpoint in several reports, including *Telecommunications: Strong Support for Extending FCC's Auction Authority Exists, but Little Agreement on Other Options to Improve Efficient Use of Spectrum*," December 20, 2005, GAO-06-236 and *Telecommunications: Options for and Barriers to Spectrum Reform*, March 14, 2006, GAO-06-526T.

[89] Licenses are designated for a specific geographic area, such as rural areas, metropolitan areas, regions, or the entire nation.

[90] FCC News, "FCC Announces Wireless Spectrum Cap to Sunset Effective January 1, 2003," November 8, 2001. Report and Order FCC-01-328. See Docket No. 01-14, *Notice of Proposed Rulemaking*, released January 23, 2001, at http://hraunfoss.fcc.gov/edocs_public/attachmatch/FCC-01-28A1.pdf.

[91] FCC RM No. 11498, October 10, 2008. Comments supporting and opposing the petition are published in this proceeding.

[92] Those supporting the RTG petition included the Organization for the Promotion and Advancement of Small Telecommunications Companies (OPASTCO), the National Telecommunications Cooperative Association, the Public Interest Spectrum Coalition, and a number of smaller (non-dominant) wireless carriers.

[93] Opponents to spectrum caps that filed comments were AT&T Inc., Verizon Wireless, CTIA—The Wireless Association, the Telecommunications Industry Association, and the Wireless Communications Association International.

[94] *Ex Parte* Submission of the United States Department of Justice, In the matter of Economic Issues in Broadband Competition: A National Broadband Plan for Our Future, GN Docket 09-51, January 4, 2010, p. 23 at http://fjallfoss.fcc.gov/ecfs/document/view?id=7020355122.

[95] Ibid., p. 24.

[96] For example, most auctions have provided bidding credits for small businesses.

[97] In 47 USC §309 (j) (3) (B), the FCC is instructed to promote "economic opportunity and competition and ensuring that new and innovative technologies are readily available to the American people by avoiding excessive concentration of licenses and by disseminating licenses among a wide variety of applicants.... "

[98] The Congressional Budget Office (CBO) reported in a 2005 study that a significant number of small companies that acquired spectrum licenses through preferential programs later transferred the licenses to larger companies: *Small Businesses in License Auctions for Wireless Personal Communications Services*, A CBO Paper, October 2005, at http://www.cbo.gov/ftpdocs/68xx/doc6808/10-24-FCC.pdf.

[99] A number of rural wireless carriers and their associations have filed comments on the increasing difficulties they face in competing for wireless customers. Comments are in a number of FCC dockets, such as RM11498, regarding spectrum caps, and WT Docket No. 09-66, on the state of wireless competition.

[100] The CBO study cited above was prepared at the request of the Senate Budget Committee to examine the impact of small-bidder preferences on federal revenue and was completed before data traffic became a significant factor in providing wireless services.

[101] 47 U.S.C. §151.

[102] P.L. 632, §3.

[103] P.L. 264, "License."

[104] An "Act to regulate radio communications," usually referred to as the Radio Act of 1912, was passed partly in response to radio problems—including interference—associated with the sinking of the *Titanic*. Hearings Before a Subcommittee of the Committee on Commerce, 62nd Congress, 2nd Session, pursuant to S. Res. 283, "Directing the Committee on Commerce to Investigate the Cause Leading to the Wreck of the White Star Liner 'Titanic,'" testimony of Guglielmo Marconi, et al.

[105] See, for example, "Mobile Broadband Evolution: the roadmap from HSPA to LTE," UMTS Forum, February 2009, Universal Mobile Telephone System Forum at http://www.umts-forum.org/.

[106] Implementation summarized in *Connecting America*, Exhibit 5-B, p. 77.

[107] Spectrum is segmented into bands of radio frequencies and typically measured in cycles per second, or hertz. Standard abbreviations for measuring frequencies include kHz—kilohertz or thousands of hertz; MHz—megahertz, or millions of hertz; and GHz—gigahertz, or billions of hertz.

[108] Founding members of the WiMAX Forum include Airspan, Alvarion, Analog Devices, Aperto Networks, Ensemble Communications, Fujitsu, Intel, Nokia, Proxim, and Wi-LAN. For additional information, see http://www.wimaxforum.org/.

[109] A discussion of backhaul technology is part of the testimony of Ravi Potharlanka, Chief Operating Officer, Fiber Tower Corp., at House of Representatives, Committee on Energy and Commerce, Subcommittee on Communications, Technology, and the Internet, "An Examination of Competition in the Wireless Industry," May 7, 2009.

[110] Unlicensed frequencies are bands set aside for devices approved by the FCC. The frequencies are effectively managed by the FCC instead of by a license-holder.

[111] "Wi-Fi Popular Now in Smartphones, Set to Boom," by Matt Hamblen, *Computerworld*, April 1, 2009.

[112] A leading advocate for replacing channel management of radio frequency with network-centric management is Preston Marshall, the source for much of the information about network-centric technologies in this report.

Mr. Marshall is Director, Information Sciences Institute, University of Southern California, Viterbi School of Engineering, Arlington, VA.

[113] A discussion of the goals of NCO is included in CRS Report RL32411, *Network Centric Operations: Background and Oversight Issues for Congress*, by Clay Wilson.

[114] DARPA announcement, February 8, 2013, at http://www.darpa.mil/NewsEvents/Releases/2013/02/08a.aspx.

In: Transformations in Telecommunications and Media
Editor: Irwin Cavazos

ISBN: 978-1-62948-413-6
© 2013 Nova Science Publishers, Inc.

Chapter 5

THE FCC'S BROADCAST MEDIA OWNERSHIP AND ATTRIBUTION RULES: THE CURRENT DEBATE[*]

Charles B. Goldfarb

SUMMARY

The Federal Communications Commission's (FCC's) broadcast media ownership rules, which place restrictions on the number of media outlets that a single entity can own or control in a local market or nationally, are intended to foster the three long-standing goals of U.S. media policy— competition, localism, and diversity of voices. The FCC is statutorily required to review these rules every four years to determine whether they continue to serve the public interest or should be modified or eliminated. One part of these rules, the FCC's attribution rules, identify criteria for determining when an entity holds sufficient ownership or control of a broadcast station that such ownership or control should be attributed to the entity for the purposes of applying the media ownership rules.

In December 2011, the FCC proposed a number of rule changes, which it has not yet adopted. It proposed eliminating its Radio/Television Cross-Ownership rule because it is no longer needed to foster the goals of diversity of voices and localism. It also proposed modifying its Newspaper/Broadcast Cross-Ownership rule to allow certain types of combinations in the 20 largest markets. It proposed a technical change in its Local Television Ownership Rule, but otherwise would continue to prohibit ownership of two stations in a local market unless one is not among the four highest-ranked stations in the market and, after the combination, there would still be eight independently owned and operating commercial full-power television stations. The FCC proposed that its Local Radio Ownership and Dual Network rules be retained as is. The FCC also sought public comment on how to define the criteria for an entity to be eligible for programs intended to promote the diversity of media ownership, and, in particular, to promote ownership by women and minorities.

---

[*] This is an edited, reformatted and augmented version of a Congressional Research Service publication, CRS Report for Congress R42436, prepared for Members and Committees of Congress, from www.crs.gov, dated January 10, 2013.

In recent years, many television stations have entered into sharing arrangements with other stations in their local market to jointly sell advertising and/or produce local news programming, typically with one station managing that shared operation and perhaps providing most or all of the staffing and other resources. The FCC sought public comment on how, for the purposes of the media ownership rules, to attribute control of a broadcast television station that has entered into such a sharing arrangement. Currently, the only sharing agreement-related attribution rule for television stations covers local marketing agreements in which one station both purchases blocks of time from another station in the same market and sells the advertising for the purchased time— that is, the broker station provides both the programming and the advertising—for at least 15% of the brokered station's broadcasting time. The FCC has enforced this as a bright-line rule. As long as (1) the block of time covered by an agreement does not exceed 15% of the brokered station's programming time, and (2) the agreement contains a certification and perhaps other language indicating that the licensee of the brokered station maintains ultimate control over station finances, personnel, and programming, the agreement will not trigger the attribution rule. Other evidence is considered immaterial. As a result, in many cases the FCC has not deemed a station to have control over another station in the same market even if such control is considered to exist, and must be reported, under generally accepted accounting practices. Such agreements create what is known in the industry as "virtual duopolies."

In late 2012, the FCC released—and made available for public comment—a report on broadcast ownership by gender, ethnicity, and race, and invited the public to comment on how its proposed ownership rule changes might affect female and minority ownership. It delayed adoption of new broadcast ownership rules until those public comments could be analyzed. It is expected to adopt new rules early in 2013.

OVERVIEW

The Federal Communications Commission's (FCC's or Commission's) broadcast media ownership rules,[1] which place restrictions on the number of media outlets that a single entity can *own or control* in a local market or nationally, are intended to foster the three long-standing goals of U.S. media policy—competition, localism, and diversity of voices. The FCC is statutorily required to review these rules every four years to determine whether they continue to serve the public interest or should be modified or eliminated.[2]

The FCC's attribution rules identify criteria for determining when an entity holds sufficient ownership or control of a broadcast station that such ownership or control should be attributed to the entity for the purposes of applying the media ownership rules.[3] According to the FCC, its attribution rules "seek to identify those interests in or relationships to licensees that confer a degree of influence or control such that the holders have a realistic potential to affect the programming decisions of licensees or other core operating functions."[4]

On May 25, 2010, the Commission adopted a Notice of Inquiry to obtain public input for the most recent review of its media ownership rules.[5] It held public hearings and commissioned and released 11 economic studies performed by outside researchers and commission staff that provided data on the impact of market structure on competition, localism, and diversity of voices.

On December 22, 2011, the Commission adopted a Notice of Proposed Rulemaking (NPRM)[6] in which it tentatively proposed loosening or eliminating some of the current constraints on cross-ownership of media outlets in local markets. For the Local Television

Ownership rule that it proposed to retain, it sought additional public comment on whether to adopt a waiver standard applicable to small markets and whether digital multicasting should be a factor in determining the television ownership limits. The NPRM also asked questions and sought public comment on how (for the purposes of the media ownership rules) to attribute control of a broadcast station when that station has entered into certain types of sharing arrangements with another station in the same market, suggesting that the FCC might consider some changes to its attribution rules that could have the effect of tightening ownership restrictions. The NPRM also asked questions and sought public comment on how to define the criteria for an entity to be eligible for programs intended to promote the diversity of media ownership—and, in particular, to increase the level of broadcast station ownership by minorities and women.

In public comments that were submitted on March 5, 2012, broadcasters argued that they face increased competition from non-broadcast media outlets (especially cable networks and Internet websites) that has reduced their audience and revenues and also that existing ownership constraints keep them from exploiting economies of scale that would allow them to offer more local news programming.[7] They therefore seek to loosen existing ownership restrictions. They also oppose expansion of attribution rules or reporting requirements related to sharing arrangements because, they claim, such requirements would impede their ability to attain efficiencies that lead to improved programming.[8]

Several consumer groups argue that the primary focus of the current review of FCC rules should be on the relationship between market structure, ownership rules, and minority and female ownership, and claim that the FCC has failed to perform analysis that would justify its proposed loosening of existing rules.[9] They allege that such analysis would demonstrate that without more stringent ownership rules ever fewer stations will be available for ownership opportunities for minorities and women.[10] Small cable companies, which must negotiate with broadcasters for permission to retransmit broadcast signals,[11] claim that ownership or contractual arrangements that allow one station to negotiate on behalf of multiple stations in a market give the broadcasters an unfair advantage and harm consumers by raising cable costs (and hence prices) and should be prohibited.[12]

Although all of the tentative rules, information requests, and questions in the NPRM have elicited comment, the proceeding has generated especially lively debate about the interaction between the FCC's Local Television Ownership rule and its attribution rules. Although no systematic database exists to confirm this, broadcasters, consumer groups, and other commenters seem to agree that the number of sharing arrangements among television stations in which one station contractually takes on some functions for one or more other stations in the same market is growing. Most (but not all) broadcasters view these sharing agreements as necessary to take advantage of scale economies they are otherwise denied by the restrictions in the ownership rules on formal combinations. Critics, however, view them as "an end around the prohibitions against controlling two top four television stations and/or controlling more than one (or two in larger markets) television stations in the same market."[13] Moreover, the critics claim that the FCC has improperly allowed sharing agreements under which one station has de facto control over the programming and core operations of another station in the same local market, thus effectively circumventing the Local Television Ownership rule.[14]

In November 2012, the FCC adopted and released a report on current ownership of broadcast stations—full power commercial, Class A, low power, commercial AM radio, and commercial FM radio—by gender, ethnicity, and race.[15] After several parties requested an additional, formal opportunity to comment on the report, the FCC established a public comment cycle, noting that the report confirmed that women and minorities continue to "own broadcast stations in disproportionately small numbers."[16] The Commission delayed adoption of its new broadcast media ownership rules until those comments, which might address the potential impact of its proposed rule changes on female and minority ownership, could be submitted and reviewed. The FCC is expected to adopt final media ownership rules early in 2013.

The FCC's on-going quadrennial review process has identified two general areas of inquiry that may be of interest to Congress. First, there is the traditional policy debate: do each of the FCC's current media ownership and attribution rules foster the long-standing policy goals of localism, diversity of voices, and competition, or should some rules be eliminated or modified or new rules added? In that regard, there is the possibility that a rule change to foster one goal might harm another goal and Congress might want to provide the FCC with guidance on how to proceed in that situation. Second, has the FCC constructed, interpreted, and enforced its attribution rules in a fashion that allows a single entity to control two of the four highest-rated stations in a market in apparent contradiction of the Local Television Ownership rule? Has the FCC, as a result, in effect changed its policy with respect to ownership without changing its rules?

Proposed Rules, Information Requests, and Questions Posed in the FCC's Notice of Proposed Rulemaking[17]

Ownership Rules

With respect to the specific ownership rules, the FCC tentatively concluded that:

- The current Radio/Television Cross-Ownership rule should be eliminated.
- The current Newspaper/Broadcast Cross-Ownership rule should be modified to include the major elements of a rule that it adopted in its previous quadrennial review but was vacated and remanded by the Court of Appeals for the Third Circuit for procedural reasons.[18] The proposed new rule would deem it presumptively consistent with the public interest for an entity to own or control both a major daily newspaper and a television or radio station in a local market if (1) the combination is in one of the 20 largest local markets, (2) for television (but not for radio), the station is not among the four highest-rated stations in the market (for radio, there is no restriction on the station's rating), and (3) after the transaction there still are at least eight independently owned and operating major media voices in the market.
- The current Local Television Ownership rule should be retained with minor modifications that take into account the transition (that has already taken place) from analog to digital transmission of broadcast television signals.
- The current Local Radio Ownership rule should be retained as is.
- The current Dual Network Rule should be retained as is.

Sharing Agreements and Attribution Rules

The media ownership rules place restrictions on the number of media properties that a single entity can own *or control* in a particular local market. Increasingly, television stations are entering into sharing agreements with other stations in their market. These agreements are legal, but may result in one station effectively controlling the programming decisions or other core operating functions of the other station. In such a situation, the first station could be viewed as controlling the second station, for the purposes of the ownership rules. The FCC has adopted attribution rules that identify certain financial or other interests in a broadcast station, including sharing agreements, that constitute control over that station and must be counted in applying the broadcast ownership rules.[19]

Currently, if two radio stations or two television stations in a local market enter into a local marketing agreement (LMA—sometimes referred to as time brokerage agreement or TBA), in which one station (the "broker" station) purchases discrete blocks of time from the other station and supplies programming and sells advertising for the purchased time, and that brokered time exceeds 15% of the weekly broadcast time of the "brokered" station, an FCC rule attributes control of the station to the broker station.[20] Also, if two radio stations (but not television stations) in a local market enter into a joint sales agreement (JSA)—an agreement for the joint sales of broadcast commercial time—in which the broker station sells more than 15% of the advertising time per week of the brokered station, a rule attributes control of the station to the broker station even though such agreements do not involve programming decisions.[21]

In 2004, the FCC tentatively concluded that JSAs have the same effect in local television markets as they do in local radio markets and therefore proposed extending the JSA attribution rule to television stations,[22] but it has never adopted a final order. Neither LMAs nor JSAs are precluded by any Commission rule or policy as long as the Commission's multiple ownership rules are not violated and the participating licensees maintain ultimate control over their facilities.

In recent years, stations have begun to enter into other types of sharing arrangements with stations in their local markets that may not be addressed in the current attribution rules. For example, under local news service (LNS) agreements multiple local broadcast television stations contribute certain news staff and equipment to a joint news gathering effort coordinated by a single managing editor. Under shared service agreements (SSAs) one television station provides operational support and programming for another station in the same market, in some cases producing the news content for one or more other stations.[23]

In its NPRM, the FCC seeks public comment on whether and how to attribute control of stations that have entered into such arrangements, with an apparent interest in agreements among television stations in a local market that explicitly involve news programming.[24] The FCC does not propose any new or modified attribution rules in the NPRM, but by asking questions and seeking comment it is signaling that it might make changes to its attribution rules in a final order.

Entities Eligible for Programs to Foster Broadcast Ownership Diversity

Congress and the FCC have a history of supporting actions—rules, policies, programs, guidelines—intended to foster diversity of broadcast station ownership, in general, and minority and female ownership, in particular, as a means to achieve the goal of diversity of voices.[25]

The courts have reinforced the notion that minority ownership is a valid concern that should be addressed by the FCC when constructing its media ownership rules,[26] but also have set restrictions on how actions can be structured to foster that goal.[27] In particular, the Supreme Court has declared that, to ensure that they satisfy the Equal Protection Clause of the Fourteenth Amendment to the United States Constitution, any government actions that employ race-based classifications are subject to strict scrutiny and may be upheld "only if they are narrowly tailored measures that further compelling governmental interests."[28]

With this in mind, when the FCC in 2008 adopted a number of rules, policies, programs, and guidelines intended to promote diversity of ownership, although it explicitly indicated its intent to foster minority and female ownership, it "decided to employ a race- and gender-neutral definition [of eligible entity] in the rules ... so as to avoid constitutional difficulties.... "[29] Specifically, it defined as eligible "any entity that would qualify as a small business consistent with the Small Business Administration (SBA) standard for its industry grouping, based on revenue."[30]

But the court ruled the FCC's revenue-based eligibility definition arbitrary and capricious, finding that the Commission failed to demonstrate that measures based on that definition would enhance the stated goal of increasing broadcast ownership by minorities and women.[31] The court remanded those measures adopted in the FCC Diversity Order that use that eligibility definition and instructed the Commission to address the issue in its 2010 Quadrennial Review of its media ownership rules.[32]

In the current NPRM, the FCC does not propose a new definition of "eligible entity." Rather, it seeks input "on how the Commission most effectively can expand upon its diversity initiatives at the same time that we address the Third Circuit's concerns and other legal considerations, including potential impediments to affording licensing preferences to minorities and women under current standards of constitutional law."[33] It asks many questions about the options available for reconsideration of the eligible entity standard to use.[34]

At this time, the Commission is merely collecting information, and will postpone any decision on eligibility criteria until its 2014 quadrennial review. The FCC concluded:

> we believe that making legally sound proposals would not be possible based on the record before us at this time. Accordingly, we plan to undertake the following actions in preparation for the 2014 broadcast ownership review to establish with the requisite foundation and clarity what additional policies can be implemented promoting greater broadcast ownership diversity, including female and minority ownership: 1) Continue to improve our data collection so that we and the public may more easily identify the diverse range of broadcast owners, including women and minorities, in all services we license; 2) Commission appropriately-tailored research and analysis on diversity of ownership; and 3) Conduct workshops on the opportunities and challenges facing diverse populations in broadcast ownership. In addition, we ask interested parties to supplement the record and provide any and all data available that can complete a picture of the current state of ownership diversity, including minority and female ownership in the broadcast industry and to justify any prospective actions the Commission may take on remand.[35]

THE RADIO/TELEVISION
CROSS-OWNERSHIP RULE

Current Rule

Under the current Radio/Television Cross-Ownership rule:[36]

- An entity may own or control up to two television stations (provided it is permitted under the Local Television Ownership rule) *and* up to six radio stations (provided it is permitted under the Local Radio Ownership rule) in a local market where at least 20 independently owned media voices would remain post-merger.
- Where entities may own a combination of two television stations and six radio stations, an entity alternatively may own one television station and seven radio stations.
- An entity may own or control up to two television stations (provided it is permitted under the local television ownership rule) *and* up to four radio stations (provided it is permitted under the local radio ownership rule) in a local market where at least 10 independently owned media voices would remain post-merger.
- A combination of one television station and one radio station is allowed regardless of the number of independently owned voices remaining in the local market.
- For the purpose of applying this rule, independently owned voices include all independently owned and operating (i) full-power broadcast television stations, (ii) broadcast radio stations, (iii) newspapers that publish at least four times a week, and (iv) cable systems (which count as a single voice).

Proposed Rule Change

In the NPRM, the FCC tentatively concludes that the Radio/Television Cross-Ownership rule is no longer necessary to promote the public interest and therefore tentatively proposes that it be repealed.[37]

The FCC does not expect that repeal would lead to significant media consolidation. But, according to the Commission, even if greater consolidation were to occur, data in the record suggest that radio/television cross-ownership does not negatively impact the amount or diversity of local news available to consumers. The FCC tentatively concludes that in the current media market, the goals of localism and diversity will be adequately protected by the Local Radio and Local Television Ownership rules without the additional cross-ownership limitation.

Broadcasters support the proposed repeal of the Radio/Television Cross-Ownership rule, arguing that it does not foster localism, diversity of voices, or competition, and citing empirical evidence that in-market television-radio combinations provide scale economies that increase local news production.[38] The National Hispanic Media Coalition (NHMC), et al., a collection of organizations representing women, people of color, and people in rural areas on telecommunications policies, opposes repeal of the rule, claiming there is not sufficient evidence on the record about radio ownership by women and people of color and that "the

Commission's conclusion runs contrary to NHMC et al.'s research in the Los Angeles and Rio Grande Valley [local markets]."[39] Most other commenters did not directly address this proposed rule.

NEWSPAPER/BROADCAST CROSS-OWNERSHIP RULE

Current Rule

The Newspaper/Broadcast Cross-Ownership rule currently in effect[40] prohibits ownership or control of a daily newspaper *and*:

- a full power television stations whose Grade A service contour[41] encompasses the entire community in which the newspaper is published; *or*
- a full power AM radio station whose predicted or measured 2 millivolt per meter contour[42] encompasses the entire community in which the newspaper is published; *or*
- a full power FM radio station whose predicted 1 millivolt per meter contour[43] encompasses the entire community in which the newspaper is published.

A broadcaster can start a new daily newspaper in a local market in which it owns a television or radio station, but cannot combine with an existing newspaper.

Despite these prohibitions, a number of newspaper/broadcast combinations exist in local markets because when the FCC first adopted the cross-ownership prohibition in 1975 it grandfathered some pre-existing combinations and, in the past decade, the FCC first granted a number of temporary waivers, and then in 2007 granted permanent waivers to five of those newspaper-broadcast station combinations.[44]

In 2007, the FCC adopted a rule that relaxed the newspaper/broadcast cross-ownership prohibition in the 20 largest local markets,[45] but the Third Circuit remanded the rule for procedural reasons, finding that the Commission had failed to comply with the notice and comment provisions of the Administrative Procedures Act.[46] The court did not address the Commission's substantive modifications to the rule.

Proposed Rule Change

In the NPRM, the FCC tentatively concludes that some newspaper/broadcast cross-ownership restrictions continue to be necessary to protect and promote viewpoint diversity, but that a blanket prohibition on such combinations is overly broad and does not allow for certain cross-ownership combinations that might be in the public interest.[47] It tentatively concludes that the opportunity to share newsgathering resources and realize other efficiencies derived from economies of scale and scope may improve the ability of commonly owned media outlets to provide local news and information,[48] and that such cross-ownership can be accommodated in the largest markets, which have many media outlets, without unduly harming viewpoint diversity.[49]

The Commission therefore proposes a rule that incorporates the central provisions in the rule that it adopted in 2007 (but was remanded by the courts on procedural grounds) and updates that rule to reflect the broadcast television transition from analog to digital transmission.[50] The proposed rule would prohibit cross-ownership subject to an exception, as follows:

(1) It would prohibit ownership or control of a daily newspaper *and*:

- a full power television station whose community of license is in the same local market (referred to as a designated market area or DMA[51]) as the entire community in which the newspaper is published; *or*
- a full power AM radio station whose predicted or measured 2 mV/m contour encompasses the entire community in which the newspaper is published; *or*
- a full power FM radio station whose predicted 1 mV/m contour encompasses the entire community in which the newspaper is published.

(2) There would be a presumption, however, that in the 20 largest DMAs such cross-ownership is consistent with the public interest so long as (a) at least eight independently owned and operating major media voices (full-power television broadcast stations and major newspapers) would remain in the DMA after the combination, and (b) if the proposed combination involves a television station, the television station is not ranked among the top four in terms of audience ratings.[52]

Under the currently proposed rule, proposed newspaper-broadcast combinations would be reviewed on a case-by-case basis without specific guidelines.[53] In contrast, the remanded 2007 rule was very complex, with every proposed newspaper-broadcast combination subject to a public interest determination that required multiple steps: there were two defined circumstances that would automatically reverse a negative presumption about a proposed combination and four factors that had to be considered to confirm or rebut a positive or negative public interest presumption about a proposed combination.[54]

Most commenters oppose the FCC proposal, either seeking elimination of the newspaper-broadcast cross-ownership rule in its entirety or seeking its retention in its entirety. The National Association of Broadcasters claims the newspaper/broadcast cross-ownership rule should be eliminated because it is unnecessary for competition or to promote viewpoint diversity, and harms localism by preventing efficiencies from newspaper-broadcast combinations that would yield more local news programming.[55] The United Church of Christ (UCC) opposes modification of the current rule because it would "facilitate further industry consolidation and reduce ownership opportunities for minorities and women."[56] Free Press, a non-profit media reform organization, supports the current rule because it "preserves news independence and ensures access to diverse and competing sources of local news, both off- and online."[57]

The Diversity and Competition Supporters, a coalition of 50 national organizations created in 2002 to advance the cause of minority ownership, states that the Newspaper-Broadcast Cross-Ownership rule has little impact on minority ownership, especially compared to the impact of the local television ownership rule.[58]

LOCAL TELEVISION OWNERSHIP RULE

Current Rule

Under the current local television ownership rule, which is sometimes referred to as the "TV duopoly" rule, an entity may own or control two television stations in the same local market (DMA) only *if either*:

- the Grade B contours[59] of the stations do not overlap, *or*
- at least one of the stations is not among the four highest-ranked stations in the DMA,[60] *and* (b) at least eight independently owned and operating commercial or non-commercial full-power broadcast television stations would remain in the DMA after the proposed combination were consummated.[61]

The second criterion is sometimes referred to as the "top four ranked/eight voices test."

An existing licensee of a failed, failing, or unbuilt television station may seek a waiver of the rule.[62] Any combination formed as a result of a failed, failing, or unbuilt station waiver may be transferred together only if the combination meets the Local Television Ownership rule or the relevant (failed, failing, or unbuilt television station) waiver standard at the time of transfer.[63] In practice, once combinations have been formed through the waiver process, the FCC has allowed them to be transferred together even if they no longer meet the waiver standard.

Historically, the "top four ranked" stations in a local market have been, almost without exception, the local affiliates of the four major English language broadcast television networks—ABC, CBS, Fox, and NBC. This is no longer the case. In recent years, in local markets with large Hispanic communities, the local Univision affiliate sometimes is among the top four ranked stations. At the same time, market rankings can change dramatically from year to year, based on the success of a national network's programming. For example, for several years NBC network programming was not very popular relative to other major network programming, and thus some local NBC affiliates were no longer in the top four rankings in their markets, but this changed in 2012 when NBC broadcast more popular programming. Thus the combination of a local NBC affiliated station with another major network-affiliated local station that might have been allowed prior to 2012 might not have been allowed in 2012.

Proposed Rule Change

In the NPRM, the FCC tentatively concludes that the Local Television Ownership rule remains necessary to promote competition in local markets.[64] With the successful transition from analog to digital broadcast transmission, however, the first criterion, which is a measure of analog signal intensity, is no longer relevant. The FCC therefore proposes elimination of that criterion—while grandfathering any existing combinations that have been approved based on that criterion.[65]

The non-overlapping Grade B criterion typically has been applicable in those unusual instances in which a DMA is so large that two stations located within that DMA could be transmitting at their full power and still be sufficiently far away from one another that their signals did not overlap. In those instances, since there would be few if any households with acceptable over-the-air reception of the signals of both stations, allowing the two stations to have common ownership likely would have negligible impact on competition, localism, or diversity of voices.

It is possible that, with the elimination of the non-overlapping Grade B criterion, there could be two stations in a DMA whose signals would be received with acceptable over-the-air reception by few if any households in the DMA, but who would no longer qualify for dual ownership (because they could not meet the "top four ranked/eight voices test"). Thus the proposed rule change could result in the prohibition of some combinations that currently would be permitted. The FCC seeks comment on how frequently such situations arise and whether and how to accommodate such a situation, while tentatively concluding that it should grandfather such ownership combinations that already exist.[66]

Request for Additional Public Comment on Waivers and on Multicasting

Some commenters raised concerns that prohibiting all television mergers in small markets could prevent competitively challenged broadcasters in those markets from realizing potential efficiencies that could be achieved through common ownership. The FCC therefore sought comment on whether allowing certain combinations in small markets, even between top-four stations, would promote additional local news.[67] For example, the FCC cited a staff analysis that suggests that markets with six or fewer stations may be less able to support four local television news operations and perhaps combinations in such markets would foster, not reduce, news programming.

With the completion of the digital television transition in June 2009, full-power television stations have the ability to use their available spectrum to broadcast not only their main program stream, but also, if they choose, additional program streams; this is known as multicasting. Some commenters state that this allows an individual licensee to broadcast multiple revenue-generating program streams without having to purchase an additional in-market station and even allows the licensee to affiliate with more than one national programming network in markets with fewer than four stations. Thus, they argue, stations do not need to resort to duopoly ownership to expand revenue opportunities that support local news programming.[68] But broadcasters argue that multicasting is not a substitute for duopoly ownership; they have had difficulty attracting popular programming for the multicast channels and cable and satellite operators are not obligated to carry those channels under the "must carry" rules.[69] Given these conflicting perspectives, the FCC seeks comment on whether multicasting replicates the potential benefits to station owners and viewers associated with owning a second in-market station (for example, efficiency gains and improved programming) and whether stations in small markets still need to own multiple stations—and perhaps even two top-four stations—to generate the revenues and exploit the cost efficiencies needed to provide local news programming.[70]

There were more comments—and more detailed comments—on the Local Television Ownership rule than on any of the other media ownership rules. The broadcasters propose that the rule be eliminated, claiming the rule is unnecessary to ensure competition and does not preserve localism or diversity and that allowing duopolies would strengthen the broadcasters' ability to serve their communities.[71]

The NAB also opposes the minor rule change proposed by the FCC—to update the rule to take into account the digital transition by eliminating the condition that the stations have overlapping Grade B contours—as overly restrictive and not reflective of technical and marketplace realities.[72]

Most other commenters either support the FCC proposal to retain the local ownership rule with a minor change to reflect the digital transition or propose that the local ownership rule be made more stringent.

For example, MMTC supports retention of the duopoly rule, claiming "duopolies have threatened minority ownership because of the fact that lenders and investors are less willing to finance a standalone station when they can finance duopolies because of their more attractive revenue models" and because "local television duopolies decrease the local programming that is available to minority consumers."[73] Free Press prefers that the FCC return to a strict "one to a market rule" but would support retention of the current rule if the FCC addressed "the problem of 'covert consolidation' by local television stations" through the use of sharing agreements.[74] UCC claims that retaining the existing local television rule would negatively affect ownership opportunities for minorities and women and thus also prefers a strict one to a market rule.[75]

As will be explained in greater detail below, the debate about the local television ownership rule frequently is tied to in-market sharing agreements and how they are treated by the FCC's attribution rules.

LOCAL RADIO OWNERSHIP RULE

Current Rule

The local radio ownership limits currently in place are those that the FCC adopted in 1996 to codify the language in Section 202(b)(1) of the 1996 Telecommunications Act (though they now use a methodology for defining local radio markets that the Commission adopted in 2003). Specifically, the current rule provides that:

- in a radio market with 45 or more full power commercial and noncommercial radio stations, an entity may own, operate, or control up to eight commercial radio stations, not more than five of which are in the same service (AM or FM);
- in a radio market with between 30 and 44 (inclusive) full power commercial and noncommercial radio stations, an entity may own, operate, or control up to seven commercial radio stations, not more than four of which are in the same service (AM or FM);
- in a radio market with between 15 and 29 (inclusive) full power commercial and noncommercial radio stations, an entity may own, operate, or control up to six commercial radio stations, not more than four of which are in the same service (AM or FM);

- in a radio market with 14 or fewer full power commercial and noncommercial radio stations, an entity may own, operate, or control up to five commercial radio stations, not more than three of which are in the same service (AM or FM), except that an entity may not own, operate, or control more than 50 percent of the stations in such market.[76]

These numerical limits are applied to geographic markets that are defined according to boundaries defined by Arbitron, a commercial audience measurement service. The Arbitron rating boundaries are based on market factors rather than on the signal transmission contours that previously were used to define markets.[77] Since Arbitron boundaries do not cover small radio markets, the FCC performed a rulemaking proceeding to determine how to define geographic markets in those small markets for which there are no Arbitron market definitions.[78]

Proposed Rule Change

In the NPRM, the FCC tentatively concludes that the current Local Radio Ownership rule remains necessary to promote competition and does not propose any changes to the rule.[79] It does seek comment, however, on whether to change the existing numerical limits and/or market tiers; on the impact of the ongoing digital radio transition on the differences between AM and FM stations; on whether to adopt a specific waiver standard; and on the impact of the Local Radio Ownership rule on minority and female ownership.[80]

The NAB proposes elimination of the radio ownership rule, claiming that competition in the audio market from internet radio, satellite radio, and mobile devices render the rule obsolete and that the localism and diversity are impeded by the rule.[81] UCC claims that retention of the current rule would negatively affect ownership opportunities for minorities and women and proposes that the radio ownership rule be tightened, with entities that could not meet the new requirements forced to divest within a reasonable period of time, thereby making stations available to women and minorities.[82]

Dual Network Rule

Current Rule

The Dual Network rule permits common ownership of multiple broadcast television networks, but prohibits a merger among ABC, CBS, FOX, and NBC,[83] which historically have been the major national networks.

In 2001, as part of an earlier review of its broadcast media ownership rules, the FCC modified the rule to allow those four major networks to own, operate, maintain, or control broadcast networks other than the four majors. With this change, Viacom, then owner of CBS, was allowed to purchase UPN, and NBC was able to purchase Telemundo, the second-largest Spanish-language television network in the United States.

Proposed Rule Change

In the NPRM, the FCC tentatively finds "that the top four broadcast networks continue to possess characteristics that distinguish them from other broadcast and cable networks and therefore still serve a unique role in the electronic media that justifies retaining a rule specific to them."[84] It tentatively concludes that "a top-four network merger would restrict the availability, price, and quality of primetime entertainment programming to the detriment of consumers ... and would substantially lessen competition for advertising dollars in the national advertising market."[85] It therefore tentatively concludes that the dual network rule should be retained as is.

Few, if any, parties commented on the Dual Network rule.

SHARING AGREEMENTS AND ATTRIBUTION RULES

The FCC's attribution rules identify criteria for determining when an entity holds sufficient ownership or control of a broadcast station that such ownership or control should be attributed to the entity for the purposes of applying the multiple ownership rules.[86] According to the FCC order in which the rules were adopted, the attribution rules "seek to identify those interests in or relationships to licensees that confer a degree of influence or control such that the holders have a realistic potential to affect the programming decisions of licensees or other core operating functions."[87]

Most significantly, under the current Local Television Ownership rule, also known as the duopoly rule, an entity may own *or control* two television stations in the same local market only if at least one of the stations is not ranked among the four highest-ranked stations in the market *and* at least eight independently owned and operating commercial or non-commercial full-power stations would remain in the market after the proposed combination was consummated. In practice, this rule applies primarily to small markets, since larger markets are likely to have more than eight independent stations. If certain sharing agreements between two stations in a market are deemed, by the FCC's attribution rules, to give one station control over the other, then that may result in the controlling station being out of compliance with the duopoly rule.

Currently, the only sharing agreement-related attribution rule for television stations covers local marketing agreements (LMAs) in which one station both purchases blocks of time from another station in the same market and sells the advertising for the purchased time—that is, the broker station provides both the programming and the advertising—for at least 15% of the brokered station's broadcasting time.[88] The FCC has enforced this as a bright-line rule with a clearly defined standard that leaves little or no room for varying interpretation and produces predictable and consistent results in its application. As long as (1) the block of time covered by an agreement does not exceed 15% of the brokered station's programming time, and (2) the agreement contains a certification and perhaps other language indicating that the licensee of the brokered station maintains ultimate control over the station's facilities, including, specifically, control over station finances, personnel, and programming, the agreement will not trigger the attribution rule.[89] (The second criterion reflects note 2j(3) to the Commission's attribution rule.[90])

In their comments, most broadcasters note approvingly that the FCC has placed very few restrictions on broadcast sharing arrangements and urge the FCC "to continue to provide broadcasters with flexibility to structure management and shared services agreements, joint sales agreements, and other innovative cost-sharing arrangements between same-market stations."[91] According to Nexstar Broadcasting:

> Although duopolies provide stations with the greatest level of efficiencies (providing the greatest benefit to revenue-challenged medium and small market stations), because the current Local TV Ownership Rule prohibits duopolies in those markets most in need of such ownership relief, many of these stations have entered into Local Service Agreements. Nexstar's LSAs allow the stations to produce and broadcast more local news and other local programming; allow the stations to invest in significant capital expenditures for Doppler weather radar, satellite trucks and high definition news equipment; allow the stations to be more active community participants; and provide public service organizations with an expanded platform to promote their causes with the community. Without these LSAs, both stations would be significantly impacted, with one station losing the ability to produce local programming without the expenditure of resources that simply are not available in these small markets.[92]

The broadcaster position is succinctly expressed by the National Association of Broadcasters (NAB):

> ... consumers benefit when stations have flexibility to cooperate and share resources with other local media outlets, whether through duopolies, LMAs, joint sales agreements or shared services agreements. Indeed, these types of arrangements help make it possible for local stations to continue providing extensive and expensive local news programming and emergency journalism, despite facing significant economic challenges.[93]

NAB argues that "sharing arrangements do not provide the opportunity to exert significant influence over another licensee's programming or core operating functions" because they typically limit the amount of provided programming to no more than 15% of the licensee's weekly schedule, the threshold level in the FCC's attribution rules.[94] Moreover, NAB claims that the sharing agreements typically are structured to specify a flat fee to be paid by the licensee in exchange for services, which creates a financial structure that ensures that the licensee retains economic incentives to control the station.[95] For these reasons, most broadcasters oppose any expansion of the current attribution rules relating to sharing arrangements.

Other commenters in the FCC's media ownership proceeding, and also in the Commission's future of media proceeding,[96] however, raise two general areas of concern about the attribution rules relating to sharing agreements that they want the Commission to address.

- Some parties express concern that stations have entered into sharing arrangements involving news gathering and news programming that are not covered by current attribution rules but that may harm localism, diversity of voices, and competition and thus merit careful FCC review and analysis.[97] They want the Commission to collect and analyze data on all types of sharing arrangements, to require stations to include all sharing agreements (not just those currently subject to the attribution rule) in their public files, and to expand attribution rules to cover a broader range of sharing arrangements. The United Church of Christ proposes a bright line, multifactor test, which is much stricter than the current attribution rule, to attribute ownership where one station exercises substantial influence over another station in the market as a result of a sharing agreement.[98]

- Some parties claim that the FCC has adopted attribution criteria, permitted waivers and exceptions to its media ownership and attribution rules, or interpreted and enforced those rules in a fashion that has resulted in many instances of two top-four stations in a local market participating in sharing arrangements in which one of the licensees effectively cedes control of its station to the other licensee.[99] They claim this has resulted in wide-scale violation of the Local Television Ownership rule and seek stricter enforcement of existing ownership and attribution rules.

Sharing Agreements Involving the Production of News Programming

Currently, the only sharing agreement-related attribution rule for television stations narrowly addresses local marketing agreements in which one station both purchases blocks of time from another station in the same market and sells the advertising for the purchased time—that is, the broker station provides both the programming and the advertising—for at least 15% of the brokered station's broadcasting time.[100] When the FCC adopted the rule in 1999,[101] it had identified 74 LMAs in which the brokering and brokered stations were located in the same market and, for these, on average the LMA covered more than 90% of the brokered station's broadcasting time.[102] From that perspective, the 15% threshold in the rule, which was a carryover from the preexisting radio rule, did not appear to provide a constraint on any station decision to participate in an agreement. One commenter, Paxson Communications, which owned independent television stations and its own small programming network, argued that control of a television station should not be attributed based on the same standard as radio because radio stations (at that time) generally were programmed entirely on a local basis (that is less the case today), while most television stations relied (and still rely) heavily on national network and syndicated programming. To more closely parallel the radio LMA attribution rule, Paxson proposed that any television attribution rule be based on the amount of locally produced programming provided by the broker station.[103]

According to Bob Pepper, who is the primary researcher and author of an annual Radio and Television Digital News Association (RTDNA) survey of broadcast station news directors that is widely recognized in the industry, just under half the non-PBS broadcast television stations in the United States produce and air original local news programming and an additional 14% broadcast local news programming produced by another station. According to the most recent *RTDNA/Hofstra Survey*,[104] those stations that produce their own local news programming aired an average of 5.3 hours of news programming on weekdays and 1.7 hours on Saturdays and Sundays. If stations are broadcasting 24/7, this would represent 17.8% of their broadcasting time. The amount of local news programming broadcast by stations that do not produce their own programming is, of course, on average much lower, with more than 600 stations not airing any local news programming at all. Thus, for most broadcast television stations, if another station in their market were to produce all of their local news programming, such programming would not reach the 15% threshold in the current attribution rule.

Although three-quarters of the news directors participating in the *RTDNA/Hofstra Survey* were "not sure" of the percentage of their station revenues generated by news, the one-quarter that did know reported that news generated a substantial portion of station revenues—the average figure reported was 46.8%, the median figure 45%.[105] The survey does not indicate whether this includes revenues received by stations for producing news programming for other stations, and if it does this figure may overstate the financial importance of news programming to a typical station, but given that the local station gets to keep all the advertising revenues generated during the time slots for which it produces it own programming rather than relying on network or syndicated programming, it would appear that for many if not most stations the approximately 15% of total programming time taken up by news is financially significant. An entity that provides that slot of programming for another station may attain a degree of influence or control such that it has a "realistic potential to affect the programming decisions of licensees or other core operating functions."

In recent years, many broadcast television stations have entered into sharing arrangements with other stations in their market for the production of news programming, sometimes with an additional arrangement for joint advertising. In its comments in the FCC's future of media proceeding,[106] the Communications Workers of America (CWA) identified agreements in at least 42 different markets. It identified three types of agreements: (1) shared service agreements (SSAs) in which one broadcast licensee has the contractual right to produce programming and operate the station of another licensee, in return for consideration; (2) joint sales agreements (JSAs) in which one broadcast licensee (the broker) is allowed to sell the advertising time for another station (the brokered station), typically in exchange for a percentage of the advertising revenues, though sometimes the brokered station pays the broker a commission and the brokered station retains the advertising revenues; and (3) local news service (LNS) arrangements in which two or more stations in a local market share news gathering resources and share the footage among the participating broadcasters. According to its submission:

- In 19 markets, two or more stations participated in local news service arrangements; in one of these markets, two (of the three) participants in the LNS also had a shared service agreement and in two of the markets, the participants also had joint sales agreements. Most LNS arrangements are in large markets (markets ranked 1-5, 7-9, 11-13, 17, 32, 37, 49, 71, and 116). In three of these, five stations participated; in six of the markets, the local stations affiliated with three (or all four) of the four major national broadcast networks participated.

- In 26 markets, two or more stations were parties to shared service agreements. In four of these markets, the stations also had joint sales agreements and in one the participants, along with a third station, had a local news service arrangement. Most SSAs are in mid-size and smaller (but not the smallest) markets—six are in markets ranked from 18 to 50, 11 in markets ranked from 51-100, and nine in markets ranked from 101-139.

CWA did not have access to the contracts to provide details,[107] but it alleges that all SSAs include LMAs and therefore the only thing currently keeping them from being subject to the

Commission's attribution rule is the 15% threshold level in the rule.[108] It therefore proposes that the Commission lower that threshold level. It claims that such a rule would not keep a top-four station in a large market from providing some news programming to a small independent station, but would prohibit two top-four stations from reducing the number of voices in a market by combining and simulcasting their local news operations.[109]

UCC proposes that the FCC collect data on broadcast joint ventures and their effect on localism, competition, and diversity.[110] CWA wants the FCC to require copies of LNS agreements and other sharing agreements be placed in a station's public file and to require that any contribution by an LNS to a report be credited on-air so viewers know the source of their news.[111]

UCC also proposes that the FCC adopt a bright line, multifactor test, which would be far more restrictive than the current bright line rule, to attribute ownership where one station exercises substantial influence over another station in the same market as a result of a sharing agreement.[112] Under its test, the "serving broadcaster" that provides services under a sharing agreement would be attributed with ownership of another license-holding station that receives services (the "licensee") under the sharing arrangement if *any* of the following circumstances exists:

- the servicing broadcaster provides all or substantially all local news programming for the licensee's station;
- the servicing broadcaster sells 15% or more of the licensee's weekly advertising time;
- the stations share management personnel;
- the licensee station maintains no separate facilities;
- the serving broadcaster reports to the Securities and Exchange Commission that the servicing broadcaster owns or operates the licensee's station;
- 50% or more of the licensee's total revenues go to the servicing broadcaster; or
- the licensee outsources its retransmission consent negotiations to the serving broadcaster.

UCC further proposes that the FCC assign attribution in situations where none of these seven bright line tests are met, but where a combination of three or more of a list of factors demonstrate "substantial influence when combined with other factors."[113]

Most broadcasters strongly oppose any restrictions on sharing arrangements for news production, claiming that these agreements facilitate greater collaboration between media outlets and permit stations to sustain labor intensive journalism, thereby offering communities greater access to local news content that could otherwise be achieved.[114] They cite many examples of sharing arrangements that result in expanded local news programming that benefit the public but fail such bright line tests.

None of the studies sponsored by the FCC address these sharing agreements and thus the FCC does not currently have empirical evidence to evaluate the conflicting claims. In the NPRM, it seeks such evidence.[115]

The American Cable Association (ACA) claims that when stations participating in an SSA or LMA jointly negotiate retransmission consent terms with the local cable operator, those stations have undue market leverage—especially if both stations are affiliated with one of the four major national networks—and are able to charge higher rates than others broadcast stations, thus raising cable costs and rates.[116] It proposes that the FCC "deem all forms of agreements that permit one television station to provide another with formal authority to negotiate retransmission consent on its behalf attributable."[117]

The FCC's Attribution Criteria and "Virtual Duopolies"

Over the past decade, the FCC appears to have quietly accepted the use of SSAs, even where the agreements give a single entity the responsibility to perform core functions for two or more top-four stations in a local market. The Commission has delegated to its Media Bureau the authority to review and rule on sharing agreements; decisions are made without a vote by the five FCC commissioners. Media Bureau decisions in one case set precedent for future cases.[118]

The Media Bureau has not constructed written guidelines, but in reaching its decisions it appears to have chosen a narrow interpretation of the language in the two relevant provisions in the attribution rules: (1) the block of programming time covered by an agreement must not exceed 15% of the brokered station's programming time, and (2) focus must be on whether the brokered station retains "ultimate control" over its programming decisions and policies, even if it has ceded certain core day-to-day functions to the broker station, and thus greater weight must be given to certifications made by the brokered station in its sharing agreements that it controls station programming, finances, and personnel than to evidence provided by outside parties.[119] In practice, the Media Bureau generally has deemed evidence provided by outside parties immaterial,[120] effectively limiting evidence to that provided to it by the parties to the agreement.

In the absence of Media Bureau guidelines, Steve Lovelady, a communications attorney who advises broadcasters, recently posted a blog that lays out "the allowable parameters for SSAs and JSAs [that] have been disclosed during private conversation with the FCC's staff when they are evaluating a proposed SSA/JSA arrangement and in the context of an assignment of FCC licenses from one party to another."[121] According to Lovelady, these "boundaries have not been generally publicized," often "aren't written down anywhere," and "can change from one day to the next and from one deal to the next," but cover the three basic elements of programming, financing, and personnel, as follows:

- To demonstrate control over programming, (1) the total amount of all provided news programming must be below the 15% threshold; (2) any network affiliation agreement for entertainment programming must be directly between the brokered station and the network, and the broker station may not negotiate network affiliation or syndicated programming agreements on behalf of the brokered station; and (3) the brokered station must have the right to reject any advertising sold by the broker station for broadcast on the brokered station.

- To demonstrate control over finances, (1) the value of the assets owned or controlled by the brokered station must be at least 20% of the total value of the station's assets, and thus the broker station cannot own or control more than 80% of the asset value of the brokered station; (2) direct loans to the broker station from the brokered station must not exceed 33% of the combined equity and debt of the brokered station, though the broker station may guarantee lease payments and loans made to the brokered station by banks and other unrelated financing sources; and (3) no more than 30% of the revenue collected by the broker station from the sale of advertising on the brokered station may be retained by the broker station, though the brokered firm may receive additional revenue from the brokered station for performing additional services.

- To demonstrate control over personnel, the brokered station (1) must retain at least two full-time employees on its payroll, one of whom must be a manager with full editorial discretion in carrying out the brokered station's policies, and (2) must be able to hire and fire its employees and pay their salaries without interference from the broker station.

- The broker station may have an option to acquire the brokered station so long as the option price is neither a fixed amount nor an amount that decreases during the life of the option if the station does well and cash flow is higher than predicted.

- The agreement cannot exceed eight years, which is the length of a broadcast television license, but need not coincide with the time period covered by the brokered station's license.

Several of these criteria are found in FCC rules that apply to all broadcast stations, not just to those stations that are party to sharing agreements,[122] but others do not appear to have been developed through a formal rulemaking proceeding with public comment, and even the 15% threshold was initially created for radio and then applied to television. The specific criteria employed may have little practical impact, in any case, if the Media Bureau relies primarily on licensee self-certification of ultimate control and considers evidence provided by outside parties immaterial.

The Commission appears to support the Media Bureau's process. It has not taken action on applications for review (dating back to January 2005) of several specific Media Bureau decisions, filed by other stations that allege the agreements give the broker station control over the brokered station and are anti-competitive and not in the public interest.[123] Several commenters now explicitly call on the FCC to grant those applications for review.[124]

The ACA submitted a list of 57 instances of multiple top-four affiliates in the same local market operating under some sort of sharing agreement, which it equated with common control of the stations.[125] It acknowledged that it did not know the terms of the sharing agreements, which in almost all cases were not publicly disclosed. Most of the identified sharing agreements were in smaller markets where common control would not be allowed under the ownership rule, but ACA did not have information needed to demonstrate which, if any, of the sharing agreements met the 15% brokered programming threshold or other requirements of the Local Marketing Agreement attribution rule. Nor do there appear to be other sources of publicly available data to fully address such questions of fact.[126]

Given the duopoly rule restriction on joint ownership in small markets, station owners who seek to exploit economies of scale in program production and to strengthen their negotiating position with advertisers and multichannel video programming distributors (MVPDs, such as cable and satellite television operators) may have the incentive to pursue sharing agreements with other stations in their market if such an agreement would not trigger an attribution rule or if the FCC chose not to enforce the attribution rule as strictly as it does direct ownership restrictions. This incentive could be particularly strong if the FCC has a history of grandfathering arrangements once it has initially allowed them to be implemented. This appears to be the situation prevailing today. The industry commonly refers to sharing arrangements in which one station effectively controls a second station as a "virtual duopoly." For example, Harry A. Jessell, editor and co-publisher of *TVNewsCheck,* who frequently comments on public policy issues from the perspective of the broadcasting industry, wrote in December 2011:

> Broadcasters in small markets interested in doubling up through shared services agreements and the like, had better act fast. The FCC will soon launch a rulemaking to determine whether such arrangements, which essentially allow the broadcasters to circumvent the ban against actual ownership of two stations in small markets, are still a good idea. And the rulemaking could very well lead to a prohibition against the so-called virtual duopolies.... As the "could" in the proceeding sentence suggests, nothing is for sure.... One thing broadcasters can count on the FCC not doing is ordering the dismantling of any existing arrangements. By its silence over the years, it has tacitly approved them. It isn't going to go back now and declare them illegal. It simply isn't in the FCC's nature to undo deals once done. I can't recall a single case, at least not in the broadcast side of FCC world. So, again, an NBC affiliate who is thinking it would be a good idea to cut a deal with the CBS affiliate and run the stations in tandem should make it a New Year's resolution to do it soon. The only guarantee, I think, is that deals done prior to a final order next year would be grandfathered in.[127]

This statement is a bit ambiguous. It might be interpreted as a warning that the FCC may take future action that would make attributable certain sharing arrangements that currently do not represent attributable control of another station. Or it might suggest that the FCC currently is allowing stations to participate in certain sharing agreements that meet the criteria in the current attribution rule and thus strictly speaking are in violation of the duopoly rule, but might not continue to look the other way in the future for new sharing agreements. Since stations have a disincentive to be subject to the attribution rule and, in any case, local programming frequently comprises less than 15% of a station's total programming, it is likely that for the vast majority of these sharing arrangements the amount of programming and advertising sales performed by the broker station is below the 15% threshold level in the FCC's attribution rule.

It may be instructive to note how one of the companies that has been an active participant in such arrangements describes its business operations.[128] Mission Broadcasting owns television stations in 14 markets and Nexstar Broadcasting also owns television stations in each of those markets. Mission has sharing agreements with Nexstar for each of its television stations. According to Mission's 2011 10-K submission to the Securities and Exchange Commission,[129] as of December 31, 2010:

- Mission owned and operated 16 broadcast television stations in 14 medium and small markets (ranked between 54 and 196 out of 210 DMAs) and had a total of 33 employees, all of which were full-time. In the two markets where it owned and operated two stations, one of the stations was a low-power station.
- Each of the stations was affiliated with a national network with the right to broadcast all programs transmitted by the network with which it is affiliated. Thirteen of those

affiliations are with Top-Four networks (ABC, CBS, Fox, or NBC). The network had the right to sell a substantial majority of the advertising time during these broadcasts. (Some of the stations received some compensation from the network based on the hours of network programming they broadcast.)

- Mission had a time brokerage agreement (TBA) for two of its stations that allowed Nexstar to program most of the stations' (non-network) broadcast time, sell the stations' (non-network) advertising time, and retain the advertising revenue generated in exchange for monthly payments to Mission.

- Mission had both a shared services agreement (SSA) and a joint sales agreement (JSA) with Nexstar for each of its 14 other stations. The SSA allowed the sharing of services including news production, technical maintenance, and security, in exchange for Nexstar's right to receive certain payments from Mission. The JSAs permitted Nexstar to sell and retain a percentage of the net revenue from the station's advertising time in return for monthly payments to Mission of the remaining percentage of net revenues. The SSAs and JSAs generally had ten year terms.

- Due to the TBAs, SSAs, and JSAs, Nexstar received substantially all the cash, after debt service costs, generated by Mission's stations. Mission anticipated that Nexstar will continue to receive substantially all of Mission's available cash.

- Nexstar guaranteed all obligations incurred under Mission's senior secured credit facility. Also, Mission was a guarantor of the senior secured credit facility entered into by Nexstar and the senior subordinated notes issued by Nexstar. If Nexstar, which was highly leveraged with debt, became unable to meet its obligations under the indentures governing its senior subordinated notes or its senior secured credit facility agreement, Mission could have been held liable for those obligations under guarantees.

- In consideration of Nexstar's guarantee of Mission's senior secured credit facility, Mission's sole shareholder had granted Nexstar purchase options to acquire the assets and assume the liabilities of each Mission television station, subject to FCC consent, for consideration equal to the greater of (1) seven times the station's cash flow less the amount of its indebtedness, or (2) the amount of its indebtedness. These option agreements (which expire on various dates between 2011 and 2018) were freely exercisable or assignable by Nexstar without consent or approval by Mission's sole shareholder. Mission expected these option agreements to be renewed upon expiration.

- The operating revenue of the Mission stations was derived primarily from broadcast advertising revenue sold and collected by Nexstar and paid to Mission under the JSAs. Mission's primary operating expense consisted of fixed monthly SSA fees paid to Nexstar for news production and technical and other services.

- Mission had net losses of $4.6 million, $2.6 million, and $7.5 million, respectively, for the years ended December 31, 2010, 2009, and 2008. Mission stated it may not be able to achieve or maintain profitability.

- Mission stated that its ability to continue as a going concern is dependent on Nexstar's pledge to continue the local services agreements.

For its part, Nexstar stated in its 2011 10-K submission,[130]

We do not own Mission or Mission's television stations. However, as a result of (a) local service agreements Nexstar has with the Mission stations, (b) Nexstar's guarantee of the obligations incurred under Mission's senior secured credit facility, (c) Nexstar having power over significant activities affecting Mission's economic performance, including budgeting for Mission's advertising revenues, advertising and hiring and firing of sales force personnel and (d) purchase options (which expire on various dates between 2011 and 2018) granted by Mission's sole shareholder which permit Nexstar to acquire assets and assume the liabilities of each Mission station, subject to Federal Communications Commission ("FCC") consent, we are deemed under U.S. GAAP to have a controlling financial interest in Mission while complying with the FCC's rules regarding ownership limits in television markets. The purchase options are freely exercisable or assignable by Nexstar without consent or approval by Mission's sole shareholder. In compliance with FCC regulations for both us and Mission, Mission maintains complete responsibility for and control over programming, finances, and personnel for its stations.

With only 33 employees serving 16 television stations in 14 different markets, Mission does not have the staffing required to produce any of its own local programming and is dependent on Nexstar for its local (news or other) programming.[131] Given the mutual guarantees they provide of the other's financial obligations, the two are financially entwined as well. It appears likely that Nexstar meets the standard enunciated in the FCC attribution order, but not in the attribution rule: it has "a realistic potential to affect the programming decisions of [Mission] or other core operating functions." In a recent article, Mr. Jessell of *TVNewsCheck* referred to Mission as a "shell" company set up by Nexstar.[132] However, the Nexstar-Mission agreements meet the criteria for not triggering an attribution finding. Each explicitly covers less than 15% of the Mission stations' programming time and each includes a certification that Mission retains ultimate control over station finances, personnel, and programming. According to Media Bureau precedents, the information about the two companies' relationships contained in their SEC filings is immaterial in determining whether Nexstar exercises control over Mission.

By relying on a 15% programming threshold bright line rule and giving great weight to certification language in the agreements, the FCC's attribution rule process avoids the potential complexities of evidence-intensive determinations of control. Erwin G. Krasnow, a communications attorney who advises broadcasters, advised in a recent article in *Radio & Television Business Report*[133] that, for a station seeking FCC approval of a local marketing agreement or time brokerage agreements, there be in the LMA:

> multiple (indeed, repetitive) provisions that say in different ways that the licensee of a broadcast station retains ultimate responsibility for all decisions of the station, including matters relating to finances, personnel and programming.... Here's a nutshell summary of what the FCC looks for in LMAs: The LMA must unequivocally reserve to the licensee/seller the unlimited right to suspend, cancel, or reject any programming furnished or recommended by the lessee/buyer. To remove any doubt, the LMA must make clear that the licensee remains responsible for the salaries of certain employees and certain other costs of the station's operations, as well as for compliance with all regulatory requirements such as maintenance of the public inspection file and political file.

There are good public policy reasons for employing a bright line attribution rule. It allows potential sharing agreement participants to make business decisions with some certainty about the regulatory treatment of the agreement. It minimizes room for varying interpretation and thus fosters predictable and consistent results in its application. It also reduces the FCC staff resources needed to review agreements or to litigate decisions. But, unless the bright line is in no way a constraining factor on market participants, the level at which it is set could have policy implications. In this case, by setting a bright line that consists of a threshold percentage of programming below the level that many television stations allocate to local programming, and by relying heavily on a self-certification process rather than a full review of all available empirical evidence, the FCC has adopted a local television ownership policy that concurrently prohibits duopolies and permits virtual duopolies in many market situations.

The Information and Analysis Needed to Construct Policy on Attribution Rules and Sharing Agreements

In response to its NPRM, the FCC has received considerable anecdotal evidence about sharing agreements. However, as only those sharing agreements that are subject to the attribution rules must be made publicly available, and as broadcasters have every incentive to structure agreements to not fit that category, the Commission has little systematic data. The wide variety of agreements now in use could provide data to perform statistical analyses of the impact of these agreements on such variables as the total level of news programming in the market, the amount of original news programming, the amount of news programming aired by broker stations and brokered stations, advertising rates, retransmission consent rates and retail cable rates. It may be possible, for example, for the FCC to perform statistical analysis to distinguish between markets in which resources are shared but advertising sales and perhaps also actual programming are performed by individual stations and markets with full LMAs.

Since issues of control typically have a financial aspect to them, the FCC also could use data provided by licensees to the Securities and Exchange Commission in their 10-K and other submissions. Sometimes this information is purely financial, for example, as cited above, Nexstar indicated in its 10-K form that it is deemed under U.S. generally accepted accounting principles to have a controlling financial interest in Mission. But sometimes the information provides more general insights about market competition. For example, in the narrative section of its 10-K providing an overview of its business, Mission states that it:

> believe[s] that medium-sized markets offer significant advantages over large-sized markets, most of which result from a lower level of competition. First, because there are fewer well-capitalized acquirers with a medium-market focus, we have been successful in purchasing stations on more favorable terms than acquirers of large market stations. Second, in many of our markets only four or five local commercial television stations exist. As a result, we achieve lower programming costs than stations in larger markets because the supply of quality programming exceeds the demand.[134]

Presumably Mission is referring to lower costs for quality entertainment programming available from national networks and syndicators, not to local programming. Still, it suggests that stations in mid-sized markets may not be less advantageously situated than those in larger markets.

The FCC also might look to information provided by television licensees in litigation, such as the antitrust lawsuit that Nexstar filed in 2011 against Granite Broadcasting for entering into exclusive affiliate agreements with two of the four major television networks in Fort Wayne, IN.[135]

Such data collection and analysis might indicate that the current bright line test is appropriate or it might lead the FCC to conclude that different attribution rules are needed to address sharing agreements.

End Notes

[1] 47 C.F.R. 73.3555.

[2] Section 629 of the FY2004 Consolidated Appropriations Act, P.L. 108-199, modified Section 202(h) of the Telecommunications Act of 1996 (P.L. 104-104), instructing the FCC to perform a quadrennial review of all of its broadcast media ownership rules, except the National Television Ownership rule.

[3] 47 C.F.R. 73.3555 Note 2.

[4] In the Matter of Review of the Commission's Regulations Governing Attribution of Broadcast and Cable/MDS Interests; Review of the Commission's Regulations and Policies Affecting Investment in the Broadcast Industry; Reexamination of the Commission's Cross-Interest Policy, MM Docket Nos. 94-150, 92-51, and 87-154, Report and Order (1999 Attribution Order), adopted August 5, 1999 and released August 6, 1999, at para. 1.

[5] In the Matter of 2010 Quadrennial Regulatory Review—Review of the Commission's Broadcast Ownership Rules and Other Rules Adopted Pursuant to Section 202 of the Telecommunications Act of 1996, MB Docket No. 09-182, Notice of Inquiry, adopted and released May 25, 2010.

[6] In the Matter of 2010 Quadrennial Regulatory Review—Review of the Commission's Broadcast Ownership Rules and Other Rules Adopted Pursuant to Section 202 of the Telecommunications Act of 1996: Promoting Diversification of Ownership in the Broadcasting Services, MB Dockets No. 09-182 and 07-294, Notice of Proposed Rulemaking (NPRM), adopted and released December 22, 2011.

[7] See, for example, In the Matter of 2010 Quadrennial Regulatory Review—Review of the Commission's Broadcast Ownership Rules and Other Rules Adopted Pursuant to Section 202 of the Telecommunications Act of 1996; Promoting Diversification of Ownership in the Broadcasting Services, MB Docket Nos. 09-182 and 07-294, Comments of the National Association of Broadcasters (NAB March 2012 Comments), March 5, 2012, at p. ii.

[8] See, for example, In the Matter of 2010 Quadrennial Regulatory Review—Review of the Commission's Broadcast Ownership Rules and Other Rules Adopted Pursuant to Section 202 of the Telecommunications Act of 1996; Promoting Diversification of Ownership in the Broadcasting Services, MB Docket Nos. 09-182 and 07-294, Comments of Sinclair Broadcast Group, Inc. (Sinclair March 2012 Comments), March 5, 2012, at p. 5.

[9] See, for example, In the Matter of 2010 Quadrennial Regulatory Review—Review of the Commission's Broadcast Ownership Rules and Other Rules Adopted Pursuant to Section 202 of the Telecommunications Act of 1996; Promoting Diversification of Ownership in the Broadcasting Services, MB Docket Nos. 09-182 and 07-294, Comments of Free Press (Free Press March 2012 Comments), March 5, 2012, at pp. 8-17.

[10] See, for example, In the Matter of 2010 Quadrennial Regulatory Review—Review of the Commission's Broadcast Ownership Rules and Other Rules Adopted Pursuant to Section 202 of the Telecommunications Act of 1996; Promoting Diversification of Ownership in the Broadcasting Services, MB Docket Nos. 09-182 and 07-24, Comments of Office of Communication of United Church of Christ, Inc., Media Alliance, National Organization for Women Foundation, Communications Workers of America, Common Cause, Benton Foundation, Media Council Hawai'i (UCC March 2012 Comments), March 5, 2012, at pp. 23-29.

[11] Section 325(b) of the Communications Act, which was enacted as part of the 1992 Cable Act, requires cable operators that seek to retransmit the signal of a broadcaster to obtain the consent of that broadcaster.

[12] See In the Matter of 2010 Quadrennial Regulatory Review—Review of the Commission's Broadcast Ownership Rules and Other Rules Adopted Pursuant to Section 202 of the Telecommunications Act of 1996; Promoting

Diversification of Ownership in the Broadcasting Services, MB Docket Nos. 09-182 and 07-24, Comments of American Cable Association (ACA March 2012 Comments), at p. 2.

[13] In the Matter of 2010 Quadrennial Regulatory Review of the Commission's Broadcast Ownership Rules, MB Docket No. 09-182, Comments of Office of Communication of United Church of Christ, Inc., Media Alliance, Common Cause, Benton Foundation (UCC 2009 Comments), November 20, 2009, at p. 13.

[14] See, for example, UCC November 2009 Comments at pp. 12-14.

[15] In the Matter of 2010 Quadrennial Regulatory Review—Review of the Commission's Broadcast Ownership Rules and Other Rules Adopted Pursuant to Section 202 of the Telecommunications Act of 1996; Promoting Diversification of Ownership in the Broadcasting Services, MB Docket Nos. 09-182 and 07-294, Report on Ownership of Commercial Broadcast Stations, adopted and released November 14, 2012.

[16] "Commission Seeks Comment on Broadcast Ownership Report," FCC Public Notice DA 12-1946, MB Docket Nos. 09-182 and 07-294, released December 3, 2012.

[17] This section presents the proposed rules, information requests, and questions as framed in the FCC's NPRM. As explained in later sections, the public comments identified related issues—such as how the FCC has been implementing, interpreting, and enforcing its existing media ownership and attribution rules—that the Commission itself did not raise in the NPRM.

[18] Prometheus Radio Project v. FCC, 652 F.3d 437 (2011) (Prometheus II). The court concluded that the Commission had failed to comply with the notice and comment provisions of the Administrative Procedures Act. The court did not address the Commission's substantive modifications to the rule.

[19] 47 C.F.R. 73.3555 Note 2.

[20] 47 C.F.R. 73.3555 Note 2(j).

[21] 47 C.F.R. 73.3555 Note 2(k).

[22] In the Matter of Rules and Policies Concerning Attribution of Joint Sales Agreements in Local Television Markets, MB Docket No. 04-256, Notice of Proposed Rulemaking, adopted July 3, 2004 and released August 2, 2004, 19 FCC Rcd 15238.

[23] NPRM at para. 199.

[24] NPRM at para. 195.

[25] See CRS Report RL34269, Minority Ownership of Broadcast Properties: A Legal Analysis, by Kathleen Ann Ruane, which provides the basis for the legal analysis presented here.

[26] Prometheus Radio Project v. FCC, 373 F.3d 372 (2004).

[27] Adarand Construction v. Peña, 515 U.S. 200, 227 (1995).

[28] Adarand, 515 U.S. at 227.

[29] In the Matter of Promoting Diversification of Ownership in the Broadcasting Services, MB Docket No. 07-294, Report and Order and Third Further Notice of Proposed Rulemaking (Diversification Order), adopted December 18, 2007 and released March 5, 2008, at para. 9.

[30] Diversification Order at para. 6.

[31] Prometheus II, 652 F.3d at 470-471.

[32] Prometheus II, 652 F.3d at 472.

[33] NPRM at para. 149.

[34] NPRM at paras. 159-167.

[35] NPRM at para. 158.

[36] 47 C.F.R. 73.3555(c).

[37] NPRM at para. 119.

[38] NAB March 2012 Comments at p. 51.

[39] In the Matter of 2010 Quadrennial Regulatory Review—Review of the Commission's Broadcast Ownership Rules and Other Rules Adopted Pursuant to Section 202 of the Telecommunications Act of 1996; Promoting Diversification of Ownership in the Broadcasting Services, MB Docket Nos. 09-182 and 07-294, Comments of National Hispanic Media Coalition (NHMC), Center for Rural Strategies, Center for Media Justice (NHMC March 2012 Comments), March 5, 2012, at pp. 35-36.

[40] 47 C.F.R. §73.3555(d).

[41] A Grade A service contour maps the geographic area that is predicted by an engineering model to receive a broadcast television signal at an intensity associated with good reception.

[42] A 2 millivolt per meter (or 2 mv/m) contour maps the geographic area that is predicted by an engineering model or measured to receive a radio signal of that intensity.

[43] A 1 millivolt per meter (or 1 mv/m) contour maps the geographic area that is predicted by an engineering model or measured to receive a radio signal of that intensity.

The FCC's Broadcast Media Ownership and Attribution Rules   113

---

[44] In the Matter of 2006 Quadrennial Regulatory Review—Review of the Commission's Broadcast Ownership Rules and Other Rules Adopted Pursuant to Section 202 of the Telecommunications Act of 1996; 2002 Biennial Review— Review of the Commission's Broadcast Ownership Rules and Other Rules Adopted Pursuant to Section 202 of the Telecommunications Act of 1996; Cross-Ownership of Broadcast Stations and Newspapers; Rules and Policies Concerning Multiple Ownership of Radio Broadcast Stations in Local Markets; Definition of Radio Markets; Ways to Further Section 257 Mandate and to Build on Earlier Studies; Public Interest Obligations of TV Broadcast Licenses, MB Dockets No. 06-121, 02-277, and 04-228 and MM Dockets No. 01-235, 01-317, 00-244, and 99-360, Report and Order and Order on Reconsideration, adopted December 18, 2007 and released February 4, 2008 (2007 Order), at para. 77.

[45] 2007 Order at para. 13 and Appendix A, pp. 84-85.

[46] Prometheus II, 652 F. 3rd 437 (2011).

[47] NPRM at para. 89.

[48] NPRM at para. 89.

[49] NPRM at para. 90.

[50] For a discussion of the transition from analog to digital broadcast television transmission, see CRS Report RL34165, The Transition to Digital Television: Is America Ready?, by Lennard G. Kruger.

[51] DMAs are geographic designations developed by Nielsen Media Research. There are 210 DMAs in the United States. A DMA is made up of all the counties that get the preponderance of their broadcast programming from a given television market. The Nielsen DMAs are both complete (all counties in the United States are assigned to a DMA) and exclusive (DMAs do not overlap). In the 1992 Cable Act, Congress amended the 1934 Communications Act to require, subject to certain exceptions, each cable system to carry the signals of all the local full power commercial television stations "within the same television market as the cable system," with that market determined by "commercial publications which delineate television markets based on viewing patterns." 47 U.S.C. §534. The DMAs represent the only nationwide commercial mapping of television audience viewing patterns. Each county in the United States is assigned to a television market based on the viewing habits of the residents of the county.

[52] It also is presumed that such cross-ownership is inconsistent with the public interest in markets smaller than the top 20 markets.

[53] NPRM at para. 103.

[54] For a detailed discussion of the remanded 2007 rule, see CRS Report RL34416, The FCC's Broadcast Media Ownership Rules, by Charles B. Goldfarb.

[55] NAB March 2012 Comments at pp. 39-49.

[56] UCC March 2012 Comments at p. 26.

[57] Free Press March 2012 Comments at p. 28.

[58] In the Matter of 2010 Quadrennial Regulatory Review—Review of the Commission's Broadcast Ownership Rules and Other Rules Adopted Pursuant to Section 202 of the Telecommunications Act of 1996; Promoting Diversification of Ownership in the Broadcasting Services, MB Docket Nos. 09-182 and 07-294, Initial Comments of the Diversity and Competition Supporters in Response to the Notice of Proposed Rule Making (DCS March 2012 Comments), March 5, 2012, at p. 41.

[59] Grade B is a measure of signal intensity associated with acceptable reception when a television signal is being transmitted using analog technology. The FCC's rules define this contour, often a circle drawn around the transmitter site of a television station, in such a way that 50 percent of the locations on that circle are statistically predicted to receive a signal of Grade B intensity at least 90 per cent of the time. Although a station's predicted signal strength increases as one gets closer to the transmitter, there will still be some locations within the predicted Grade B contour that do not receive a signal of Grade B intensity.

[60] Defined as follows in 47 C.F.R. 73.3555(b)(1)(i): "At the time of application ... among the top four stations in the DMA, based on the most recent all-day (9 a.m.-midnight) audience share, as measured by Nielsen Media Research or by any comparable professional accepted audience rating service."

[61] 47 C.F.R. 73.3555(b), adopted in In the Matter of Review of the Commission's Regulations Governing Television Broadcasting; Television Satellite Stations Review of Policy and Rules, MM Docket Nos. 91-221 and 87-8, Report and Order (Local TV Ownership Report and Order), adopted August 5, 1999 and released August 6, 1999, at para. 8.

[62] A "failed" station is one that has been dark for at least four months or is involved in court-supervised involuntary bankruptcy or involuntary insolvency proceedings. Under the standard for "failing" stations, a waiver is presumed to be in the public interest if the applicant satisfies each of the following criteria: (1) one of the merging stations has had all-day audience share of 4% or lower; (2) the financial condition of one of the merging stations is poor; (3) and the merger will produce public interest benefits. Under the standard for

"unbuilt" stations, a waiver is presumed to be in the public interest if an applicant meets each of the following criteria: (1) the combination will result in the construction of an authorized but as yet unbuilt station; and (2) the permittee has made reasonable efforts to construct, and has been unable to do so. (47 C.F.R. 73.3555, Note 7 (1) and Local TV Ownership Report and Order at para. 86.)

[63] Local TV Ownership Report and Order at paras. 77, 81, 86.

[64] NPRM at para. 26.

[65] NPRM at paras. 36-39.

[66] NPRM at para. 39.

[67] NPRM at para. 53.

[68] NPRM at para. 56.

[69] NPRM at para. 56. Under Section 534 of the Communications Act (enacted as part of the 1992 Cable Act), cable companies are required to carry the primary signals of local full-power commercial television stations, but this requirement does not extend to the carriage of non-primary multicast signals.

[70] NPRM at para. 57.

[71] See, for example, In the Matter of 2010 Quadrennial Regulatory Review—Review of the Commission's Broadcast Ownership Rules and Other Rules Adopted Pursuant to Section 202 of the Telecommunications Act of 1996; Promoting Diversification of Ownership in the Broadcasting Services, MB Docket Nos. 09-182 and 07-294, Comments of Nexstar Broadcasting, Inc. (Nexstar March 2012 Comments) at pp. 11-26.

[72] NAB March 2012 Comments at pp. 29-30.

[73] DCS March 2012 Comments at p. 40.

[74] Free Press March 2012 Comments at pp. 44-50.

[75] UCC March 2012 Comments at pp. 26-27.

[76] Section 202(b) also provides that the Commission may permit a party to exceed these limits "if the Commission determines that [it] will result in an increase in the number of radio broadcast stations in operation." 1996 Act, §202(b)(2), 110 Stat. at 10-11.

[77] 2003 Order at para. 239.

[78] 2003 Order at para. 239.

[79] NPRM at para. 61.

[80] NRPM at para. 62.

[81] NAB March 2012 Comments at pp. 32-38.

[82] UCC March 2012 Comments at pp. 27-29.

[83] 47 C.F.R. 73.658(g).

[84] NPRM at para. 137.

[85] NPRM at para. 137.

[86] 47 C.F.R. 73.3555 Note 2.

[87] 1999 Attribution Order at para. 1.

[88] 47 C.F.R. 73.3555 Note 2j(2).

[89] See, for example, Letter from Barbara A. Kreisman, Chief, Video Division, Media Bureau, Federal Communications Commission, to Malara Broadcast Group of Duluth Licensee LLC, Re: Application for Assignment of License of KDLH-TC, Duluth, Minnesota (Facility ID # 4691), File No. BALCT-20040504ABU, DA 04-3908, released December 14, 2004.

[90] 47 C.F.R. 73.3555 Note 2j(3) states "Every time brokerage agreement of the type described in this Note shall be undertaken only pursuant to a signed written agreement that shall contain a certification by the licensee or permittee of the brokered station verifying that it maintains ultimate control over the station's facilities including, specifically, control over station finances, personnel and programming.... "

[91] In the Matter of 2010 Quadrennial Regulatory Review—Review of the Commission's Broadcast Ownership Rules and Other Rules Adopted Pursuant to Section 202 of the Telecommunications Act, MB Docket No. 09-182, Comments of Gray Television, Inc. on the FCC's May 25, 2010 Notice of Inquiry (Gray June 2010 Comments), June 12, 2010, at p. 13.

[92] Nexstar March 2012 Comments at p. vi.

[93] In the Matter of 2010 Quadrennial Regulatory Review—Review of the Commission's Broadcast Ownership Rules and Other Rules Adopted Pursuant to Section 202 of the Telecommunications Act of 1996, MB Docket No. 09-182, Comments of the National Association of Broadcasters (NAB July 2010 Comments), July 12, 2010, at p. 84.

[94] NAB March 2012 Comments at pp. 64-65.

[95] NAB March 2012 Comments at p. 66.

[96] "FCC Launches Examination of the Future of Media and Information Needs of Communities in a Digital Age; Comment Sought," GN Docket No. 10-25, DA 10-100, released January 21, 2010, hraunfoss.fcc.gov/edocs_public/attachmatch/DA-10-100A1.pdf.

[97] See, for example, UCC November 2009 Comments at pp. 4-6, and In the Matter of FCC Launches Examination of the Future of Media and Information Needs of Communities in a Digital Age, GN Docket No. 10-25, Comments of Communications Workers of America and Media Council Hawai'i (CWA Future of Media Comments), May 7, 2010, at pp. 9-17.

[98] UCC March 2012 Comments at pp. 15-20.

[99] See, for example, UCC November 2009 Comments, at pp. 12-14, and In the Matter of 2010 Quadrennial Regulatory Review—Review of the Commission's Broadcast Ownership Rules and Other Rules Adopted Pursuant to Section 202 of the Telecommunications Act of 1996, MB Docket No. 09-182, American Cable Association Comments (ACA July 2010 Comments), July 12, 2010, Appendix A ("57 Identified Instances of Common Control of Multiple Big 4 Network Stations in the Same Market") and Appendix B (William P. Rogerson, "Joint Control or Ownership of Multiple Big 4 Broadcasters in the Same Market and its Effect on Retransmission Consent Fees," May 18, 2010, at pp. 5-7).

[100] 47 C.F.R. 73.3555 Note 2j.

[101] 1999 Attribution Order at paras. 66-99.

[102] 1999 Attribution Order at paras. 74 and 75.

[103] 1999 Attribution Order at para. 78 and footnote 173.

[104] RTDNA 2011 TV and Radio News Staffing and Profitability Survey (RTDNA 2011 Survey), "Part II: Record Amount of Local News Produced on TV," unpaginated, http://www.rtdna.org/pages/media_items/2011-tv-and-radionews-staffing-and-profitability-survey2033.php?id=2033.

[105] RTDNA 2011 Survey, "Part I: More Jobs, Higher Profits in TV News," unpaginated.

[106] CWA Future of Media Comments at p. 2 and appendix.

[107] Under the current rule, only those LMAs that meet the 15% threshold have to be placed in the station's public file and filed with the FCC; thus there is no public information on other sharing arrangements.

[108] In the Matter of 2010 Quadrennial Regulatory Review—Review of the Commission's Broadcast Ownership Rules and Other Rules Adopted Pursuant to Section 202 of the Telecommunications Act of 1996; RFQ for Media Ownership Studies, MB Docket Nos. 09-192 and 09-182, Comments of Communications Workers of America, The Newspaper Guild/CWA, National Association of Broadcast Employees and Technicians/CWA (CWA July 2010 Comments), July 12, 2010, at p. 32.

[109] CWA July 2010 Comments at p. 32.

[110] UCC November 2009 Comments at pp. 4-6.

[111] CWA July 2010 Comments at pp. 33-35.

[112] UCC March 2012 Comments at pp. 15-20.

[113] UCC March 2012 Comments at pp. 19-20.

[114] See, for example, Gray June 2010 Comments at p. 14.

[115] NPRM at paras. 204-208.

[116] ACA July 2010 Comments at p. 2.

[117] ACA March 2012 Comments at p. 3.

[118] See, for example, Letter from Barbara A. Kreisman, Chief, Video Division, Media Bureau, Federal Communications Commission, to Piedmont Television of Springfield License LLC, et al., Re: KSPR(TV), Springfield, Missouri, Application for Assignment of License, File No. BALCT-20061005ADY, ID No. 35630, DA 07-3746, released July 30, 2007, citing its decision in an earlier case.

[119] See, for example, Letter from Barbara A. Kreisman, Chief, Video Division, Media Bureau, Federal Communications Commission, to Malara Broadcast Group of Duluth Licensee LLC, et al. (Media Bureau Letter to Malara), Re: Application for Assignment of License of KDLH-TC, Duluth, Minnesota (Facility ID # 4691), File No. BALCT20040504ABU, DA 04-3908, released December 14, 2004.

[120] See, for example, Media Bureau Letter to Malara, at footnote 12, which states that "we view as immaterial to our resolution of this case statements allegedly made by Granite in an SEC filing."

[121] Steve Lovelady, "SSAs and JSAs—Some Unwritten Rules," CommLawBlog, Fletcher, Heald & Hildreth, March 26, 2012, http://www.commlawblog.com/2012/03/articles/broadcast/ssas-and-jsas-some-unwritten-rules/print.html.

[122] For example, under 47 C.F.R. 73.3555 Note 2i(1)(A), an entity would have an attributable interest in a station if its equity and debt interests, in the aggregate, exceed 33% of the total asset value, defined as the aggregate of all equity plus all debt, of the station, whether or not the entity was in a sharing agreement with the station. Similarly, the FCC main studio rule requires all broadcast television and radio stations to maintain a main

studio staffed by two full time employees, one of whom must be supervisory or managerial. (See Gregg P. Skall, "Main Studio Rule and Staffing," Womble, Carlyle, Sandridge & Rice, http://www.wcsr.com/resources/pdfs/telecommuncationmemos1328.pdf.)

[123] See, for example, In re Application of Piedmont Television of Springfield License LLC (Assignor) and Perkin Media, LLC (assignee) For Consent to the Voluntary Assignment of the License for Station KSPR(TV), Springfield, MO, File No. BALCT-20061005ADY, Facility ID No. 35630, Application for Review, Koplar Communications International, Inc., August 29, 2007, and UCC November 2009 Comments at pp. 13-14.

[124] See, for example, CWA July 2010 Comments at p. 33 and UCC November 2009 Comments at p. 14.

[125] ACA July 2010 Comments at p. 9 and Appendix A, table entitled "57 Identified Instances of Common Control of Multiple Big 4 Network Stations in the Same Market."

[126] For example, ACA identifies 17 markets—ranked, by size, numbers 54, 74, 80, 100, 116, 131, 134, 138, 143, 146, 147, 149, 152, 165, 169, 170, and 198—in which Nexstar Broadcasting was the controlling entity in a sharing arrangement. Nexstar, itself, identifies 23 markets in which it owns and operates a station and has entered into a Management Service Agreement (MSA) to provide services to another station in the market owned and operated by an independent third party. (See http://www.nexstar.tv/index.php?option=com_content&view=article&id=301&Itemid=2.) Nexstar provides a brief description of the services it offers in each market. In the smallest of these markets, it has entered into both a joint sales agreement and shared service agreements in San Angelo, TX, Utica, NY, Abilene, TX, and Terre Haute, IN. It initiated a newscast for its partner in Terre Haute. It provides non-programming services in Billings, MT, but neither station offers local news programming. Nexstar does not provide information on the percentage of programming that it provides to its partners, however, so it is not possible to determine if control should be attributed in these small markets.

[127] Harry A. Jessell, "Now's The Time To Make Virtual Duopolies," TVNewsCheck, December 9, 2011, http://www.tvnewscheck.com/article/2011/12/09/55959/nows-the-time-to-make-virtual-duopolies.

[128] CRS has selected this example because it exemplifies how the FCC addresses attribution when a licensee has participated in sharing agreements and also has made publicly available in its submissions to the Securities and Exchange Commission substantial information on station finances, personnel, and programming. CRS has no opinion and makes no judgment about the company's activities or FCC's actions.

[129] Mission Broadcasting, Inc. Form 10-K, United States Securities and Exchange Commission, March 16, 2011, Commission File Number 333-62916-02, at various pages.

[130] Nexstar Broadcasting Group Inc. Form 10-K, United States Securities and Exchange Commission, March 16, 2011.

[131] As explained earlier, the FCC main studio rule requires a station to maintain a main studio staffed by two full time employees, one of whom must be supervisory or managerial. Thus the FCC rules require Mission to have at least 32 employees, two for each station.

[132] Reporting on the UCC and ACA (and other) proposals for the FCC to adopt much more stringent attribution rules, which he characterized as an "assault on broadcasters' virtual duopolies," Mr. Jessell began as follows: "The FCC's duopoly rules prohibit one broadcaster from owning two stations in markets with fewer than eight total stations or two top-four stations (in most cases the Big Four affiliates) in markets of any size. But over the years, many broadcasters have gotten around the rules through shared service agreements and other contractual arrangements that stop short of ownership, but allow them to operate two or even three stations in a market. Station groups like Sinclair and Nexstar have gone so far as to set up shell companies or licensees for the purpose of acquiring stations that they can operate in tandem with ones they own." Harry A. Jessell, "Advocacy Groups, Cable Target Virtual Duops," TVNewsCheck, March 7, 2012, http://www.tvnewscheck.com/article/2012/03/07/57926/advocacy-groups-cable-target-virtual-duops.

[33] Erwin G. Krasnow, Esq., "Lotsa Money for Communications Attorneys," Radio & Television Business Report, March 14, 2012, http://rbr.com/lotsa-money-for-communications-attorneys/.

[34] Mission Broadcasting, Inc. Form 10-K, United States Securities and Exchange Commission, March 16, 2011, Commission File Number 333-62916-02, at p. 2.

[35] "Nexstar Broadcasting Files Antitrust Lawsuit against Granite Broadcasting and Its Fort Wayne Indiana Station, WISE-TV," Business Wire, July 25, 2011.

In: Transformations in Telecommunications and Media
Editor: Irwin Cavazos

ISBN: 978-1-62948-413-6
© 2013 Nova Science Publishers, Inc.

Chapter 6

THE CORPORATION FOR PUBLIC BROADCASTING: FEDERAL FUNDING AND ISSUES[*]

Glenn J. McLoughlin and Mark Gurevitz

SUMMARY

The Corporation for Public Broadcasting (CPB) receives virtually all of its funding through federal appropriations; overall, about 15% of all public television and radio broadcasting funding comes from the federal appropriations that CPB distributes. CPB's appropriation is allocated through a distribution formula established in its authorizing legislation and has historically received two-year advanced appropriations. Congressional policymakers are increasingly interested in the federal role in supporting CPB due to concerns over the federal debt, the role of the federal government funding for public radio and television, and whether public broadcasting provides a balanced and nuanced approach to covering news of national interest.

It is also important to note that many congressional policymakers defend the federal role of funding public broadcasting. They contend that it provides news and information to large segments of the population that seek to understand complex policy issues in depth, and in particular for children's television broadcasting, has a significant and positive impact on early learning and education for children.

On June 20, 2012, the Corporation for Public Broadcasting released a report, Alternative Sources of Funding for Public Broadcasting Stations. The report was undertaken in response to the conference report accompanying the Military Construction and Veterans Affairs and Related Appropriations Act of 2012 (incorporated into the Consolidated Appropriations Act, FY2012, H.R. 2055, P.L. 112-74). The CPB engaged the consulting firm of Booz & Company to explore possible alternatives to the federal appropriation to CPB. Among its findings, the report stated that ending federal funding for public broadcasting would severely diminish, if not destroy, public broadcasting service in the United States.

On September 28, 2012, President Obama signed a Continuing Resolution (CR) of federal funding for FY2013 into law (H.J.Res. 117, P.L. 112-175). It maintains CPB's

---

[*] This is an edited, reformatted and augmented version of a Congressional Research Service publication, CRS Report for Congress RS22168, prepared for Members and Committees of Congress, from www.crs.gov, dated February 25, 2013.

advanced appropriations for FY2013 at $445 million from October 1, 2012, through October 1, 2013.

BACKGROUND

The Corporation for Public Broadcasting (CPB) was incorporated in 1967 as a private nonprofit corporation under the authority of the Public Broadcasting Act of 1967 (P.L. 90-129). CPB funding promotes public television and radio stations and their programs. These CPB-funded stations reach virtually every household in the United States. CPB is the largest single source of funding for public television and radio programming.

Most CPB-funded television programs are distributed through the Public Broadcasting Service (PBS) created in 1969 by CPB. CPB-funded radio programs are distributed primarily through National Public Radio (NPR), created in 1970 by CPB, and Public Radio International (PRI).

The number of radio and television public broadcasting stations supported by CPB increased from 270 in 1969 to 1,300 in 2012,[1] of which 368 are television stations. Public broadcasting stations are run by universities, nonprofit community associations, state government agencies, and local school boards.

CORPORATION FOR PUBLIC BROADCASTING

CPB is a nonprofit private corporation and is guided by a nine-member board of directors. These directors are appointed by the President with the advice and consent of the Senate. The directors serve for staggered six-year terms. The current chairman is Patty Cahill, elected by the board of directors in September 2012. CPB's principal function is to receive and distribute government contributions (or federal appropriations) to fund national programs and to support qualified public radio and television stations based on legislatively mandated formulas. The bulk of these funds, including the matching funds received from nonfederal sources, are used to provide Community Service Grants (or CSGs) to stations that meet specified eligibility criteria. CPB exercises minimum control of program content and other activities of local stations, and is prohibited from owning or operating any of the primary facilities used in broadcasting. In addition, it may not produce, disseminate, or schedule programs. The current president and CEO of CPB is Patricia de Stacy Harrison, appointed by the board of directors in June 2005.

Approximately 15% of all public television and radio broadcasting funding comes from the federal appropriations that CPB distributes.[2] However, among individual public broadcasting stations, the amount of federal dollars that contributes to a station's annual budget depends on the funds it receives from nonfederal sources; the number and extent of broadcast transmitters required to service its coverage area; the extent to which a station is serving rural areas and minority audiences; and whether or not it is a television or radio station.

While federal funding for CPB primarily comes from the Departments of Labor-Health and Human Services-Education appropriations bill as a separate entry under the "Related Agencies" section of that bill, it may receive other sources of funding from the federal government. For example, on October 15, 2010, CPB and PBS received notification of a Ready to Learn grant of nearly $72 million from the Department of Education's Office of Innovation and Improvement.

CPB and PBS will use this money to fund research, development, and deployment of transmedia content to improve the math and literacy skills of children ages 2-8, especially those living in poverty.[3]

PUBLIC TELEVISION: PBS

PBS was created by CPB in 1969 to operate and manage a nationwide (now satellite) program distribution system interconnecting all the local public television stations, and to provide a distribution channel for national programs to those public television stations. Although PBS does not produce programs for its members, it aggregates funding for the creation and acquisition of programs by and for the stations, and distributes programs through its satellite distribution system. Paula Kerger became the sixth and current president and CEO of PBS in March 2006.

PUBLIC RADIO: NPR AND PRI

For radio, a different division of responsibilities was established. CPB created National Public Radio (NPR) in 1970 as a news-gathering, production, and program-distribution company governed by its member public radio stations. Unlike its public television counterpart, NPR is authorized to produce radio programs for its members as well as to provide, acquire, and distribute radio programming through its satellite program distribution system. NPR Inc., located in Washington, DC, provides these administrative operations. Public Radio International (PRI) was founded in 1983 as an independent, not-for-profit corporation to act as another distributor of public radio content, in competition with National Public Radio and other existing distributors. Gary Knell is currently the president and CEO of NPR.

On October 20, 2010, Juan Williams, a news analyst working as an independent contractor reporter for NPR, was fired by executives in the office of NPR News for comments Mr. Williams made on the Fox News Channel, as well as apparently for previous incidents that violated the terms of Mr. Williams's contract.[4] In an appearance on The O'Reilly Factor show, Mr. Williams stated that he gets "nervous" when he sees someone in "Muslim garb" on an airplane. Two days after his appearance, Mr. Williams was notified by telephone that his contract with NPR was being terminated. In a memorandum to NPR staff on October 24, 2010, Vivian Schiller, then president and chief executive at NPR, stated that she "regrets that NPR executives did not meet with (Mr.) Williams in person to discuss the situation." In addition, Ms. Schiller also stated that Mr. Williams's comment was just the most recent in a series of objectionable remarks Mr. Williams has made while offering commentary on Fox News. This decision was supported by the leadership of NPR news.[5] However, almost immediately there was a strong reaction from some among the media and public about the process and fairness of this firing. This in turn has raised questions about any federal funding that supports NPR policies and programs.

On January 6, 2011, NPR Inc. announced that its board of directors had accepted several recommendations to provide greater clarity and transparency for its code of ethics regarding NPR employees. These include reviewing and updating of policies and training with respect to the role of NPR journalists appearing on other media outlets, reviewing and defining their roles (including those of news analysts) in a changing news environment, and encouraging a broad range of viewpoints to reflect the diversity of NPR's national audiences. At the same time these recommendations were announced, Ellen Weiss, vice president of news for NPR, resigned; it was also announced that Ms. Schiller would not receive a bonus for 2010. On March 9, 2011, Vivian Schiller resigned as president and CEO of NPR, over continued scrutiny and criticism over NPR's handling of an incident regarding Ronald Schiller (no relation) in a taped interview.

In response, some policymakers are revisiting the issue of whether federal appropriations should continue to support any part of the public broadcasting system. They are addressing the overall federal appropriations for CPB (discussed below), as well as any direct funding the NPR Foundation receives from federal sources. H.R. 68 and H.R. 69 (Representative Lamborn, and discussed more fully later in this report) were introduced on January 5, 2011. These bills would have eliminated federal appropriations for NPR and CPB, respectively, in FY2013.

In both FY2010 and FY2011, CPB distributed its community service grant (CSG) money to over 900 public radio stations. NPR directly received $3 million in discretionary grants designated for "special projects" ($2.96 million in FY2010 and $0 in FY2011).[6] In addition, there appears to be at least another source of federal funding for NPR. NPR Inc., which oversees the NPR system, states that annually NPR receives direct funding in the range of $1.5 million to $3 million from two federal agencies and CPB, and that this funding accounts for less than 2% of its annual budget. In FY2009 NPR Inc. received $1.6 million and in FY2010, $2.5 million. The sources of this funding are the National Endowment of the Arts, CPB, and the Department of Education.[7]

FEDERAL FUNDING

The Obama Administration requested a $445 million appropriation for CPB in its FY2013 budget request (which would be disbursed in FY2015). The vehicle that is used to provide appropriations to the CPB is the Departments of Labor-Health and Human Services-Education bill. After several Continuing Resolutions for the FY2012 federal budget, the final CR was signed into law by President Obama on December 23, 2011 (P.L. 112-74). While the final bill sustained the CPB advanced appropriations approved in the FY2010 budget, it was included in an across-the-board federal recission of 0.189%. The final appropriated funding for CPB stands at $444.1 million for FY2012.

From the last year of available information, the U.S. public broadcasting system—comprised of the national public radio and television stations—reported total income of $2.8 billion in FY2011.[8] According to the CPB, for public broadcasting revenue by source, CPB funds made up 15.1% of the total; another 2.9% came from federal grants and contracts. The remaining 82% was raised from nonfederal sources (including individuals, businesses, foundations, state and local governments, and educational institutions). The largest single income source (27.4% in FY2011) came from membership.

Federal appropriations which go through CPB to the individual public radio and television stations generally are designated as unrestricted federal funds. For those public radio stations which applied for grants, the amount of grants awarded by CPB totaled $68.6 million in FY2012. However, member stations also pay NPR fees for content and programming; some contend that federal grant money is supporting part of the revenue streams back to NPR Inc. A history of CPB appropriations is presented in *Table 1*. Additional information on both NPR and PBS funding may be obtained at their respective websites (http://www.npr.org and http://www.pbs.org, respectively).

Table 1. CPB Federal Appropriations
($ in millions)

Fiscal Year	Administration Request	House Appropriation	Senate Appropriation	Final Appropriation
1969	$9	[a]	$6	$5
1970	15	[a]	15	15
1971	22	[a]	27	23
1972	35	35	35	35
1973	45	45	45	35
1974	45	[a]	55	50
1975	60	60	65	62
1976	70	78.5	78.5	78.5
TQ[b]	17	17.5	17.5	17.5
1977	70	96.7	103	103
1978	80	107.1	121.1	119.2
1979	90	120.2	140	120.2
1980	120	145	172	152
1981	162	162	162	162
1982	172	172	172	172
1983	172	172	172	137
1984	110	110	130	137.5
1985	85	130	130	150.5
1986	75	130	130	159.5
1987	186	[a]	238	200
1988	214	[a]	214	214
1989	214	214	238	228
1990	214	238	248	229.4[c]
1991	214	[a]	302.5[d]	298.9[d]
1992	242.1	314.1d	340.5[d]	327.3[d]
1993	306.5[d]	306.5d	341.9	318.6[d]
1994	260	253.3	284	275
1995	275	271.6	310	285.6
1996	292.6	292.6	320	275
1997	292.6	[a]	330	260
1998	296.4	240	260	250
1999	275	250	250	250

Table 1. (Continued)

Fiscal Year	Administration Request	House Appropriation	Senate Appropriation	Final Appropriation
2000	325	300	300	300
2001	340	340	340	340
2002	350	340	350	350
2003	365	365	365	362.8
2004	e	365	395	377.8
2005	e	380	395	386.8[f]
2006	e, g	335	400	396[h]
2007	e, g	400	400	400
2008	e, g	400	400	393[i]
2009	e, g	None	400	400
2010	e, g	420	420	420
2011	e	430	430	429.[l]
2012	440	440	450	444.[l]
2013	460	460	460	445[l]
2014	451[j]	None	445	445[m]
2015	445[k]			

Source: Compiled by the Congressional Research Service from information from the Corporation for Public Broadcasting (http://www.cpb.org).

[a] Allowance not included in House Bill because of lack of authorizing legislation.

[b] Transition Quarter funding, during which federal budget year changed from July to September.

[c] Reduced FY1990 by Sequestration.

[d] Includes funds appropriated for the Satellite Replacement Fund.

[e] From FY2002-FY2011, the Bush Administration declined to request two-year advance funding for CPB. Similarly, the President's budget request did not provide separate funding for digital or, where applicable, interconnection replacement, but would have permitted CPB to use a portion of its general appropriation to fund both.

[f] FY2005 funding ($390 million) reduced by 0.80% across-the-board rescission in P.L. 108-447.

[g] From FY2006-FY2010, the Bush Administration proposed rescissions to CPB's already-enacted two-year advanced funding. The proposed rescissions: $10 million from FY2006; $53.5 million from FY2007; $50 million from FY2008; $200 million from FY2009 and $220 million from FY2010.

[h] FY2006 funding ($400 million) reduced by 1% across-the-board rescission in P.L. 109-148.

[i] FY2008 funding ($400 million) reduced by 1.747% across-the-board rescission in P.L. 110-161.

[j] Fiscal Year 2012 Appendix Budget of the U.S. Government.

[k] Fiscal Year 2013 Appendix Budget of the U.S. Government.

[l] P.L. 112-10 Senate Committee on Appropriations FY 2011 CR: Labor, HHS, Education Summary.

[m] P.L. 112-74 Consolidated Appropriations Act, (H.R. 2055).

There was significant legislative interest and activity regarding federal funding for CPB from the end of the 111th Congress through the 112th Congress. During the 111th Congress, Representative Lamborn (CO) introduced H.R. 5538, a bill that would have eliminated federal funding for CPB after FY2012.[9] This bill was referred to the House Committee on Energy and Commerce. During the "lame duck" period of the 111th Congress in November 2010, Representative Lamborn sought to have his bill considered for floor action in the House, but this action was defeated by a vote of 239-171. In response, Representative Earl

Blumenauer (OR) defended public broadcasting by stating that "National Public Radio is one of the few areas where the American public can actually get balanced information."[10]

On January 5, 2011, Representative Lamborn introduced H.R. 68 (To amend the Communications Act of 1934 to prohibit Federal funding for the Corporation of Public Broadcasting after FY2013) and H.R. 69 (To prohibit Federal funding of certain public radio programming, to provide for the transfer of certain public debt, and for other purposes). The first bill, like its predecessor H.R. 5538, would have eliminated federal appropriations for CPB when its two-year advanced funding ends. The second bill would have prohibited federal funding to organizations incorporated for specified purposes related to (1) broadcasting, transmitting, and programming over noncommercial educational radio broadcast stations, networks, and systems; (2) cooperating with foreign broadcasting systems and networks in international radio programming and broadcasting; (3) assisting and supporting such noncommercial educational radio broadcasting pursuant to the Public Broadcasting Act of 1967; or (4) acquiring radio programs from such organizations. In effect, it would have prohibited any individual public radio station from using federal funding to engage in transactions with NPR Inc. Both bills were referred to the House Committee on Energy and Commerce.

On January 11, 2011, NPR Inc. responded to the two bills by stating, in part: "The proposal to prohibit public radio stations from using CPB grants to purchase NPR programming interjects federal authority into local station program decision-making. Furthermore, restrictions on the authority of CPB—a Congressionally chartered, independent, nonprofit organization—to make competitive grants to NPR, or any other public broadcasting entity, is misguided."[11]

Other legislation was introduced addressing federal support for public broadcasting. On January 7, 2011, Representative Kevin Brady introduced H.R. 235 (Cut Unsustainable and Top-Heavy Spending Act of 2011 or the CUTS ACT), which provided that all unobligated balances held by the CPB that consist of federal funds be rescinded and no federal funds appropriated hereafter shall be obligated or expended.

On January 24, 2011, Representative Jim Jordan introduced H.R. 408 (Spending Reduction Act of 2011), which would have reduced federal spending by $2.5 trillion through FY2021 in part by eliminating the CPB.

On March 15, 2011, Representative Lamborn introduced H.R. 1076, a bill to prohibit the funding of National Public Radio and restrict the use of federal funds for member stations to acquire NPR broadcasting content. The House Rules Committee passed H.Res. 174, which permitted H.R. 1076 to go directly to the floor and, without any points of order or amendments, be open to one hour of debate before a full vote in the House of Representatives. H.R. 1076 passed the House 228-192, and was referred to the Senate. No further action was taken on this bill.

Other proposals in the 112th Congress addressed federal funding for public broadcasting. On January 20, 2011, the Republican Study Group, a conservative caucus comprised of 100 Members of Congress, released its list of proposed budget cuts, including elimination of CPB's appropriations starting in FY2012. At the same time, Representative Ryan (WI), the chairman of the House Committee on the Budget for the 112th Congress, proposed a new continuing resolution that would have set the rest of the FY2011 budget at FY2008 levels (excluding defense, homeland security, and veterans programs).

ISSUES

In an age of multiple cable channel options, digital radio, and computerized digital streaming, some ask whether there is a need for federal appropriations to support public broadcasting. The array of commercial all-news radio and radio talk shows, many of which are also streamed on the Internet, provides sources of news and opinion. Supporters of public broadcasting argue that public radio and television broadcasters, free of commercial interruption, provide perhaps the last bastion of balanced and objective information, news, children's education, and entertainment in an era of a changing media landscape. Still, others contend that public broadcasting has lost much of its early impact since the media choices have grown so much over the last several decades and that the federal role in public broadcasting should be re-evaluated as well.

Supporters of public broadcasting contend that public radio and public television provide education and news to many underserved parts of the American population. Public broadcasters may provide this service to an underserved and less commercially attractive population that commercial broadcasters do not address. For example, PBS broadcasting for children includes lessons in reading, counting, and spelling as part of its content, subjects not normally found on commercial broadcasts. According to NPR Inc., approximately 90% of public radio stations provide local newscasts, airing both newscast and non-newscast content (primarily in weekday drive times and especially during morning drive-time). About half of all public radio stations carry local news during the weekends, says NPR, and 74% of stations are producing and inserting stories into their programming.

On June 20, 2012, the CPB released a report, *Alternative Sources of Funding for Public Broadcasting Stations.* The report was undertaken in response to language in the Military Construction and Veterans Affairs and Related Agencies Appropriations Act of 2012 directing the CPB to provide a report to congressional appropriations committees on alternative sources of federal funding for public broadcasting stations. (H.R. 2055, P.L. 112-74). The report, undertaken by the consulting firm of Booz & Company, provides several alternative or new funding options for public broadcasting stations, with possible benefits as well as liabilities for each option. Five options considered by Booz & Company are television advertising, radio advertising, retransmission consent fees, paid digital subscriptions, and digital game publishing. In addition, fourteen current sources of revenue streams already employed by public broadcasting, ranging from merchandise licensing to mobile device applications, were also analyzed as options to replace federal funding for public broadcasting. Booz & Company found "there is simply no substitute for the federal investment" in public broadcasting and that "Ending federal funding for public broadcasting would severely diminish, if not destroy, public broadcasting service in the United States."[12] The report concludes that if the existing public broadcasting structure were commercialized, the new revenue streams would not offset the loss of federal funding; and that many public broadcasters would have to deviate from their statutory service mission or compete for advertising with established commercial broadcasters in a difficult economic environment.

Still, some critics contend that the report substantiates criticisms of the public broadcasting model: required to compete with commercial television and radio broadcasters that also provide news and entertainment, many public broadcasters could not adapt to a changing media world that provides multiple sources of information and entertainment. For these critics, if many public broadcasters struggle to operate with budget deficits even with federal funding available, what does that say about the need and viability of these stations in a multi-media world, or the ability of their audiences to sustain this business model going forward?

Several important issues are facing congressional policymakers as they address federal appropriations for all forms of public broadcasting. On the most fundamental level, many question the 1967 law that created the national public broadcasting system and whether the federal government should be in the "business" of providing general appropriations to CPB every year since 1969. They ask: is this still a relevant and appropriate role of the federal government? On a second level, some may contend that in an era of spiraling federal deficits, in which many (if not all) federal expenditures are being re-examined, appropriations for CPB should be reduced if not eliminated. These questions revolve around whether federal funding for public broadcasting should be continued at its current level; whether the funding should be modified or reduced; whether the arrangement between the federal funding process and public broadcasting should be changed; or whether federal funding for public broadcasting should be eliminated.

Public broadcasting retains its strong supporters. Most federal appropriations go through CPB to directly support member stations of NPR, PRI, PBS, and other independent affiliates. Since according to NPR, federal funding to supplement administrative functions amounts to less than 2% of its annual budget, some may question whether such a small amount is worthy of congressional action to eliminate. Independent of the 2011controversy regarding NPR, over the last several years some in Congress have questioned whether the federal appropriations for CPB should be reduced or eliminated as well. Underlying this position are concerns that the federal role, once so clear in 1967, has been eclipsed in a multi-media Internet age; concerns that the size and scope of the federal government budget deficit requires significant cutbacks in many areas; and allegations that public broadcasting is not objective, balanced, or free of an ideological slant.

As indicated in *Table 1,* CPB has consistently received increasing federal appropriations since 1969. Some would contend that this demonstrates a general consensus among congressional policymakers that there is a federal role in public broadcasting. In addition, public support of public radio and television broadcasting generally has been consistent as well. Supporters of a public broadcasting network system contend that local programming content is not determined by NPR Inc., or PBS, and that most content is local serving community needs. Balanced against concerns about the role of the federal government in public broadcasting, as well as strong pressure to reduced federal spending, these issues will likely continue to be of interest to federal policymakers.

End Notes

[1] http://www.cpb.org/aboutpb/faq/stations.html.

[2] http://www.cpb.org/stations/reports/revenue/2007PublicBroadcastingRevenue.pdf.

[3] Corporation for Public Broadcasting, "The Corporation for Public Broadcasting and PBS Receive Ready To Learn Grant Funding from the U.S. Department of Education," press release, October 15, 2010, http://www.cpb.org/pressroom/release.php?prn=840.

[4] Mike Riksen, Vice President, Policy & Representation, letter to the Congressional Research Service, December 6, 2010. p. 1.

[5] Ibid., and as cited at http://www.npr.org/blogs/thetwo-way/2010/10/25/130805049/npr-ceo-apologizes-for-hand lingof-williams-termination. Information on NPR's code of ethics may be viewed at http://www.npt. org/about/aboutnpr/ ethics/. Last viewed December 23, 2010.

[6] Also described by NPR as "occasional projects." It is important to note that the original numbers provided to CRS were $3.9 million in FY2009 and $4.1 million in FY2010, for a total of $8 million. Telephone Conversation, Office of Congressional Affairs, Corporation for Public Broadcasting, October 21, 2010. The purposes of these grants were to develop HD radio technology for those who are "print- and hearing-impaired"; to distribute next generation public broadcasting technology to individual stations; to support collaborations between public radio and public broadcasting stations for content related to news events of 2008 and 2009; and for other projects. Mike Riksen, Vice President, Policy & Representation, letter to the Congressional Research Service, December 6, 2010. p. 2.

[7] Personal Communication, Anna Christopher, Communications Office, NPR, October 26, 2010.

[8] http://www.cpb.org/stations/reports/revenue/2011PublicBroadcastingRevenue.pdf.

[9] Full title: "To amend the Communications Act of 1934 to prohibit Federal funding for the Corporation for Public Broadcasting after fiscal year 2012."

[10] Frances Symes, CQ Today Online News, "House Rejects GOP Bid to Reduce Federal Funding for NPR and Local Stations." http://www.cq.com/doc/news.

[11] NPR Statement Regarding Proposed Legislation, H.R. 68/69, NPR., Inc., Washington, DC, January 11, 2011.

[12] Corporation for Public Broadcasting, Alternative Sources of Funding for Public Broadcasting Stations, June 20, 2012, pp. 2, 3.

In: Transformations in Telecommunications and Media
Editor: Irwin Cavazos
ISBN: 978-1-62948-413-6
© 2013 Nova Science Publishers, Inc.

Chapter 7

ONLINE VIDEO DISTRIBUTORS AND THE CURRENT STATUTORY AND REGULATORY FRAMEWORK: ISSUES FOR CONGRESS[*]

Charles B. Goldfarb and Kathleen Ann Ruane

SUMMARY

Digital and Internet protocol technologies have spawned a number of online video distributors (OVDs) whose "over-the-top" video services are in some ways akin to, and in some ways different from, traditional cable and satellite video programming distribution services. However, most of the statutory and regulatory framework for video predates the commercial Internet and was developed within a policy debate that could not consider digital technology and online services. As a result, many statutory provisions apply only to cable companies or satellite carriers, or only to "multichannel video programming distributors" (MVPDs)—a category that includes cable and satellite operators, but as currently interpreted by the Federal Communications Commission excludes online video distributors.

Congress has begun to consider this issue. At both the June 27, 2012, House Energy and Commerce Subcommittee on Communications and Technology hearing on "The Future of Video" and the July 24, 2012, Senate Commerce, Science, and Transportation Committee hearing on "The Cable Act at 20," questions were posed about which of the existing statutory provisions and regulatory rules, if any, should be applied to the new service providers, which provisions and rules should be modified in light of the new technologies and new market realities, and even whether changed circumstances are so great that major statutory reform is needed.

Statutory provisions and regulatory rules affecting media and communications typically are shaped by negotiations among the many stakeholders present at the time the statutes and regulations are being developed.

The resulting framework creates obligations, prohibitions, privileges, and even rights for the various stakeholders. The industry players construct business models based on these. But statutes and regulations that are tailored to existing technologies may create

---

[*] This is an edited, reformatted and augmented version of Congressional Research Service, Publication No. R42722, dated January 14, 2013.

impediments to the deployment of new technologies, especially if they create privileges or rights that are technology-specific.

Some observers have raised concerns that the current statutory and regulatory framework no longer fosters the long-standing U.S. media policy goals of competition, diversity of voices, localism, and innovation because it does not extend to online video distributors the privileges and rights—and also the obligations and prohibitions—that are applicable to traditional video distributors.

For example, competition and innovation may be harmed if online video distributors are denied the access to programming that MVPDs enjoy through the program access and retransmission consent rules or if they are denied the guaranteed low cost compulsory copyright license that cable companies and satellite carriers enjoy. At the same time, localism may be harmed if online video distributors that rebroadcast broadcast television signals are not required to carry their subscribers' local broadcast stations and are not required to black out distant broadcast signals that duplicate the network and syndicated programming on local stations.

OVERVIEW

Most of the statutory and regulatory framework for video predates the commercial Internet and was developed within a policy debate that could not consider digital technology and online services. Notably, the regime under which cable and satellite operators negotiate with programmers—the retransmission consent and program access rules—was created by the 1992 Cable Act[1] and the compulsory copyright licenses that allow cable operators and satellite carriers to retransmit the performance of works embodied on broadcast signals were created by the 1976 Copyright Act[2] and 1988 Satellite Home Viewer Act,[3] respectively. Today, digital and Internet protocol technologies have spawned a number of online video distributors (OVDs) whose "overthe-top" video services are in some ways akin to, and in some ways different from, traditional cable and satellite video programming distribution services. But the existing statutes do not cover online video distribution.

At both the June 27, 2012, House Energy and Commerce Subcommittee on Communications and Technology hearing on "The Future of Video" and the July 24, 2012, Senate Commerce, Science, and Transportation Committee hearing on "The Cable Act at 20," questions were posed about which of the existing statutory provisions and regulatory rules, if any, should be applied to the new service providers, which provisions and rules should be modified in light of the new technologies and new market realities, and even whether changed circumstances are so great that major statutory reform is needed. In the 112th Congress, the Next Generation Television Marketplace Act, which was introduced by Representative Scalise (H.R. 3675) and Senator DeMint (S. 2008), would have, among other things, made major changes to the statutory framework for video.[4]

Statutory provisions and regulatory rules affecting media and communications typically are shaped by negotiations among the many stakeholders present at the time the statutes and regulations are being developed. The resulting framework creates obligations, prohibitions, privileges, and even rights for the various stakeholders. The industry players construct business models based on these. One of the challenges when developing and implementing federal media policy is to construct a statutory framework and regulatory regime flexible enough to accommodate technological change.

It is difficult, and potentially counterproductive, to construct statutes and regulations based on surmises as to where future technologies will lead media markets. At the same time, statutes and regulations that are tailored to existing technologies may create impediments to the deployment of new technologies, especially if they create privileges or rights that are technology-specific.

To address this, Congress has enacted certain statutes that include definitions or terminology that intentionally leave room for interpretation by the relevant expert agencies (in this case, the Federal Communications Commission (FCC) and the Copyright Office) to accommodate new technologies over time.

The FCC interpretations typically take into account the three longstanding U.S. media policy goals of competition, localism, and diversity of voices, and the more recently enunciated goal of innovation.[5] It is not unusual for an unhappy party to go to court to challenge the FCC's interpretation or to attempt to force the agency to make an interpretation. Several such court cases relating to new ways to deliver video services are moving forward today. But even as the FCC, the Copyright Office, and the courts attempt to interpret existing statutes, Congress may choose to perform its own statutory review in light of technological and market changes.

As a general rule, new providers of communications services seek access to the privileges and rights in the existing statutory and regulatory framework, and seek to avoid the obligations and prohibitions.

To the extent that the new entrants potentially threaten incumbents' business models, the incumbents seek to minimize new entrant access to privileges and rights and to maximize new entrant obligations and prohibitions. Perhaps the most notable example of this is the attempt by Sky Angel, an online video distributor that was denied access to several cable networks owned by Discovery Communications, to use one such privilege—the FCC's Program Access complaint process—to obtain access to programming, and the arguments made by cable companies and cable network programmers that Sky Angel does not have legal standing to bring such a complaint.[6]

At the same time, entities may have the incentive to deploy new technologies that would allow them to "invent around" those obligations, prohibitions, privileges, and rights in the existing regulatory framework that are unfavorable to them. For example, copyright holders have the right to control public (but not private) performances of their works.[7] A cable or satellite company must obtain a copyright license to retransmit the copyrighted material on a broadcast television program, but a consumer can use an antenna to receive over-the-air, without obtaining a license and without payment, the same copyrighted work for her own private use.

A start-up company, Aereo, now has deployed thousands of mini-antennas, which, though maintained at Aereo's premises, are assigned to individual subscribers as their personal antennas; these antennas send broadcast television signals to individual subscribers over the Internet. Aereo claims the broadcast programs received by its subscribers are private performances and not subject to copyright licensing requirements. The broadcasters claim otherwise and have sued Aereo for copyright infringement and sought a preliminary injunction.[8]

On July 11, 2012, a federal district court judge denied the injunction, concluding "that plaintiffs have failed to demonstrate they are likely to succeed in establishing that Aereo's system results in a public performance."[9] Aereo therefore will be able to continue to offer its service as the copyright infringement case moves forward.

As the congressional committees with jurisdiction—the Senate and House commerce and judiciary committees—review the existing communications and copyright statutory framework, they may have to consider two complicating factors. First, there has been market entry by online video distributors with such a wide variety of service offerings, business models, technologies deployed, consumer equipment targeted, and even ownership relationships with traditional video distributors that it might not be optimal, or even possible, for Congress to construct statutory and regulatory provisions applicable to all entities. Rather, Congress may determine that some online video distributors (OVDs) are sufficiently similar to traditional video distributors (such as cable companies and satellite carriers) that they should be treated the same, while other OVDs have offerings and business models that are far removed from those of traditional video distributors and should be treated differently.

Second, several industry stakeholders have alleged that the United States is prohibited from taking certain actions, or must take certain actions, in order to conform to an article that has been included in a number of bilateral and multilateral trade agreements for which Congress has passed implementing language. The article states, "neither Party may permit the retransmission of television signals (whether terrestrial, cable, or satellite) on the Internet without the authorisation of the right holder or right holders of the content of the signal and, if any, of the signal."[10]

The Walt Disney Company claims that extending the compulsory copyright license currently available to cable companies and satellite carriers to online video distributors would violate the article because under a compulsory license the right holders could not control authorization.[11]

The Copyright Office, which "continues to oppose an Internet statutory license that would permit any website on the Internet to retransmit television programming without the consent of the copyright owner," claims "An Internet statutory license would require renegotiating the relevant FTAs [free trade agreements] with other countries."[12] The local broadcast stations affiliated with the ABC, CBS, and NBC broadcast networks claim that online video distributors must be made subject to the retransmission consent rules "in order for the United States to honor these free trade commitments."[13] However, this article generally has been incorporated in the trade agreements at the behest of American, not foreign, companies; statutory or regulatory actions taken in furtherance of domestic media policy goals that are not in accord with the article may be opposed by the entities that sought the provision, but may not be likely to generate trade complaints against the United States.

THE CURRENT STATUTORY AND REGULATORY FRAMEWORK

Obligations, Prohibitions, Privileges, and Rights Relating to All Multichannel Video Programming Distributors (MVPDs)

In 1992, Congress responded to a perceived lack of competitive alternatives to the local cable company and concerns about cable company vertical integration into programming by passing the Cable Television Consumer Protection and Competition Act (1992 Cable Act).[14]

That act created a number of legal obligations, prohibitions, privileges, and rights relating to a newly defined group of entities—multichannel video programming distributors (MVPDs). MVPDs primarily are cable and satellite operators but, as discussed in the Appendix to this report, the definition is sufficiently broad and non-specific to potentially include other video distributors.[15]

Several of the obligations, prohibitions, privileges, and rights created by the 1992 Cable Act fundamentally affect the way that MVPDs may interact with local broadcast stations and with broadcast and cable programming networks, and vice versa. Notably, the 1992 Cable Act added provisions to the Communications Act that

- require an MVPD to obtain the consent of a local broadcast station before retransmitting that station's signal to its subscribers (retransmission consent requirements).[16] This, in effect, creates a new property right for broadcasters that allows a broadcast station to demand compensation from an MVPD for the retransmission of its *signal*.[17]
- require both the local broadcast station and the MVPD to participate in retransmission consent negotiations in good faith.[18] On one hand, this restrains an MVPD from refusing to agree to pay reasonable compensation. On the other hand, this benefits an MVPD because it implies a right to retransmit a local station's signal that cannot be circumvented by bad faith retransmission consent negotiations by the station.
- prohibit a cable or satellite operator that has an attributable interest in a programming network from denying a competing MVPD access to that programming (program access rules).[19] This allows an unaffiliated MVPD to seek redress at the FCC if a cable programming network in which a competing cable or satellite company has an interest refuses to make that network programming available for reasonable compensation.

The statutory and regulatory framework for MVPDs also includes rules to foster the commercial availability to consumers of converter boxes and other equipment needed to access MVPD services;[20] program carriage rules that prohibit a vertically integrated MVPD that owns a programming network from discriminating against a non-affiliated programming vendor when choosing which programming networks it will carry and the placement of those networks in its channel line-up;[21]

Equal Employment Opportunity requirements;[22] closed captioning and emergency information requirements to better serve persons with hearing and visual disabilities;[23] and various technical rules.

There are legal questions about whether some new online video distribution services qualify as MVPDs and are thus subject to all or some of the MVPD-specific obligations, prohibitions, privileges, and rights in the Communications Act, as well as questions about the public policy implications of applying—or not applying—some or all of the MVPD-specific rules to the new entrants.

Obligations, Prohibitions, Privileges, and Rights Relating Only to Cable Companies or Satellite Carriers

Several provisions in the Communications Act, the Copyright Act, or FCC rules create obligations, prohibitions, privileges, and rights that specifically pertain to cable companies or to satellite carriers, but not to MVPDs in general, nor to video distributors that are not cable or satellite operators. In particular:

- To foster localism, the FCC adopted network non-duplication and syndicated exclusivity rules that require cable companies to black out programming on the signals of distant (non-local) broadcast stations they import that duplicates the programming on a local stations if the local station has a contract with the program supplier to be the exclusive broadcaster of that programming within a geographic area.[24] The FCC initially adopted these rules for cable television without specific guidance from Congress. Later, Congress explicitly instructed the FCC to apply its network non-duplication and syndicated exclusivity rules to the signals of certain stations retransmitted by satellite carriers.[25] Although these rules do not apply to video programming distributors that are not cable or satellite operators, those distributors may have contractual arrangements with programmers with similar geographic restrictions.
- Also in support of localism, the 1992 Cable Act gave each broadcast station the right to demand that every cable operator in its local market carry its signal, though without compensation ("must carry" rules).[26] A satellite carrier is not required to carry the signals of its subscribers' local broadcast stations, but if it chooses to carry even one local station in a market it must carry all the stations in that market.[27] But these requirements do not apply to MVPDs or other video distributors that are not, in the first case, a cable operator or, in the second case, a satellite carrier.
- Sections 111, 119, and 122 of the Copyright Act create free or relatively low-cost compulsory copyright licenses for cable companies and for satellite carriers for the public performance or display of a work embodied in a broadcast television transmission, if the cable and satellite providers are "in compliance with the rules, regulations, or authorizations of the Federal Communications Commission governing the carriage of television broadcast station signals."[28] These privileges allow cable operators and satellite carriers that abide by the geographic restrictions in the FCC rules to pay lower fees to copyright holders when retransmitting broadcast television signals than they likely would had they been required to negotiate copyright license payments. These compulsory licenses are not available to online video distributors or other video distributors or MVPDs that are not cable companies or satellite carriers.[29]

There are serious questions about the public interest implications of making—or not making— some new online and other video distribution services subject to all or some of the cable-specific or satellite-specific obligations, prohibitions, privileges, and rights.

Table 1 shows the differential treatment in the current statutory and regulatory framework of video distributors depending on whether they are MVPDs, cable companies, satellite carriers, or other (non-video) programming distributors.

Table 1. Treatment of MVPDs, Cable Companies, Satellite Carriers, and non-MVPD Video Programming Distributors Under the Current Statutes and Regulations

Obligation, Prohibition, Privilege, or Right	Multichannel Video Programming Distributors (all MVPDs)	Cable Operators (a subset of MVPDs)	Satellite Carriers (a subset of MVPDs)	Video Programming Distributors Not Classified as MVPDs (including OVDs)
Obligation to obtain retransmission consent from local broadcast stations	Yes	Yes	Yes	No
Implicit right to retransmit broadcast station signals (from requirement that broadcaster negotiate retransmission consent in good faith)	Yes	Yes	Yes	No
Access to below market rate compulsory copyright license for works in retransmitted broadcast signals	No	Yes	Yes	No
Access to programming of programmers that have an attributable interest in a competing MVPD (Program Access Rules)	Yes	Yes	Yes	No
If affiliated with a programmer, obligation to carry similar programming of independent programmer (Program Carriage Rules)	Yes	Yes	Yes	No
Prohibitions under Network Non-Duplication and Syndicated Exclusivity Rules and/or Unserved Household Rule from retransmitting broadcast signals in certain geographic areas	No, although MVPDs have similar prohibitions in their contracts with programmers	Yes	Yes	No, although non-MVPD distributors may have similar prohibitions in their contracts with programmers
Obligations under Must Carry or Carry One/Carry All Rules	No	Yes, subject to Must Carry Rules	Yes, subject to Carry One/Carry All Rule	No
Obligation to meet FCC's Equal Employment Opportunity Rules	Yes	Yes	Yes	No
Obligation to allow customers to use or attach competitively provided navigation devices (set-top boxes)	Yes	Yes	Yes	No
Obligation to provide closed captioning for the hearing impaired	Yes	Yes	Yes	Some may meet definition of "distributor of video programming for residential reception that delivers such programming directly to the home and is subject to the jurisdiction of the Commission"

Table 1. (Continued)

Obligation, Prohibition, Privilege, or Right	Multichannel Video Programming Distributors (all MVPDs)	Cable Operators (a subset of MVPDs)	Satellite Carriers (a subset of MVPDs)	Video Programming Distributors Not Classified as MVPDs (including OVDs)
Obligation to provide emergency video programming accessible to persons with hearing and visual disabilities	Yes	Yes	Yes	Some may meet definition of "distributor of video programming for residential reception that delivers such programming directly to the home and is subject to the jurisdiction of the Commission"

Source: Congressional Research Service, based on interpretations by the Federal Communications Commission and the Copyright Office.

THE OVERARCHING POLICY ISSUE WITH RESPECT TO ONLINE VIDEO DISTRIBUTORS: ACCESS TO PROGRAMMING

Under current law, online video distributors, and other video distributors that do not fit the statutory categories of MVPD, cable company, or satellite carrier, do not have the privileges and rights firms in those categories enjoy under the Communications Act and the Copyright Act.

The provision in the Communications Act requiring broadcasters to negotiate retransmission consent terms "in good faith" with MVPDs implies a right to MVPDs for access to the broadcast signal if they offer reasonable compensation. Online video distributors that are not deemed MVPDs do not enjoy that implied right.

The program access provision in the Communications Act allowing an unaffiliated MVPD to seek redress at the FCC if it faces discriminatory access to a programming network affiliated with a competing MVPD is not available to an online video distributor that is not deemed an MVPD.

The provisions in Sections 111, 119, and 122 of the Copyright Act that create statutory compulsory public performance licenses for cable companies and satellite carriers that retransmit the signals of broadcast stations give cable and satellite providers the right to retransmit the broadcast signals as long as they pay the statutory fees and meet the other requirements, such as abiding by the FCC's network non-duplication and syndicated exclusivity rules. Moreover, these fees are generally viewed as being lower than the rates likely to prevail in market negotiations. In contrast, online video distributors and other non-cable and non-satellite distributors must negotiate public performance copyright permission with the copyright holders, under Section 106(4) of the Copyright Act, and the copyright holders are under no obligation to grant permission.

Although copyright holders often have the incentive to distribute their works as widely as possible, there may be situations when more exclusive distribution represents a more profitable business strategy. If the copyright holders believe that granting these rights to video distributors that compete against the incumbent cable and satellite video distributors could undermine the existing industry business model that has been very lucrative for the copyright holders and distributors alike, they have the right not to grant such rights. For example, if a copyright holder participates in the "TV Everywhere" strategy of making its programming available for use on all receiving devices (television sets, computers, game consoles, tablets, cell phones) but only if the consumer subscribes to the pay television service of a participating distributor, then that copyright holder might not have the incentive to grant permission to unaffiliated online distributors. This is in some way analogous to a manufacturer prohibiting discount stores from carrying its product line in order to maintain profit margins (that might be reinvested into product innovation).

In addition, to the extent that a broadcast network also has its own online video distribution service or has a financial interest in an online video distribution service (for example several networks have ownership shares in Hulu), it may not have the incentive to grant copyright permission to competing over-the-top services.

Many of the copyright holders for broadcast network programming are large programmers that also hold the copyrights for cable network programming. They may have the same incentives to withhold copyright permission from over-the-top providers for their cable network programming as they have for their broadcast network programming.

Given that there is certain "must have" progra mming that multichannel video distributors must provide in order to successfully attract customers, if the copyright holders of such programming choose to deny online video distributors access to that programming, successful competitive entry could be difficult.[30] That concern was the basis for Congress enacting the Program Access requirements in the first place.

THE POLICY ISSUES RAISED BY THE ENTRY OF ONLINE VIDEO DISTRIBUTORS

Some observers have raised concerns that the current statutory and regulatory framework no longer fosters the policy goals of competition, diversity of voices, localism, and innovation because it does not extend to online video distributors the privileges and rights—and also the obligations and prohibitions—that are applicable to traditional video distributors. For example, competition and innovation may be harmed if online video distributors are denied the access to programming that MVPDs enjoy through the program access and retransmission consent rules or if they are denied the guaranteed low cost compulsory copyright license that cable companies and satellite carriers enjoy. At the same time, localism may be harmed if online video distributors are not subject to the geographic restrictions in the network non-duplication and syndicated exclusivity rules that cable and satellite providers must obey.

As explained above, the current statutory and regulatory framework is characterized by provisions that are specific to certain statutorily-defined subcategories of video programming distributors—multichannel video programming distributors (MVPDs), cable companies, or satellite carriers. Congress may choose to review these provisions to determine whether, with some definitional modifications, the statute can accommodate online video distributors and better meet the policy goals. In so doing, it may find that, given the new technologies and new market dynamics, more extensive changes are needed to the statutory and regulatory framework, perhaps even a full re-evaluation of the retransmission consent/must carry regime and the compulsory copyright license. Similarly, in light of its longstanding concern about vertically integrated media companies leveraging market power from one market to another, Congress may choose to analyze the ability of vertically integrated Internet service providers to discriminatorily favor their own online video services over competing non-affiliated services.

Any review may have to address two complicating factors. There has been wide-scale market entry by online video distributors. But among those entrants there is such a wide variety of service offerings, business models, technologies deployed, consumer equipment targeted, and even ownership relationships with traditional MVPDs that it might not be optimal, or even possible, for Congress to construct statutory and regulatory provisions applicable to all OVDs. Rather, Congress may determine that some OVDs are sufficiently similar to traditional MVPDs that they are close substitutes that should be treated as MVPDs, while other OVDs have offerings and business models that are far removed from those of traditional MVPDs and should be treated differently.

Second, as explained earlier, a number of free trade agreements include language that arguably constrains the ability of the United States to expand the compulsory copyright license to online video distributors and requires online distributors to get permission from broadcasters to retransmit their signals. Although these provisions generally have been incorporated in the trade agreements at the behest of American, not foreign, companies, and thus statutory or regulatory actions taken in furtherance of domestic media policy goals that are not in accord with these trade provisions may not generate trade complaints against the United States, any such actions are likely to be strongly opposed by the entities that sought the provisions.

Refining or Modifying Constructs in the Current Statutory and Regulatory Framework

In the 1992 Cable Act, Congress defined an "MVPD" as "a person such as, but not limited to, a cable operator, a multichannel multipoint distribution service, a direct broadcast satellite service, or a television receive-only satellite program distributor, who makes available for purchase by subscribers or customers, multiple channels of video programming."[31]

Congress could not predict the direction that video distribution technology would take. Given the many technological and market developments since 1992, Congress may wish to review the concept of an MVPD, both in terms of what constitutes an MVPD and what obligations, prohibitions, privileges, and rights should apply to MVPDs and the entities dealing with MVPDs—in particular with respect to online video distributors.

Should a Video Distributor Have to Offer Multiple Channels of Pre-Scheduled, "Linear" Programming to Qualify as an MVPD Eligible for Program Access Privileges?

There are many OVDs offering a wide variety of video services using different business models. Some offer multiple channels of pre-scheduled, "linear" programming of the sort long offered by television broadcast stations and incumbent MVPDs; others offer video-on-demand and pay-perview services of the sort offered by many incumbent MVPDs but not by broadcasters; and yet other OVDs offer both types of services. Depending on the OVD service to which they subscribe, consumers may be able to view the video on their television sets, game consoles, computers, or mobile devices.

By common usage, and by statutory definition, a central element of an MVPD has been the provision of "multiple channels of video programming." Most of the MVPD-specific provisions in the 1992 Cable Act appear to apply to channels of pre-scheduled, linear programming offered by broadcast networks and cable networks, rather than on-demand offerings. Retransmission consent applies to the retransmission of a broadcast signal, but on-demand services do not retransmit a broadcaster signal, they provide a stored program. Similarly, the program access rules apply to "satellite broadcast programming" and "satellite cable programming,"[32] which by definition appear to be limited to pre-scheduled, linear programming retransmitted to MVPD subscribers, thus not including programming stored for future use by subscribers.

Yet, if a vertically integrated programmer-MVPD refuses to make its non-linear programming available to an unaffiliated video distributor, or does so only in a discriminatory fashion, that could have the same anti-competitive effect as a vertically integrated programmer-MVPD refusing to make its linear programming available, especially as the market appears to be shifting away from linear programming offerings toward on-demand offerings. Is there a policy reason to apply the program access rules to pre-scheduled, linear programming but not to on-demand programming?

The FCC appears to recognize this issue. It has sought public comment on how to address a hypothetical future scenario in which "traditional MVPDs...eventually make available video programming for purchase or rental exclusively on an on-demand basis?"[33] Congress may want to provide guidance.

This suggests an additional question. If a company, such as Verizon, has a traditional MVPD service offering, such as its FiOS, but also offers an over-the-top IP-based service, such as Verizon's joint venture with Redbox, to what extent, if at all, should the obligations, prohibitions, privileges, and rights associated with MVPD status apply to the second service?[34]

Should Online Video Distributors be Subject to the Geographic Restrictions in the FCC's Network Non-duplication and Syndicated Exclusivity Rules or the "Unserved Household" Rules in Order to Qualify as MVPDs?

In order to protect local broadcast stations from duplicative programming imported from distant broadcast stations, as early as the 1960s the FCC adopted network non-duplication and syndicated exclusivity rules that require cable companies to black out duplicative network programming or syndicated programming.[35] Later, Congress explicitly instructed the FCC to apply its network non-duplication and syndicated exclusivity rules to the signals of certain stations retransmitted by satellite carriers.[36] In addition, the Satellite Home Viewer Act of 1988 (P.L. 100-667) modified the Copyright Act by adding Section 119, which created a statutory compulsory license for satellite carriers for the retransmission of copyrighted material on broadcast signals to "unserved households"—broadly speaking, those households that are not able to receive that programming from local broadcast stations. These rules set geographic restrictions on the retransmission of broadcast station signals by cable companies and satellite carriers. The statutory compulsory licenses covering the retransmission of broadcast signals by cable operators (in Section 111 of the Copyright Act) and by satellite carriers (in Sections 119 and 122 of the Copyright Act) all restrict these licenses to cable and satellite providers that abide by the FCC's network non-duplication and syndicated exclusivity rules.

If a video distributor is deemed an MVPD, then it attains the access to broadcast programming and cable network programming provided by the retransmission consent and program access rules. A satellite carrier does not need to obtain a broadcaster's retransmission consent to provide its subscribers the signals of a "superstation"[37] if it is abiding by the FCC's network non-duplications and syndicated exclusivity rules. It also does not have to obtain a broadcaster's retransmission consent to provide its subscribers the signals of a non-local network station so long as the subscribers qualify as "unserved households."[38] These provisions suggest that Congress has sought to place geographic limits on satellite retransmission of broadcast signals. If Congress continues to be concerned about the geographic reach of retransmitted signals, it might want to set the same or similar geographic restrictions on online video distributors that retransmit broadcast station signals.

It is noteworthy that when the then-register of copyrights, Marybeth Peters, testified to Congress that she opposed extending the compulsory license for retransmission of broadcast signals to Internet distributors, she stated that "Our principal concern is the extent to which Internet retransmissions of broadcast signals can be controlled geographically."[39]

She further stated that the Section 111 license for cable companies

> was tailored to a heavily-regulated industry subject to requirements such as must-carry, programming exclusivity, and signal quota rules—issues that have also arisen in the context of the satellite compulsory satellite license. Congress has properly concluded that the Internet should be largely free of regulation, but the lack of such regulation makes the Internet a poor candidate for a compulsory license that depends so heavily on such restrictions.[40]

Ms. Peters also expressed concern that the compulsory licenses for cable operators and satellite carriers harmed copyright holders because they were set at a fee that was below market level. But if these compulsory licenses remain in effect for cable and satellite providers, and if Internet providers were required to abide by—and could demonstrate an ability to abide by—the same geographic and other rules imposed on cable and satellite video distributors, then differential (and inferior) treatment of online video distributors would have anticompetitive implications. If an online video distributor agreed to abide by all the obligations and prohibitions currently imposed on MVPDs, should it be granted the privileges and rights given those MVPDs?

In comments submitted to the FCC, Syncbak, Inc. claims its service "is designed to permit distribution of traditional television program networks via the Internet while observing and respecting all of the obligations imposed on MVPDs by programmers under traditional licensing and distribution agreements, and by the government, through law and regulation."[41] It further claims:

- with proprietary servers in local broadcast stations and proprietary hardware and software for use in homes and on mobile devices, it can and does control broadcast signal distribution with more granularity than the great majority of incumbent MVPDs; and
- before permitting a viewer to tune to a particular channel or program, it can authenticate every aspect of the session, including the user's right to view the particular channel or program where and when used.

If these claims can be confirmed—if Syncbak's investment in servers, set-top boxes, and other equipment give it the same capability as traditional MVPDs to control geographically the retransmission of broadcast signals—it is possible that the public policy concerns raised by the Copyright Office more than a decade ago about Internet distribution of video programming may no longer be an issue for those providers with these capabilities.

This same issue was raised in a recent interview of Bob Saunders, president of the broadband DSL cable provider Skitter, which seeks to stream broadcast signals and other programming via the Internet to paying subscribers in rural communities.[42] Saunders claims that Skitter is willing to pay retransmission consent compensation to the local broadcasters whose signals it retransmits and to deploy the equipment (including set-top boxes) needed to ensure that it only retransmits those signals within the local broadcast market, and thus should qualify for the cable compulsory copyright license. But he further argues that other online video distributors, such as ivi.tv, which he alleges do not deploy all the equipment needed to meet the geographic restrictions on cable retransmission, should not qualify for the compulsory copyright license: "Why would they have the rights to be treated as a cable TV system when they haven't made the multimillion dollar investment in the facility that cable TV requires? We made the multimillion dollar investment." Similarly, M3X Media, Inc., which uses Internet Protocol Television to provide multiple channels of live, real time video programming (as well as video on demand) to connected devices and private networks via its Internet platform, argues that "online MVPD 'facilities' must be capable of meeting legitimate geographical requirements associated with the program access [rules]."[43]

These OVDs, of course, would benefit from a requirement that is constructed in a fashion that they could meet but that they allege some of their OVD competitors could not meet. Congress is likely to consider whether the current geographic requirements should remain or be modified.

As explained earlier, the United States is party to a number of free trade agreements that appear to prohibit Internet video distributors from obtaining a compulsory copyright license for the public performance of works on broadcast signals. Interestingly, however, for at least some of these agreements there is a signed side letter to the principal agreement that would allow this provision to be renegotiated if either party believes "that there has been a significant change in the reliability, robustness, implementability, and practical availability of technology to effectively limit the reception of Internet retransmission to users located in a specified geographic area."[44] Given the claims of Syncbak, Skitter, and M3X Media, it is possible that the technological capabilities needed to restrict Internet retransmissions to the geographic areas specified in U.S. statute and regulation already exist.

Should MVPDs be Required to Meet Certain Facilities Requirements?

Since the break-up of the old AT&T monopoly in 1984, there has been a continuing debate— should U.S. telecommunications and media policy foster facilities-based competition or allow multiple service providers to share the same network infrastructure. In 1992, satellite operators were in their infancy, telephone companies were not providing multichannel video distribution services, and there were no broadband networks capable of streaming video. At that time, Congress sought to foster the competitive entry of facilities-based video providers because there was no other potential alternative to cable service. Today, most American households have at least two facilities-based alternatives to cable—the two national satellite carriers—and many also have a telephone-based alternative. Most Americans also have access to broadband networks capable of streaming full channels of programming or on-demand programming.

If online video distribution over the Internet provides an efficient alternative to traditional distribution modes, what is the policy justification for treating it differently simply because the online distributor is not end-to-end facilities-based, but rather requires its subscribers also to use an Internet service provider? Successful online distribution entry does not appear to be the result of some market distortion. Broadband network providers are moving toward usage-based pricing mechanisms that presumably will recover from online video subscribers the costs they impose on the broadband networks they use.[45] Thus there would not appear to be an issue of subscribers of online services being artificially subsidized to the detriment of traditional video distribution services.

Today, Internet Protocol Television (IPTV) is increasingly the technology of choice. It is being deployed both by online video distributors that rely on the Internet to reach their subscribers and by more traditional distributors, such as AT&T with its U-verse, that use their own private networks to reach their subscribers. In this environment, is there a public policy basis for requiring that MVPDs be end-to-end facilities-based and, if so, what facilities or capabilities should be required?

On their face, the MVPD-related provisions in current law do not appear to address policy issues that are affected by whether the MVPD is facilities-based—unless, perhaps, certain facilities or transmission paths are necessary to provide the geographic restrictions required under various statutory and regulatory provisions discussed in the previous section. For example, is it necessary that the transmission path provided be fixed and certain?

Syncbak claims that while all OVDs rely on Internet links for some portion of the transmission path to their subscribers, only some OVDs design and build technical platforms using proprietary hardware and software to create and manage the end-to-end transmission path for each video service they provide, thus ensuring that they meet all the required geographic restrictions in the current statutory and regulatory framework.[46] If the geographic restrictions are in the public interest, then perhaps the relevant standard might be end-to-end management of the transmission path, rather than the provision of an end-to-end facility.

Should Online Video Distributors be Subject to the "Must Carry" Rules That Apply to Cable Companies or the "Carry One, Carry All" Rule That Applies to Satellite Carriers in Order to Qualify as MVPDs?

Approximately 85% of U.S. television households subscribe to MVPD service provided by cable, satellite, or telephone companies. Those households tend not to maintain rooftop antennas and as a result many of them cannot receive broadcast signals over the air. If a local broadcast station is not carried by the MVPDs in its service area, then many households in the service area will not be able to receive its signal. To support localism, the 1992 Cable Act allows a broadcast station that is concerned the cable companies in its service area may not voluntarily carry its signal to require such carriage. This is called "must carry." Similarly, if a satellite carrier wants to offer its subscribers in a local market even one local broadcast television station signal, it must offer those subscribers the signals of all the local television stations ("carry one, carry all"). These rules do not apply to video distributors other than cable or satellite video providers.

There are very few OVDs that currently carry the signals of their local broadcast stations. Thus, the lack of any "must carry" or "carry one, carry all" rule for them currently has minimal impact on local broadcasters. But some day OVDs might enjoy greater market penetration. If OVDs gain MVPD status and the associated access to broadcast and cable network programming, that time may not be far away.

Would Treating Online Video Distributors as MVPDs Harm the Relationship Between Program Networks and Content Creators?

Discovery Communications, the programmer against which Sky Angel filed the program access complaint described in the **Appendix** to this report, argues that

Classifying OVDs as MVPDs would have serious repercussions for the video programming market. Giving OVDs program access rights would place programmers in the difficult position of being forced to acquire online distribution rights for all or substantially all of their programming, from content providers who may not wish to sell those rights—either at all or at prices that make sense for programmers to acquire those rights.[47]

This appears to be a subset of a larger business issue that the industry already is facing—how to negotiate distribution rights in the online world. For example, earlier this year Time Warner Cable and Viacom settled a lawsuit regarding Time Warner Cable's streaming of Viacom video content on its iPad application.[48] A programmer that assembles content (some of which it may have created itself) into a full network schedule of programming[49] of course seeks to establish a profit maximizing distribution strategy; this typically involves setting contractual limits on how networks or distributors may use the content.

In the July 24, 2012, House Commerce Committee hearing, Preston Padden, a long-time industry attorney now at the Colorado School of Law, testified that owners of hundreds of cable program networks already act as "rights aggregators," assembling the performance rights to all of the programs shown on their network by negotiating licenses with the copyright holders that include provisions for sublicensing the rights to the MVPDs that carry the program network.[50] With the proliferation of options for consumers to receive their desired video programming, this has become an increasingly complex task, but clearly a feasible one. For example, Scripps Network Interactive Inc., which owns a number of "lifestyle" networks such as HGTV, Food Network, and the Travel Channel, recently announced a multiyear distribution agreement with Comcast that covers linear, on-demand, mobile, and online platforms on many devices, as well as advanced advertising services.[51] To do this, Scripps had to reach licensing agreements with all the underlying content providers.

If all (or some subset of) online video distributors were to be categorized as MVPDs, and thus were to gain statutory rights to access the programming on cable and broadcast networks (through the program access and retransmission consent rules), those program networks might have to modify their contracts with their content providers if those contracts did not already include or allow sublicensing to online distributors. But business contracts typically have clauses relating to modifications necessitated by change in law. As discussed earlier, a major concern of content creators and programmers is to retain control over the geographic distribution of their programming. This may not be a problem, however, because the program access provision in the Communications Act states that "Nothing in this section shall require any person who is engaged in the national or regional distribution of video programming to make such programming available in any geographic area beyond which such programming has been authorized or licensed for distribution."[52]

Discovery's fundamental concern appears to be that providing some or all online video distributors MVPD status could undermine the current profit-maximizing distribution strategy of content creators or network programmers, that is, could harm their prevailing business models, such as TV Everywhere. But as explained at the outset of this report, those business models typically are based on the obligations, prohibitions, privileges, and rights in the statutory and regulatory framework that were created by negotiations among the then-existing providers. To the extent these are technology-specific, they may impede the introduction of new technologies. It may be difficult to justify retaining such impediments unless they demonstrably foster the public policy goals of localism, diversity of voices, competition, and innovation.

Should Online Video Distributors Have Access to a Compulsory Copyright License for the Performance of Works on Broadcast Signals Analogous to Those for Cable and Satellite Video Distributors?

Congress enacted compulsory copyright license provisions for the public performance of works on retransmitted broadcast signals for cable companies (Copyright Act of 1976, P.L. 94-553) and for satellite carriers (Satellite Home Viewer Act of 1988, P.L. 100-667) in large part because cable and satellite were then in their relative infancy and it was viewed as burdensome for them to have to negotiate license agreements with the multitude of copyright holders with works on the broadcast signals. Cable and satellite are now well established, but the compulsory license provisions remain intact. Online video distributors are in their infancy, and compete against cable and satellite video distributors, but currently do not have access to compulsory copyright licenses. Extending compulsory licenses to OVDs would foster the policy goals of competition and innovation.

But most copyright holders, and the Copyright Office, have long criticized compulsory copyright licenses for generating less revenues than would be forthcoming through market negotiations and thus acting as a disincentive for creative works. They oppose any expansion of compulsory licensing and instead would prefer to have all compulsory copyright license provisions eliminated.

In recent years, the United States has entered into a number of bilateral and multilateral free trade agreement that include, largely at the urging of American copyright holders, an article under which "neither party may permit the retransmission of television signals (whether terrestrial, cable, or satellite) on the Internet without the authorisation of the right holder or right holders, if any, of the content of the signal and of the signal."[53] If the compulsory license provision were extended to online video distributors, the copyright holders would not have the authority to keep online video distributors from retransmitting broadcast signals with their creative works. This would violate the terms of the FTA but it is not clear whether the trading partner would have many copyright holders pushing their government to bring a trade complaint. U.S. content creators likely would argue that enactment of U.S. legislation that violates the intellectual property terms of the agreement would undermine the ability of U.S. firms to enforce their intellectual property rights in the other country.

Should the Broadcast-Related Copyright Laws be Reviewed in Light of Technological Developments That are Changing the Way Broadcast Programming Can be Retransmitted?

The public performance of copyrighted works on broadcast television signals is subject to copyright licensing, but the private performance of such works is not. [54] A cable or satellite operator must obtain a copyright license to retransmit the copyrighted material on broadcast signals to its subscribers, but a consuming household does not have to pay a license fee to capture the same broadcast programming directly over the air with use of a rooftop antenna. Video distributors therefore have the incentive to develop and deploy technologies that mimic the household's rooftop antenna or in some other fashion allow the consuming household to directly capture the broadcast signal. A start-up company, Aereo, now has deployed thousands of mini-antennas, which, though maintained at Aereo's own premises, are assigned to individual subscribers as their personal antennas; these antennas send the broadcast television signal to individual subscribers over the Internet. Aereo therefore claims that it is not a video distributor, but rather a technology rental company, and the programs received by its subscribers are private performances and not subject to copyright licensing requirements. The broadcasters claim otherwise and have sued Aereo for copyright infringement and sought a preliminary injunction.[55] On July 11, 2012, a federal district court judge denied the injunction, concluding "that plaintiffs have failed to demonstrate they are likely to succeed in establishing that Aereo's system results in a public performance."[56] Aereo therefore will be able to continue to offer its service as the copyright infringement case moves forward.

No matter how the courts ultimately rule in the Aereo case, Congress may want to review copyright law as it pertains to the performance of work on retransmitted broadcast signals. Several provisions in the Communications Act refer to provisions in the Copyright Act, and vice versa. Thus uncertainty in one area can create even wider uncertainty. As an example, at least one broadcast industry observer has noted that it is imperative for local broadcasters that online video distributors be classified as MVPDs so that the broadcasters can control the Internet distribution of their programming and be assured of a stream of retransmission consent revenues, in case the court were to rule that OVDs like Aereo are not subject to copyright licensing.[57]

Should the Basic Statutory Framework of Compulsory Copyright Licenses and Retransmission Consent Be Reconsidered?

Under the current statutory framework, local broadcast stations enjoy intellectual property rights that generate two revenue streams when their signals are retransmitted by cable and satellite video distributors. To the extent they hold the copyright for the works incorporated in their local television programming, they receive compulsory license fees under Sections 111, 119, and 122 of the Copyright Act. These compulsory rates, however, are almost certainly below rates that would prevail in an open market. In addition, though, under Section 325(b)(1) of the Communications Act, local broadcast stations can demand compensation from cable and satellite operators (and other MVPDs) before consenting to the retransmission of their signals, and such compensation is determined by market negotiations. Typically, if the local broadcast station is affiliated with a national broadcast network, it shares its retransmission consent revenues with that network. Similarly, the national network is likely to hold some of the copyright for the work on its programming and thus also will receive some compulsory copyright fees.

In the 112th Congress, Representative Scalise and Senator DeMint proposed, in the Next Generation Television Marketplace Act (H.R. 3675 and S. 2008) and in hearings of the House and Senate Commerce Committees,[58] that the three compulsory license provisions in the Copyright Act and the retransmission consent provision in the Communications Act be eliminated. Instead, any video distributor seeking to retransmit a broadcast television signal would have to negotiate with the copyright holders for the public performance rights to the works incorporated in that signal.[59]

This proposal could have several potential benefits. It would eliminate technology-specific provisions in the Copyright Act and Communications Act that may be hindering entry by providers using new video distribution technologies. It also would allow the marketplace to set the copyright value of the works, thus rewarding those content creators (and programmers) who provide the programming most valued by consumers.

But these changes in compulsory licensing and retransmission consent also could harm some video distributors by removing two protections they currently enjoy and by imposing on them the costs associated with negotiating payment with the large number of copyright holders whose works are incorporated in broadcast signals. Cable and satellite operators (though not other video distributors) now are guaranteed a below market price copyright license and thus cannot be denied access to the copyrighted material. Similarly, all MVPDs currently have the implicit right created by the "good faith negotiations" clause in the retransmission consent provision in the Communications Act[60] to retransmit broadcast signals at market negotiated prices. Under the proposed legislation, broadcasters could no longer demand compensation for the retransmission of their signals, but they and/or other copyright holders could refuse to negotiate a license agreement with a video distributor. A copyright holder might find it advantageous to employ a business model that restricts program access to a limited number of video distributors. One possible business strategy would be to limit licensing to incumbent MVPDs, through a TV Everywhere-type distribution strategy that expands the array of devices consumers can use to access programming but limits the number of distributors given access to the programming.

Martin Luther King III, of Bounce TV, a new African American-oriented broadcast network, reportedly has criticized the proposal as a threat to diversity of voices because it would be damaging to economically fragile broadcast stations that are serving minority communities. He reportedly claimed the elimination of retransmission consent would harm small-market broadcasters that "lack the leverage to negotiate equitable carriage agreements with the giant pay services" and that "broadcasters will lose the financial flexibility to fund diverse programming, both on a broadcaster's prime channel as well as the digital sub-channels that are allowing Bounce TV to entertain and inform previously neglected African American viewers." [61]

Empirically, it is uncertain how the replacement of the current combination of retransmission consent and compulsory copyright license with a negotiated copyright license would affect the flow of revenues from video distributors to content providers, broadcast networks, and local broadcast stations. It would seem likely, however, that the elimination of the property right for retransmission of a broadcast signal and the strengthening of the property right for copyright holders would result in a more revenue flowing to content providers and less to broadcast stations and networks.

The Scalise-DeMint legislation in the 112th Congress would have conformed with the provision in a number of free trade agreements that does not permit the retransmission of broadcast signals without the authorization of the right holders of the content on the signal, since it would eliminate any possibility of a compulsory copyright license for Internet video distributors. But it would not have conformed with the provision that does not permit the retransmission of broadcast signals without the authorization of the right holders of the signal, since it would eliminate retransmission consent.

APPENDIX. THE LEGAL ISSUE CURRENTLY BEFORE THE FCC AND THE COURTS

The Current Definition of MVPD

As noted above, multichannel video programming distributors (MVPDs) are subject to many requirements by the Communications Act, as amended, but also receive many statutory rights and privileges as a result of their status as MVPDs. The question of which entities are MVPDs, therefore, becomes important, and, in the rapidly changing market for the delivery of video programming, potentially ambiguous. The FCC is presently engaged in proceedings to determine whether certain online video programming distributors could be considered MVPDs under the current statute. The question has arisen because an online video programming distributor known as Sky Angel has attempted to claim program access rights, described above, in the course of a dispute over access to programming owned by Discover Communications (owner of the Discovery Channel, and other cable programming channels).

Program Access Statute and Definitions

The Communications Act, as amended by the 1992 Cable Act, prohibits cable operators and cable-affiliated program networks from engaging in unfair methods of competition that could hinder or prevent other MVPDs from providing satellite cable or satellite broadcast programming to their customers.[62] The statute also directed the FCC to promulgate rules to prohibit discrimination by satellite and cable programming vendors when those entities negotiate with MVPDs to provide those entities access to programming.[63] The regulations require owners of programming that are affiliated with cable operators to make that programming available to all MVPDs on a non-discriminatory basis.[64]

MVPDs that believe that these regulations were violated in the course of attempting to obtain rights to programming from an owner affiliated with a cable operator may file a complaint with the FCC to enforce the rules.[65] Congress directed the FCC to provide expedited review for these complaints.[66] In implementing this requirement, the FCC set a goal of resolving typical program access complaints within five months of their receipt by the Media Bureau of the FCC.[67]

However, the FCC reserved the right to extend that time period when the complaint presented more complex issues.[68]

Before moving on to the Sky Angel complaint, it is important to note, again, that the rights to access programming apply only to MVPDs under current law. The statute defines a multichannel video programming provider (MVPD) as "[A] person such as, but not limited to, a cable operator, a multichannel multipoint distribution service, a direct broadcast satellite service, or a television receive-only satellite program distributor, who makes available for purchase, by subscribers or customers, multiple channels of video programming."[69]

The terms "channel" and "video programming" are also statutory terms that may differ in important ways from the common understanding of those words. Channel is defined as "a portion of the electromagnetic frequency spectrum which is used in a cable system and which is capable of delivering a television channel," as further defined by FCC regulation.[70] FCC regulation defines a television channel as "a band of frequencies in the 6MHz wide in the television broadcast band and designated either by number or by extreme lower or upper frequencies."[71] Video programming is defined as programming provided by or generally considered comparable to programming provided by a television broadcast station.[72] Therefore, an MVPD must be a person who makes available for purchase multiple channels of video programming as those terms are defined by statute and regulation. Ambiguity has arisen regarding whether programming providers that deliver programming online, but do not provide the transmission pathway for that programming, in other words, do not provide broadband Internet access along with programming, can be considered MVPDs under the current definition.

Sky Angel Case

Sky Angel provides programming to its subscribers over the Internet.[73] The subscriber must purchase a broadband Internet connection from a third-party provider of that service. Sky Angel then provides the subscriber with equipment to watch Sky Angel's programming over the subscriber's existing broadband connection. The user experience, according to Sky Angel, is similar to cable programming providers in the way in which the programming is displayed, except Sky Angel does not provide subscribers access to local broadcasting stations over its service.

In 2007, Sky Angel entered into an agreement with Discovery Communications for access to its programming.[74] In July 2010, Discovery terminated the agreement, although the agreement was not set to expire until 2014. After failure to negotiate a new deal for program access with Discovery, Sky Angel filed a program access complaint with the FCC on March 24, 2010.[75] Sky Angel alleged that Discovery had discriminatorily denied access to programming to Sky Angel. In conjunction with its complaint, Sky Angel also filed a "standstill" request to prevent Discovery from denying Sky Angel access to Discovery's programming while the program access complaint was being considered by the FCC.[76] Discovery opposed both the standstill petition and the program access complaint, on the grounds that Sky Angel is not an MVPD and does not qualify for the protections the 1992 Cable Act provides to MVPDs.

Media Bureau Decision

On April 21, 2010, the Media Bureau issued its decision denying the standstill petition.[77] The Bureau found that Sky Angel had not shown a likelihood of success on the merits of the question of whether it is an MVPD entitled to avail itself of the program access complaint process. While Sky Angel had asserted that it was an MVPD, it had not demonstrated to the Media Bureau's satisfaction, at that point in the proceedings, how Sky Angel met the elements of being an MVPD. In particular, Sky Angel argued that it provides multiple "channels" of video programming to its subscribers. However, the Media Bureau noted that Sky Angel appeared to be referring to "channels" in the sense that the company provides different streams of programming, but ignored the technical definition of channel that appears in the Communications Act.

As noted above, the definition of "channel" in the act is "a portion of the electromagnetic frequency spectrum which is used in a cable system and which is capable of delivering a television channel," as further defined by FCC regulation. [78] The Media Bureau read this definition of channel to include not just the programming stream, but also a transmission pathway for that programming.[79] Sky Angel does not provide the transmission pathway for its programming; the subscriber's broadband Internet provider does. Furthermore, the Bureau claimed that the illustrative list of entities considered to be MVPDs in the current statutory definition all provide both programming and the transmission pathway for that programming to their subscribers.[80] As a result, the Media Bureau found that Sky Angel was unlikely to succeed on the merits of its program access complaint because Sky Angel had not met its burden of demonstrating that it was an MVPD.

It is important to note that though Sky Angel's standstill petition was denied, no decision was issued regarding Sky Angel's original program access complaint. The Media Bureau stated specifically that its decision should not be read as a statement regarding the Commission's ultimate decision on the issue of whether Sky Angel is an MVPD.[81] Sky Angel's program access complaint remains pending before the FCC.

Mandamus Petition

Sky Angel filed its program access complaint on March 24, 2010, and the Media Bureau issued its denial of the standstill petition a month later, but the program access complaint remained pending. At the beginning of 2012, it had been nearly two years since the FCC had taken any action regarding Sky Angel's program access complaint.

As a result, on February 27, 2012, Sky Angel filed a petition for a writ of mandamus in the United States Court of Appeals for the District of Columbia Circuit.[82] The petition asked the court to order the FCC to issue a final ruling on the merits Sky Angel's program access complaint within 30 days of the court's decision.

Mandamus is an "extraordinary remedy reserved for extraordinary circumstances."[83] Courts generally will only interfere with agency proceedings by issuing a writ of mandamus where the court has found a clear violation of an agency's statutory duties, including an unreasonable delay on the part of an agency to act where the governing statute requires action.[84] Here, Sky Angel argued that the FCC's failure to act on its program access petition for nearly two years (at the time the petition was filed) was an unreasonable delay so egregious as to warrant a court order to force action. The delay, according to their argument, had harmed and continues to harm Sky Angel and the public interest in competition among distributors of video programming, particularly those that would challenge the dominance of incumbent cable operators.

Furthermore, Sky Angel pointed to the statutory requirement that the FCC provide expedited adjudicatory proceedings to program access complaints, and the FCC's own implementing regulations of that section that state that program access complaints would be resolved within 5 months, as evidence that the agency's delay in ruling on the complaint was unreasonable.[85] Importantly, Sky Angel is not asking the court to rule on whether it is an MVPD. Sky Angel is arguing only that the FCC's delay in addressing its complaint is unreasonable and is asking the court to order the FCC to render a decision.

On March 30, 2012, about one month after Sky Angel filed its petition, the FCC issued a notice seeking public comments on the definitions of multichannel video programming distributor and channel in the Communications Act.[86] The proceeding is intended to clarify the definitions in order for the FCC to proceed on Sky Angel's program access complaint. Following the issuance of the notice, the FCC filed its opposition to Sky Angel's petition for a writ of mandamus on April 5, 2012.[87] The FCC argued that Sky Angel had not shown that the FCC's delay was unreasonable.

The FCC argued that its self-imposed five month timeframe for ruling on program access complaint applied only to typical complaints and that Sky Angel's complaint is not typical, because it poses novel questions of law and policy. The FCC claimed that it has always reserved the right to spend more time reviewing "cases that 'involve numerous issues requiring legal, economic, and accounting expertise.'"[88] In the Commission's opinion, Sky Angel's complaint presents one of the more complicated cases that would require more time to resolve. Furthermore, answering the questions presented by the complaint could have effects that resonate throughout the Internet video distribution industry. The FCC pointed out that it had issued a public notice to begin to address these questions.[89] For that reason, among others, the FCC asked the court to deny Sky Angel's mandamus petition. The court has yet to issue its decision.[90]

Current FCC Proceedings to Clarify the Definition of MVPDs

As noted above, the FCC has issued a public notice seeking comments regarding the scope of the statutory definitions of MVPD and channel, as well as other statutory terms. The way in which the FCC answers these questions could have far-reaching effects on the video programming distribution industry. The questions posed by the FCC ultimately meld complicated questions of law and policy. However, the decision the FCC renders necessarily will be constrained by the language of a statute that was written before technology employed by companies by Sky Angel was contemplated.

The FCC posed a number of questions regarding whether Internet video distributors could be considered to be MVPDs under the current statutory language.[91] As discussed above, the Media Bureau found that the definition of channel in the Communications Act appeared to encompass both the stream of programming and the transmission pathway for that programming.[92] The FCC asked for comment on this interpretation specifically.[93] The FCC pointed out that the definition of channel relied on by the Media Bureau was adopted by Congress as part of the 1984 Cable Act and refers to portions of the electronic spectrum, *which is used in a cable system*.[94] The Commission noted that this could be read to mean that the definition of channel in the statute was only meant to define channels as they applied to cable systems.[95] The definition of MVPD was not added to the statute until 1992, it lists entities that are not cable systems as examples of MVPDs, and it uses the term "channels," rather than the singular to define the video programming that should be offered. These facts could support an argument that the word "channels" in the definition of MVPD was not intended to be understood to have the statutory meaning previously given to the word "channel" in the 1984 Cable Act.

The Commission asked commenters whether Congress intended to incorporate the previous statutory definition of "channel" when it used the plural of the word as part of the definition of MVPDs. In other words, the Commission asked whether Congress meant that MVPDs had to be offering "channels" of programming in the colloquial sense of simply offering multiple streams of programming, or if MVPDs must offer "channels" in the statutory sense, which would seem to include not only the offering of streams of programming, but also the pathway for that programming. The Commission noted that it has previously held that an entity does not need to own the facilities over which it distributes video programming in order to be considered an MVPD, and may use a third party's facilities for distribution.[96] However, each of the examples in the non-exhaustive list of MVPDs contained in the statutory definition offers both a transmission pathway and multiple video programming networks.[97] The Commission queried to what extent the distribution pathway and programming streams must be packaged together in order to qualify an entity to be an MVPD. In other words, assuming that in order to be an MVPD an entity must offer a pathway in conjunction with the programming, the Commission has asked to what extent those two attributes must be packaged together for an entity to be an MVPD.

The Commission then went on to request comment on whether the definitions of MVPD and channel could be interpreted to include entities that offer multiple video programming networks without regard for whether the entity also offered a transmission path for the programming.[98] This is the precise question presented by the Sky Angel complaint. Sky Angel offers multiple programming streams, but its customers' broadband Internet providers offer the transmission pathway. The Commission inquired as to whether the agency has leeway under the statutory language to interpret the offering of "multiple channels of video programming" in the more common understanding of the term, such that a company, like Sky Angel, that is offering a user experience not unlike the experience of flipping through cable "channels" could be considered an MVPD.

The Commission also inquired about the extent to which the statutory term "video programming" might constrain the types of entities that may be considered to be MVPDs.[99] The definition of "video programming" was added to the Communications Act, by the 1984 Cable Act. Video programming is defined as "programming provided by, or generally considered comparable to programming provided by, a television broadcast station."[100] The Commission noted that programming comparable to that of a television broadcast station might be understood to mean prescheduled, "linear" programming; however, the Commission had previously found that videoon-demand, which is not linear, constituted "video programming."[101] The Commission asked for comment regarding whether an entity that offered programming only on an on-demand basis could "make available" programming as defined by the act.

The Commission has yet to reach a decision on these important questions. Any decision the FCC makes regarding whether online video programming providers are MVPDs may be subject to court challenge. Such proceedings may take months or years to resolve. Congress has the option to clarify the status of online video distributors under the Communications Act, keeping in mind the complex policy issues described in this report.

End Notes

P.L. 102-385.
[2] P.L. 94-553.
[3] P.L. 100-667.
[4] The bill would have eliminated the compulsory copyright license for cable and satellite, most of the requirements relating to cable and satellite company carriage of broadcast signals, the retransmission consent requirements, the broadcast network affiliation rules, most of the broadcast media ownership rules, and the network non-duplication, syndicated exclusivity, and sports blackout rules. In their place, stakeholders would have been expected to rely on market negotiations.
[5] Section 706 of the 1996 Telecommunications Act (P.L. 104-104), titled "Advanced Telecommunications Incentives," instructs the FCC and state regulatory commissions "to encourage the deployment on a reasonable and timely basis of advanced telecommunications capability to all Americans" and instructs the FCC to "take immediate action to accelerate deployment of such capability by removing barriers to infrastructure investment and by promoting competition in the telecommunications market."
[6] This case is discussed in greater detail in the Appendix to this report.
[7] Section 106(4) of the Copyright Act (17 U.S.C. §106(4)) states that the owner of copyright "has the exclusive rights," in the case of motion pictures and other audiovisual works "to perform the copyrighted work publicly."

[8] *American Broadcasting Cos. et al. v. Aereo Inc.*, U.S. District Court, Southern District of New York, No. 12-01540 and *WNET et al. v. Aereo Inc.*, U.S. District Court, Southern District of New York, No. 12-01543.

[9] *American Broadcasting Cos. et al. v. Aereo Inc.*, U.S. District Court, Southern District of New York, No. 12-01540 and *WNET et al. v. Aereo Inc.*, U.S. District Court, Southern District of New York, No. 12-01543, Opinion of Alison J. Nathan, District Judge, July 11, 2012, at p. 37. If the broadcasters are unsuccessful in this case and lose copyright revenues, they are likely to seek alternate compensation, most likely by seeking to make Aereo and similarly situated video distributors subject to retransmission consent requirements. (See Harry A. Jessell, "Stations' Online Future Hinges on 'MVPD,'" *TVNewsCheck,* April 20, 2012, http://www.tvnewscheck.com/article/2012/04/20/58897/stationsonline-future-hinges-on-mvpd.)

[10] United States-Australia Free Trade Agreement, Art. 17.4(10)(b). This language, or similar language, is included in the Dominican Republic-Central America Free Trade Agreement and in the free trade agreements with Bahrain, Morocco, and South Korea.

[11] See Testimony of Preston Padden, Executive Vice President, The Walt Disney Company, Before the U.S. Copyright Office "Sec. 109 Hearings on the Operation of, and Continued Necessity for, the Cable and Satellite Statutory Licenses," November 12, 2010.

[12] Satellite Home Viewer Extension and Reauthorization Act Section 109 Report: A Report of the Register of Copyrights, June 2008, at p. 189, http://www.copyright

[13] *Before the Federal Communications Commission, In the Matter of Interpretation of the Terms "Multichannel Video Programming Distributor" and "Channel" As Raised in Pending Program Access Complaint Proceeding,* MB Docket No. 12-83, Reply Comments of ABC Television Affiliates Association, CBS Television Network Affiliates Association, and NBC Television Affiliates, June 13, 2012, at pp. 26-27.

[14] P.L. 102-385. See, in particular, the findings in Section 2(a) and statement of policy in Section 2(b).

[15] The Federal Communications Commission has characterized the definition as "broad in its coverage" and "unclear" in its scope. See "Media Bureau Seeks Comment on Interpretation of the Terms "Multichannel Video Programming Distributor" and "Channel" as Raised in Pending Program Access Complaint Proceeding," MB Docket No. 12-83, DA 12-507, Public Notice released March 30, 2012, at p. 2, fn. 2.

[16] 47 U.S.C. §325(b)(1).

[17] This is separate from holding the copyright for works within the programming carried on that signal.

[18] 47 U.S.C. §325(b)(3)(C)(ii) and (iii).

[19] 47 U.S.C. §548(b).

[20] 47 U.S.C. §549.

[21] 47 U.S.C. §536(a).

[22] 47 U.S.C. §554. Although the language in this provision refers only to cable television, the language in the rule adopted by the FCC to implement the provision explicitly includes all MVPDs (47 C.F.R. 76.71) and this rule has been upheld by the courts. Most recently, on June 25, 2012, the FCC Media Bureau chief released a Memorandum Opinion and Order and Notice of Apparent Liability for Forfeiture against Grande Communications of San Marcos, TX, an MVPD, finding that the company willfully and repeatedly violated the FCC's EEO rules and imposed a $10,000 forfeiture.

[23] 47 U.S.C. §613 and 47 C.F.R. §§79.1 and 79.2.

[24] 47 C.F.R. §§76.92-76.130. As explained in "A Short History of the Program Exclusivity Rules," appended to *In the Matter of Petition for Rulemaking to Amend the Commission's Rules Governing Retransmission Consent,* MB Docket No. 10-71, Opposition of the Broadcaster Associations, May 18, 2010, the first FCC program exclusivity rule was adopted in 1965.

[25] 47 U.S.C. §339(b).

[26] 47 U.S.C. §534.

[27] 47 U.S.C. §338(a)(1).

[28] 17 U.S.C. §§111, 119, and 122.

[29] See, for example, Statement of Marybeth Peters, Register of Copyrights, before the Subcommittee on Courts and Intellectual Property, House Committee on the Judiciary, June 15, 2000, available at http://www.copyright regstat61500.html.

[30] Access to "must have" programming may be less important for an OVD that does not seek to compete directly with traditional cable and satellite video providers, but instead pursues a business plan based on offering a low-priced alternative that does not include sports or other high-cost programming.

[31] 47 U.S.C. §522(13).

[32] "Satellite broadcast programming" is defined as broadcast video programming when such programming is retransmitted by satellite and the entity retransmitting such programming is not the broadcaster or an entity

performing such retransmission on behalf of and with the specific consent of the broadcaster (47 U.S.C. §548(i)(3) and 47C.F.R. §76.1000(f)). "Satellite cable programming" is defined as video programming which is transmitted via satellite and which is primarily intended for direct receipt by cable operators for their retransmission to cable subscribers, except that such term does not include satellite broadcast programming (47 U.S.C. §605(d)(1) and 47 C.F.R. §76.1000(h)).

[33] "Media Bureau Seeks Comment on Interpretation of the Terms 'Multichannel Video Programming Distributor' and 'Channel' as Raised in Pending Program Access Complaint Proceeding," *FCC Public Notice*, MB Docket No. 12-83, DA 12-507, released March 30, 2012, at para. 14.

[34] See, for example, *In the Matter of Public Notice on Interpretation of the Terms "Multichannel Video Programming Distributor" and "Channel" as Raised in Pending Program Access Complaint Proceeding*, MB Docket No. 12-83, Comments of Verizon, submitted May 14, 2012, at p. 23, in which Verizon strongly argues that the two services should be treated separately, with its FiOS video service treated as an MVPD, but its joint venture with Redbox left unregulated.

[35] See "A Short History of the Program Exclusivity Rules," appended to *In the Matter of Petition for Rulemaking to Amend the Commission's Rules Governing Retransmission Consent*, MB Docket No. 10-71, Opposition of the Broadcaster Associations, May 18, 2010.

[36] 47 U.S.C. §339(b).

[37] The Communications Act identifies a class of "nationally distributed superstations" (47 U.S.C. §339(d)(2)) that is limited to six stations that were in operation prior to May 1, 1991. These are independent broadcast television stations whose broadcast signals are picked up and redistributed by satellite to local cable television operators and to satellite television operators all across the United States. These nationally distributed superstations in effect function like cable program networks rather than local broadcast television stations or broadcast television networks. The nationally distributed superstations are WTBS, Atlanta; WOR and WPIX, New York; WSBK, Boston; WGN, Chicago; and KTLA, Los Angeles.

[38] 47 U.S.C. §325(b)(2).

[39] Statement of Marybeth Peters, Register of Copyrights, before the Subcommittee on Courts and Intellectual Property, House Committee on the Judiciary, June 15, 2000, available at http://www.copyright

[40] Ibid.

[41] *In the Matter of Public Notice on Interpretation of the Terms "Multichannel Video Programming Distributor" and "Channel" as Raised in Pending Program Access Complaint Proceeding*, MB Docket No. 12-83, Comments of Syncbak, Inc., submitted May 14, 2012, at pp. 3-4.

[42] Harry A. Jessell, "Skitter Chief: 'We Would Be Great Partners,'" *TVNewsCheck*, July 11, 2012, http://www.tvnewscheck.com/article/60695/skitter-chief-we-would-be-great-partners.

[43] *In Re: Media Bureau Seeks Comment on Interpretation of the Terms 'Multichannel Video Programming Distributor" and "Channel" as Raised in Pending Program Access Complaints Proceeding*, MB Docket No. 12-83, Comments of M3X Media, Inc., at p. 7. The program access provision in the Communications Act, at 47 U.S.C. §548(c)(3)(A), states: "Geographic limitations. Nothing in this section shall require any person who is engaged in the national or regional distribution of video programming to make such programming available in any geographic area beyond which such programming has been authorized or licensed for distribution."

[44] See, for example, Letter dated May 18, 2004, from Mark Vaile, Australian Minister for Trade, to Robert B. Zoelick, U.S. Trade Representative, and Letter dated November 22, 2006, from Jorge Humberto Botero, Colombian Minister of Commerce, Industry, and Tourism, to John K. Veroneau, Deputy United States Trade Representative. See, also, the discussion in Robert Burrell and Kimberlee Weatherall, "Exporting Controversy? Reactions to the Copyright Provisions of the U.S.-Australia Free Trade Agreement: Lessons for U.S. Trade Policy," *University of Illinois Journal of Law, Technology, and Policy*, 2008, at pp. 290-291.

[45] There is a related policy issue of whether the ISPs that are imposing these usage based charges are discriminating in favor of their own video distribution services by not including usage of their own video streaming services in the usage calculations of their ISP customers.

[46] *In the Matter of Public Notice on Interpretation of the Terms "Multichannel Video Programming Distributor" and "Channel" as Raised in Pending Program Access Complaint Proceeding*, MB Docket No. 12-83, Comments of Syncbak, Inc., submitted May 14, 2012, at pp. 3-4.

[47] Comments of Discovery Communications, LLC, at pp. 8-9.

[48] See, for example, Amy Chozick, "Viacom and Time Warner Cable Reach Deal on App for Streaming Programs," *New York Times*, May 16, 2012.

[49] This is sometimes referred to as linear programming.

[50] Testimony of Preston Padden, Senior Fellow, Silicon Flatirons Center, Colorado School of Law, before the Committee on Commerce, Science and Transportation, United States Senate, July 24, 2012.

[51] See Sarah Barry James, "2 opposite approaches to battling programming cost increases," *SNL Kagan,* July 23, 2012, http://www.snl.com/InteractiveX/article.aspx?CDID=A-15327953-12077&KPLT=2.

[52] 47 U.S.C. §548(c)(3)(A).

[53] U.S.-Australia Free Trade Agreement, Article 17.4(10)(b). Similar articles appear in other FTAs.

[54] Section 106(4) of the Copyright Act (17 U.S.C. §106(4)) states that the owner of copyright "has the exclusive rights," in the case of motion pictures and other audiovisual works "to perform the copyrighted work publicly."

[55] *American Broadcasting Cos. et al. v. Aereo Inc.*, U.S. District Court, Southern District of New York, No. 12-01540 and *WNET et al. v. Aereo Inc.*, U.S. District Court, Southern District of New York, No. 12-01543.

[56] *American Broadcasting Cos. et al. v. Aereo Inc.*, U.S. District Court, Southern District of New York, No. 12-01540 and *WNET et al. v. Aereo Inc.*, U.S. District Court, Southern District of New York, No. 12-01543, Opinion of Alison J. Nathan, District Judge, July 11, 2012, at p. 37. If the broadcasters are unsuccessful in this case and lose copyright revenues, they are likely to seek alternate compensation, most likely by seeking to make Aereo and similarly situated video distributors subject to retransmission consent requirements. See Harry A. Jessell, "Stations' Online Future Hinges on 'MVPD,'" *TVNewsCheck,* April 20, 2012, http://www.tvnewscheck.com/article/2012/04/20/58897/stationsonline-future-hinges-on-mvpd.

[57] Harry A. Jessell, "Stations' Online Future Hinges on 'MVPD,'" *TVNewsCheck,* April 20, 2012, http://www.tvnewscheck.com/article/2012/04/20/58897/stations-online-future-hinges-on-mvpd.

[58] House Energy and Commerce Subcommittee on Communications and Technology hearing on "The Future of Video," June 27, 2012, and Senate Commerce, Science, and Transportation Committee hearing on "The Cable Act at 20," July 24, 2012.

[59] In the absence of a statutory compulsory copyright license, Section 106(4) of the Copyright Act (17 U.S.C. §106(4)) gives the owner of copyright the exclusive right to authorize the public performance of the copyrighted work.

[60] 47 U.S.C. §325(b)(3)(C)(ii).

[61] Dave Seyler, "Bounce TV sees DeMint dereg bill as existential threat," *Radio & Television Business Report,* August 8, 2012, reporting on a letter from Martin Luther King III, Bounce TV, to Jay Rockefeller, chairman, Senate Commerce Committee.

[62] 47 U.S.C. §548(b) ("It shall be unlawful for a cable operator, a satellite cable programming vendor in which a cable operator has an attributable interest, or a satellite broadcast programming vendor to engage in unfair methods of competition or unfair or deceptive acts or practices, the purpose or effect of which is to hinder significantly or to prevent any multichannel video programming distributor from providing satellite cable programming or satellite broadcast programming to subscribers or consumers.")

[63] 47 U.S.C. 548(c)(2)(B).

[64] 47 C.F.R. 76.1001-1002.

[65] 47 U.S.C. §548(d).

[66] 47 U.S.C. §548(f)(1).

[67] Implementation of the Cable Television Consumer Protection and Competition Act of 1992, 22 FCC Rcd 17791 (2007).

[68] *Id.* at 15843.

[69] 47 U.S.C. 522 (13).

[70] 47 U.S.C. 522(4).

[71] 47 C.F.R. 73.681.

[72] 47 U.S.C. 522(20).

[73] Sky Angel U.S., LLC, http://www.skyangel.com/?aid=00010&gclid=CMqA_e_g57ECFQjc4Aod6VoAeA.

[74] See Sky Angel U.S., LLC v. Discovery Communications LLC, *et al.*, Program Access Complaint, MB Docket No. 12-80, File No. CSR-8605-P (March 24, 2010). [Sky Angel Complaint]

[75] *Id.*

[76] Sky Angel U.S., LLC, Emergency Petition for Temporary Standstill (March 24, 2010).

[77] In the Matter of Sky Angel U.S., LLC Emergency Petition for Temporary Standstill, Order, DA 10-679 (April 21, 2010). [Media Bureau Order]

[78] 47 U.S.C. 522(4).

[79] Media Bureau Order, *supra* note 16, at para. 7.

[80] This assertion has been question in comments by a number of industry participants. For example, DirecTV pointed out in its comments that the illustrative list in the definition includes "television receive-only satellite program distributors." These providers, according to DirecTV, "do not provide any transmission path for the delivery of video programming." Comments of DirecTV, LLC, In the Matter of Interpretation of the Terms "Multichannel Video Programming Distributor" and Channel" as Raised in Pending Program Access Complaint Proceeding, FCC MB Docket No. 12-83 (May 14, 2012). If this assertion is found to be true, it may undermine the Media Bureau's interpretation that in order to be an MVPD an entity must offer a transmission pathway.

[81] *Id*. at para. 10.

[82] In re Sky Angel U.S., LLC, Pet. for Writ of Mandamus, Case No. 12-1119, D.C. Cir. (February 27, 2012). [Mandamus Petition].

[83] In re American Rivers and Idaho Rivers United, 372 F.3d 413, 418 (D.C. Cir. 2004).

[84] Telecomm. Research & Action Ctr. v. FCC, 750 F.2d 70, 79-80 (D.C. Cir. 1984).

[85] Mandamus Petition at 14.

[86] Public Notice DA 12-507 (Media Bur. March 30, 2012), available at http://hraunfoss.fcc.gov/edocs_public/attachmatch/DA-12-507A1.pdf. [Public Notice]

[87] In re Sky Angel U.S., LLC, Opposition of the Federal Communications Commission to Sky Angel's Pet. for a Writ of Mandamus, Case No. 12-1119, D.C. Cir. (April 5, 2012).

[88] *Id*. at 16.

[89] Public Notice, *supra* note 24.

[90] FCC, List of Pending Appellate Cases, http://transition.fcc.gov/Daily_Releases/Daily_Business/2012/db0802/DOC315545A1.pdf (August 1, 2012).

[91] Public Notice, *supra* note 24.

[92] Media Bureau Order, *supra* note 16, at para. 7.

[93] Public Notice, *supra* note 24.

[94] Cable Communications Policy Act of 1984, P.L. 98-549 §2 (emphasis added).

[95] Public Notice, *supra* note, 24 at para. 7.

[96] *Id*. at para. 9 (citing In the Matter of Implementation of Section 302 of the Telecommunications Act of 1996 Open Video Systems Second Order on Recon., 11 FCC Rcd 20227, 20301 (1996)).

[97] As noted *supra* note 76, this assertion has been questioned in comments by a number of industry participants. DirecTV pointed out in its comments that the illustrative list in the definition includes "television receive-only satellite program distributors." These providers, according to DirecTV, "do not provide any transmission path for the delivery of video programming." Comments of DirecTV, LLC, In the Matter of Interpretation of the Terms "Multichannel Video Programming Distributor" and Channel" as Raised in Pending Program Access Complaint Proceeding, FCC MB Docket No. 12-83 (May 14, 2012). It is thus possible that the Commission may decide that the non-exhaustive list does include entities that do not provide both programming and a transmission pathway to customers.

[98] Public Notice, *supra* note 24, at para. 11.

[99] *Id*. at para. 13.

[00] 47 U.S.C. §522(20).

[01] Public Notice, *supra* note 24, at para. 13.

In: Transformations in Telecommunications and Media   ISBN: 978-1-62948-413-6
Editor: Irwin Cavazos   © 2013 Nova Science Publishers, Inc.

Chapter 8

INTERNET GOVERNANCE AND THE DOMAIN NAME SYSTEM: ISSUES FOR CONGRESS[*]

Lennard G. Kruger

SUMMARY

The Internet is often described as a "network of networks" because it is not a single physical entity, but hundreds of thousands of interconnected networks linking hundreds of millions of computers around the world. As such, the Internet is international, decentralized, and comprised of networks and infrastructure largely owned and operated by private sector entities. As the Internet grows and becomes more pervasive in all aspects of modern society, the question of how it should be governed becomes more pressing.

Currently, an important aspect of the Internet is governed by a private sector, international organization called the Internet Corporation for Assigned Names and Numbers (ICANN), which manages and oversees some of the critical technical underpinnings of the Internet such as the domain name system and Internet Protocol (IP) addressing. ICANN makes its policy decisions using a multistakeholder model of governance, in which a "bottom-up" collaborative process is open to all constituencies of Internet stakeholders.

National governments have recognized an increasing stake in ICANN policy decisions, especially in cases where Internet policy intersects with national laws addressing such issues as intellectual property, privacy, law enforcement, and cybersecurity. Some governments around the world are advocating increased intergovernmental influence over the way the Internet is governed. For example, specific proposals have been advanced that would create an Internet governance entity within the United Nations (U.N.). Other governments (including the United States), as well as many other Internet stakeholders, oppose these proposals and argue that ICANN's multistakeholder model, while not perfect and needing improvement, is the most appropriate way to govern the Internet.

---

[*] This is an edited, reformatted and augmented version of Congressional Research Service, Publication No. R42351, dated April 23, 2013.

Currently, the U.S. government, through the National Telecommunications and Information Administration (NTIA) at the Department of Commerce, enjoys a unique influence over ICANN, largely by virtue of its legacy relationship with the Internet and the domain name system. A key issue for the 113th Congress is whether and how the U.S. government should continue to maximize U.S. influence over ICANN's multistakeholder Internet governance process, while at the same time effectively resisting proposals for an increased role by international governmental institutions such as the U.N. An ongoing concern is to what extent will future intergovernmental telecommunications conferences (such as the December 2012 World Conference on International Telecommunications or WCIT) constitute an opportunity for some nations to increase intergovernmental control over the Internet, and how effectively will NTIA and other government agencies (such as the State Department) work to counteract that threat? H.R. 1580, introduced on April 16, 2013, states that "[I]t is the policy of the United States to preserve and advance the successful multistakeholder model that governs the Internet."

The ongoing debate over Internet governance will likely have a significant impact on how other aspects of the Internet may be governed in the future, especially in such areas as intellectual property, privacy, law enforcement, Internet free speech, and cybersecurity. Looking forward, the institutional nature of Internet governance could have far-reaching implications on important policy decisions that will likely shape the future evolution of the Internet.

WHAT IS INTERNET GOVERNANCE?

There is no universally agreed-upon definition of "Internet governance." A more limited definition would encompass the management and coordination of the technical underpinnings of the Internet—such as domain names, addresses, standards, and protocols that enable the Internet to function. A broader definition would include the many factors that shape a variety of Internet policy-related issues, such as such as intellectual property, privacy, Internet freedom, e-commerce, and cybersecurity.

One working definition was developed at the World Summit on the Information Society (WSIS) in 2005:

> Internet governance is the development and application by governments, the private sector and civil society, in their respective roles, of shared principles, norms, rules, decision-making procedures, and programmes that shape the evolution and use of the Internet.[1]

Another definition developed by the Internet Governance Project (IGP)[2] delineates three aspects of the Internet that may require some level of governing: *technical standardization*, which involves arriving at and agreeing upon technical standards and protocols; *resource allocation and assignment* which includes domain names and Internet Protocol (IP) addresses; and *human conduct on the Internet*, encompassing the regulations, rules, and policies affecting areas such as spam, cybercrime, copyright and trademark disputes, consumer protection issues, and public and private security. With these three categories in mind, the IGP definition is:

> Internet governance is collective decisionmaking by owners, operators, developers, and users of the networks connected by Internet protocols to establish policies, rules, and

dispute resolution procedures about technical standards, resource allocations, and/or the conduct of people engaged in global internetworking activities.[3]

HOW IS THE INTERNET CURRENTLY GOVERNED?

The nature of the Internet, with its decentralized architecture and structure, makes the practice of governing a complex proposition. First, the Internet is inherently international and cannot in its totality be governed by national governments whose authority ends at national borders. Second, the Internet's successful functioning depends on the willing cooperation and participation by mostly private sector stakeholders around the world. These stakeholders include owners and operators of servers and networks around the world, domain name registrars and registries, regional IP address allocation organizations, standards organizations, Internet service providers, and Internet users.

Given the multiplicity and diversity of Internet stakeholders, a number of organizations and entities play varying roles. It is important to note that all of the Internet stakeholders cited above participate in various ways within the various fora, organizations, and frameworks addressing Internet governance and policy.

Key organizations in the private sector include the following:

Internet Corporation for Assigned Names and Numbers (ICANN)—ICANN was created in 1998 through a Memorandum of Understanding with the Department of Commerce (see the following section of this report, "Role of U.S. Government"). Directed by an internationally constituted Board of Directors, ICANN is a private, not-for-profit organization based in Marina Del Ray, CA, which manages and oversees the critical technical underpinnings of the Internet such as the domain name system and IP addressing (see the **Appendix** for more background information on ICANN). ICANN implements and enforces many of its policies and rules through contracts with *registries* (companies and organizations who operate and administer the master database of all domain names registered in each top level domain, such as .com and .org) and accredited *registrars* (the hundreds of companies and organizations with which consumers register domain names). Policies are developed by Supporting Organizations and Committees in a consensus-based "bottom-up" process open to various constituencies and stakeholders of the Internet. As such, ICANN is often pointed to as emblematic of the "multistakeholder model" of Internet governance.

Internet standards organizations—As the Internet has evolved, groups of engineers, researchers, users, and other interested parties have coalesced to develop technical standards and protocols necessary to enable the Internet to function smoothly. These organizations conduct standards development processes that are open to participants and volunteers from around the world. Internet standards organizations include the Internet Engineering Task Force (IETF), the Internet Architecture Board (IAB), the Internet Society (ISOC), and the World Wide Web Consortium (W3C).

Governmental entities involved in Internet governance include the following:

Governmental Advisory Committee (GAC)—As part of ICANN's multistakeholder process, the GAC provides advice to the ICANN Board on matters of public policy, especially in cases where ICANN activities and policies may interact with national laws or international agreements related to issues such as intellectual property, law enforcement, and privacy. Although the ICANN Board is required to consider GAC advice and recommendations, it is not obligated to follow those recommendations. Membership in the GAC is open to all national governments who wish to participate. Currently, there are 113 nations represented, and the GAC Chair is presently held by Canada, with Vice Chairs held by Kenya, Sweden, and Singapore.

Internet Governance Forum (IGF)—The IGF was established in 2006 by the United Nation's World Summit on the Information Society (WSIS). The purpose of the IGF is to provide a multistakeholder forum which provides an open discussion (in yearly meetings) on public policies related to the Internet. Open to all stakeholders and interested parties (governments, industry, academia, civil society), the IGF serves as an open discussion forum and does not have negotiated outcomes, nor does it make formal recommendations to the U.N. In December 2010, the U.N. General Assembly renewed the IGF through 2015 and tasked the U.N.'s Commission on Science and Technology for Development (CSTD) to develop a report and recommendations on how the IGF might be improved. A Working Group on Improvements to the Internet Governance Forum was formed by the U.N., which includes 22 governments (including the United States) and the participation of Internet stakeholder groups.

Other International Organizations—Other existing international organizations address Internet policy issues in various ways. The International Telecommunications Union (ITU) is the United Nations specialized agency for communications and information technology. The World Intellectual Property Organization (WIPO) is another specialized agency of the U.N., which addresses a wide range of intellectual property issues, including those related to Internet policy. The Organisation for Economic Co-operation and Development (OECD) provides a forum for governments to work together to address economic issues, including the recent development of Internet policymaking principles. While none of these organizations have direct control or authority over the Internet, their activities can have influence over future directions of global Internet policy.

National governments—National governments have acted to address various Internet policy issues within their own borders. Many of the national laws and regulations pertain to user behavior on the Internet. For example, in the United States, laws have been passed addressing such issues as cybersecurity and cybercrime, Internet gambling, Internet privacy, and protection of intellectual property on the Internet. Governments have also established internal Internet policy coordinating bodies (e.g., the National Telecommunication and Information Administration's Internet Policy Task Force and the European Commission's Information Society).

ROLE OF U.S. GOVERNMENT

The United States government has no statutory authority over ICANN or the domain name system. However, because the Internet evolved from a network infrastructure created by the Department of Defense, the U.S. government originally owned and operated (primarily through private contractors) many of the key components of network architecture that enabled the domain name system to function. In the early 1990s, the National Science Foundation (NSF) was given a lead role in overseeing domain names used in the civilian portion of the Internet (which at that time was largely comprised of research universities). By the late 1990s, ICANN was created, the Internet had expanded into the commercial world, and the National Telecommunications and Information Administration (NTIA) of the Department of Commerce (DOC) assumed the lead role.

A 1998 Memorandum of Understanding between ICANN and the DOC initiated a process intended to transition technical DNS coordination and management functions to a private-sector not-for-profit entity. While the DOC plays no role in the internal governance or day-to-day operations of ICANN, the U.S. government, through the DOC/NTIA, retains a role with respect to the DNS via three separate contractual agreements. These are:

- a 2009 Affirmation of Commitments (AoC) between DOC and ICANN;[4]
- a contract between ICANN and DOC to perform various technical functions such as allocating IP address blocks, editing the root zone file, and coordinating the assignment of unique protocol numbers; and
- a cooperative agreement between DOC and VeriSign to manage and maintain the official DNS root zone file.

By virtue of those three contractual agreements, the United States government—through DOC/NTIA—exerts a legacy authority over ICANN, and arguably has more influence over ICANN and the DNS than other national governments.

While NTIA is the lead agency overseeing domain name issues, other federal agencies maintain a specific interest in the DNS that may affect their particular missions. For example, the Federal Trade Commission (FTC) seeks to protect consumer privacy on the Internet, the Department of Justice (DOJ) addresses Internet crime and intellectual property issues, and the Department of Defense and Department of Homeland Security address cybersecurity issues. However none of these agencies have legal authority over ICANN or the running of the DNS.

Affirmation of Commitments

On September 30, 2009, DOC and ICANN announced agreement on an Affirmation of Commitments (AoC) to "institutionalize and memorialize" the technical coordination of the DNS globally and by a private-sector-led organization.[5] The AoC replaced the previous Memorandum of Understanding and subsequent Joint Project Agreement between DOC and ICANN. It has no expiration date and would conclude only if one of the two parties decided to terminate the agreement.

Under the AoC, ICANN committed to remain a not-for-profit corporation "headquartered in the United States of America with offices around the world to meet the needs of a global community." According to the AoC, "ICANN is a private organization and nothing in this Affirmation should be construed as control by any one entity." Specifically, the AoC called for the establishment of review panels which will periodically make recommendations to the ICANN Board in four areas: ensuring accountability, transparency, and the interests of global Internet users (panel includes the Administrator of NTIA); preserving security, stability, and resiliency; impact of new generic top level domains (gTLDs); and WHOIS policy.[6]

On December 31, 2010, the Accountability and Transparency Review Team (ATRT) released its recommendations to the Board for improving ICANN's transparency and accountability with respect to Board governance and performance, the role and effectiveness of the GAC and its interaction with the Board, public input and policy development processes, and review mechanisms for Board decisions.[7] At the June 2011 meeting in Singapore, the Board adopted all 27 ATRT recommendations. According to NTIA, "the focus turns to ICANN management and staff, who must take up the challenge of implementing these recommendations as rapidly as possible and in a manner that leads to meaningful and lasting reform."[8]

DOC Contracts with ICANN and VeriSign

A contract between DOC and ICANN authorizes the Internet Assigned Numbers Authority (IANA) to perform various technical functions such as allocating IP address blocks, editing the root zone file, and coordinating the assignment of unique protocol numbers. Additionally, a cooperative agreement between DOC and VeriSign (a company that operates the .com and .net registries) authorizes VeriSign to manage and maintain the official root zone file that is contained in the Internet's root servers that underlie the functioning of the DNS.[9] By virtue of these legal agreements, the DOC must approve changes or modifications made to the root zone file (changes, for example, such as adding a new top level domain).[10]

Debate among Internet stakeholders was ongoing over the renewal of the IANA contract between DOC and ICANN, which was due to expire on September 30, 2012. The IANA contract renewal provided a further arena for the larger debate over Internet governance. NTIA's draft Statement of Work (SOW) detailing work requirements for the IANA contract[11] included a provision requiring that requests to IANA for new gTLDs be accompanied by documentation demonstrating how the proposed new gTLD "reflects consensus among relevant stakeholders and is supportive of the global public interest."[12] ICANN and many others in the domain name community submitted comments to NTIA, expressing strong opposition to the proposal that requests to IANA for new gTLDs be accompanied by documentation demonstrating global public support and consensus. According to ICANN, such a step would undermine ICANN's multistakeholder model by revising the gTLD implementation and policy processes already adopted through the bottom-up decision-making process.[13]

NTIA's final contract solicitation, released on November 10, 2011, lessened the IANA contractor requirements for adding new gTLDs, stating that when adding new gTLDs to the root zone, the contractor must provide "specific documentation demonstrating how the process provided the opportunity for input from relevant stakeholders and was supportive of the global public interest."[14] The IANA contract solicitation issued by NTIA specified that the contractor must be a wholly U.S. owned and operated firm or a U.S. university or college; that all primary operations and systems shall remain within the United States; and that the U.S. government reserves the right to inspect the premises, systems, and processes of all facilities and components used for the performance of the contract.

On July 2, 2012, NTIA announced the award of the new IANA contract to ICANN for up to seven years (through September 2019). The new contract included a separation between the policy development of IANA services and the implementation by the IANA functions contractor. The contract also featured "a robust company-wide conflict of interest policy; a heightened respect for local national law; and a series of consultation and reporting requirements to increase transparency and accountability."[15]

U.S. government authority and control over IANA and the management of the root zone file is a long-standing point of contention internationally. For example, while the European Commission approved many aspects of the new IANA contract, it sounded the following caution:

> The Commission believes greater respect should be given by the IANA contractor to respecting applicable law (such as EU personal data protection laws). The Commission will continue to take the initiative for such provisions in future IANA contracts as part of its efforts to ensure sustainable multi-stakeholder governance of the Internet, in the service of public interest, as a matter of both principle and efficient practice. In that context, it noted with regret that non-US companies are not allowed to compete for the forthcoming IANA contract.[16]

DEBATE OVER FUTURE MODEL OF INTERNET GOVERNANCE

Given its complexity, diversity, and international nature, how should the Internet be governed? Some assert that a multistakeholder model of governance is appropriate, where all stakeholders (both public and private sectors) arrive at consensus through a transparent bottom-up process. Others argue that a greater role for national governments is necessary, either through increased influence through the multistakeholder model, or under the auspices of an international body exerting intergovernmental control.

To date, ICANN and the governance of the domain name system has been the focal point of this debate. While ICANN's mandate is to manage portions of the technical infrastructure of the Internet (domain names and IP addresses), many of the decisions ICANN makes affect other aspects of Internet policy, including areas such as intellectual property, privacy, and cybersecurity. These are areas which many national governments have addressed for their own citizens and constituencies through domestic legislation, as well as through international treaties.

As part of the debate over an appropriate model of Internet governance, criticisms of ICANN have arisen on two fronts. One criticism reflects the tension between national governments and the current performance and governance processes of ICANN, whereby governments feel they lack adequate influence over ICANN decisions that affect a range of Internet policy issues. The other criticism is fueled by concerns of many nations that the U.S.

government holds undue legacy influence and control over ICANN and the domain name system.

The debate over multistakeholderism vs. intergovernmental control initially manifested itself in 2005 at the World Summit on the Information Society (WSIS), which was a conference organized by the United Nations. More recently, this debate has been rekindled in various international fora, partially sparked by two ICANN actions in 2011: the approval of the .xxx top-level domain and the approval of a process to allow an indefinite number of new generic top level domains (gTLDs).

2005 World Summit on the Information Society (WSIS)

Following the creation of ICANN in 1998, many in the international community, including foreign governments, argued that it was inappropriate for the U.S. government to maintain its legacy authority over ICANN and the DNS. They suggested that management of the DNS should be accountable to a higher intergovernmental body. The United Nations, at the first phase of the WSIS in December 2003, debated and agreed to study the issue of how to achieve greater international involvement in the governance of the Internet, and the domain name system in particular. The study was conducted by the U.N.'s Working Group on Internet Governance (WGIG). On July 14, 2005, the WGIG released its report,[17] stating that no single government should have a preeminent role in relation to international Internet governance. The report called for further internationalization of Internet governance, and proposed the creation of a new global forum for Internet stakeholders. Four possible models were put forth, including two involving the creation of new Internet governance bodies linked to the U.N. Under three of the four models, ICANN would either be supplanted or made accountable to a higher intergovernmental body. The report's conclusions were scheduled to be considered during the second phase of the WSIS held in Tunis in November 2005. U.S. officials stated their opposition to transferring control and administration of the domain name system from ICANN to any international body. Similarly, the 109th Congress expressed its support for maintaining existing U.S. control over ICANN and the DNS (H.Con.Res. 268 and S.Res. 323).[18]

The European Union (EU) initially supported the U.S. position. However, during the September 2005 preparatory meetings, the EU seemingly shifted its support towards an approach which favored an enhanced international role in governing the Internet. Conflict at the WSIS Tunis Summit over control of the domain name system was averted by the announcement, on November 15, 2005, of an Internet governance agreement between the United States, the EU, and over 100 other nations. Under this agreement, ICANN and the United States maintained their roles with respect to the domain name system. A new international group under the auspices of the U.N. was formed—the Internet Governance Forum (IGF)—which would provide an ongoing forum for all stakeholders (both governments and nongovernmental groups) to discuss and debate Internet policy issues.

Creation of the .xxx Domain and New gTLDs

Starting in 2010 and 2011, controversies surrounding the roll-out of new generic top level domains (gTLDs) and the addition of the .xxx TLD led some governments to argue for increased government influence on the ICANN policy development process.[19]

.xxx

Since 2000, ICANN has repeatedly considered whether to allow the establishment of a gTLD for adult content. On June 1, 2005, ICANN announced that it had entered into commercial and technical negotiations with a registry company (ICM Registry) to operate a new ".xxx" domain, which would be designated for use by adult websites. With the ICANN Board scheduled to consider final approval of the .xxx domain on August 16, 2005, the Department of Commerce sent a letter to ICANN requesting that adequate additional time be provided to allow ICANN to address the objections of individuals expressing concerns about the impact of pornography on families and children and opposing the creation of a new top level domain devoted to adult content. ICANN's Governmental Advisory Committee (GAC) also requested more time before the final decision.

On March 30, 2007, the ICANN Board voted 9-5 to deny the .xxx domain. ICM Registry subsequently challenged ICANN's decision before an Independent Review Panel (IRP), claiming that ICANN's rejection of ICM's application for a .xxx gTLD was not consistent with ICANN's Articles of Incorporation and Bylaws. On February 19, 2010, a three-person Independent Review Panel ruled primarily in favor of ICM Registry, finding that its application for the .xxx TLD had met the required criteria.

Subsequently, on June 25, 2010, at the ICANN meeting in Brussels, the Board of Directors voted to allow ICM's .xxx application to move forward, and at the December 2010 ICANN meeting, the ICANN Board passed a resolution stating that while "it intends to enter into a registry agreement with ICM Registry for the .xxx TLD," the Board would enter into a formal consultation with the Governmental Advisory Committee on areas where the Board's decision was in conflict with GAC advice relating to the ICM application.[20]

While not officially or formally in opposition to the approval of .xxx, the GAC advised ICANN that "there is no active support of the GAC for the introduction of a .xxx TLD" and that "while there are members, which neither endorse nor oppose the introduction of a .xxx TLD, others are emphatically opposed from a public policy perspective to the introduction of an .xxx TLD."[21] The GAC listed a number of specific issues and objections that it wished ICANN to resolve.

A February 2011 letter from ICANN to the GAC acknowledged and responded to areas where approving the .xxx registry agreement with ICM would conflict with GAC advice received by ICANN.[22] The Board acknowledged that ICANN and the GAC were not able to reach a mutually acceptable solution, and ultimately, on March 18, 2011, the Board approved a resolution giving the CEO or General Counsel of ICANN the authority to execute the registry agreement with ICM to establish a .xxx TLD. The vote was nine in favor, three opposed, and four abstentions.

The decision to create a .xxx TLD was not viewed favorably by many governments.[23] In an April 6, 2011, letter to the Department of Commerce, the European Commissioner for the Digital Agenda asked that the introduction of .xxx be delayed.[24] In its response, NTIA said it "share[s] your disappointment that ICANN ignored the clear advice of governments worldwide, including the United States, by approving the new .xxx domain."[25] However, NTIA stated why it would not (and did not) interfere with the addition of .xxx:

> While the Obama Administration does not support ICANN's decision, we respect the multi-stakeholder Internet governance process and do not think that it is in the long-term best interest of the United States or the global Internet community for us unilaterally to reverse the decision. Our goal is to preserve the global Internet, which is a force for innovation, economic growth, and the free flow of information. I agree with you that the Board took its action without the full support of the community and accordingly, I am dedicated to improving the responsiveness of ICANN to all stakeholders, including governments worldwide.[26]

gTLD Expansion

Top Level Domains (TLDs) are the suffixes that appear at the end of an address (after the "dot"). Prior to ICANN's establishment in 1998, the Internet had eight generic top level domains (gTLDs), including .com, .org, .net, and .gov. In 2000 and 2004, ICANN held application rounds for a limited number of new gTLDs—currently there are twenty-two. Some are reserved or restricted to particular types of organizations (e.g., .museum, .gov, .travel) and others are open for registration by anyone (.com, .org, .info). Applicants for new gTLDs are typically commercial entities and non-profit organizations who seek to become ICANN-recognized registries that will establish and operate name servers for their TLD registry, as well as implement a domain name registration process for that particular TLD.

The growth of the Internet and the accompanying growth in demand for domain names have focused the debate on whether and how to further expand the number of gTLDs. Beginning in 2005, ICANN embarked on a long consultative process to develop rules and procedures for introducing and adopting an indefinite number of new gTLDs into the domain name system. A new gTLD can be any word or string of characters that is applied for and approved by ICANN. Between 2008 and 2011, ICANN released seven iterations of its gTLD Applicant Guidebook (essentially the rulebook for how the new gTLD program will be implemented). On June 20, 2011, the ICANN Board of Directors voted to approve the launch of the new gTLD program, under which potentially hundreds of new gTLDs could ultimately be approved by ICANN and introduced into the DNS. Applications for new gTLDs were to be accepted from January 12 through April 12, 2012.

The rollout of new gTLDs was controversial. Advocates (including the domain name industry) argued that a gTLD expansion will provide opportunities for Internet innovation and competition. On the other hand, many trademark holders pointed to possible higher costs and greater difficulties in protecting their trademarks across hundreds of new gTLDs. Similarly, governments expressed concern over intellectual property protections, and along with law enforcement entities, also cited concerns over the added burden of combating various cybercrimes (such as phishing and identity theft) across hundreds of new gTLDs. Throughout ICANN's policy development process, governments, through the Governmental Advisory Committee, advocated for additional intellectual property protections in the new gTLD process. The GAC also argued for more stringent rules that would allow for better law enforcement in the new domain space to better protect consumers. Although changes were made, strong opposition from many trademark holders[27] led to opposition from some parts of the U.S. government towards the end of 2011. For example:

- On December 8, 2011, the Senate Committee on Commerce, Science and Transportation held a hearing on the ICANN's expansion of TLDs. Subsequently, on December 28, 2011, a letter from Senator John Rockefeller, chairman of the Senate Committee on Commerce, Science and Transportation, to the Secretary of Commerce and the Administrator of NTIA, stated his concern that "this expansion of gTLDs, if it proceeds as planned, will have adverse consequences for the millions of American consumers, companies, and non-profit organizations that use the Internet on a daily basis" and that at the hearing, "witnesses speaking on behalf of more than a hundred companies and non-profit organizations explained that ICANN's current plan for gTLD expansion will likely cause millions of dollars in increased costs related to combating cybersquatting." In the letter, Senator Rockefeller requested that NTIA "should consider asking ICANN to either delay the opening of the application period or to drastically limit the number of new gTLDs it approves next year."[28] A subsequent December 22, 2011, letter to ICANN from Senators Klobuchar and Ayotte, also registered concern over the TLD expansion and asked ICANN to further address law enforcement, trademark, and consumer concerns before launching the program.[29]

- On December 14, 2011, the House Committee on Energy and Commerce, Subcommittee on Communications and Technology, held a hearing on ICANN's top level domain program. Subsequently on December 21, 2011, a bipartisan group of Committee Members sent a letter to ICANN requesting that the expansion of the gTLDs be delayed, noting that "many stakeholders are not convinced that ICANN's process has resulted in an acceptable level of protection."[30] The Energy and Commerce Committee Members argued that "a short delay will allow interested parties to work with ICANN and offer changes to alleviate many of them, specifically concerns over law enforcement, cost and transparency that were discussed in recent Congressional hearings."[31]

- A December 16, 2011, letter to the Secretary of Commerce from Representative Bob Goodlatte, chairman of the House Subcommittee on Intellectual Property, Competition, and the Internet, and Representative Howard Berman, ranking Member of the House Committee on Foreign Affairs, urged DOC to take all steps necessary to encourage ICANN to undertake further evaluation and review before the gTLD expansion is permitted to occur. The letter asked DOC to determine whether the benefits of the expansion outweigh the costs and risks to consumers, businesses, and the Internet, and that if the program proceeds, that ICANN should initially limit the expansion to a small pilot project which can be evaluated.[32] Previously, the Subcommittee on Intellectual Property, Competition, and the Internet had held a May 4, 2011, hearing on oversight of the gTLD program.

- A December 16, 2011, letter from the Federal Trade Commission (FTC) to ICANN argued that a "rapid, exponential expansion of gTLDs has the potential to magnify both the abuse of the domain name system and the corresponding challenges we encounter in tracking down Internet fraudsters." The FTC urged ICANN to implement the new gTLD program as a pilot program and substantially reduce the number of gTLDs that are introduced in the first application round, strengthen ICANN's contractual compliance program, develop a new ongoing program to monitor consumer issues that arise during the first round of implementing the new gTLD program, conduct an assessment of each new proposed gTLD's risk of consumer harm as part of the evaluation and approval process, and improve the accuracy of WHOIS data, including by imposing a registrant verification requirement. The FTC added that "ICANN should address these issues before it approves any new gTLD applications. If ICANN fails to address these issues responsibly, the introduction of new gTLDs could pose a significant threat to consumers and undermine consumer confidence in the Internet."[33]

- A December 27, 2011, letter to ICANN from the Senate and House Judiciary Committees expressed concerns over the new gTLD program and urged ICANN to "strengthen protections for consumers and trademark holders who risk being harmed by the proliferation of domain names on the web." The letter also urged ICANN to work closely with the law enforcement community "to ensure that the program's rollout does not adversely impact their efforts to fight fraud and abuse on the Internet."[34]

At the December 2011 House and Senate hearings, ICANN stated its intention to proceed with the gTLD expansion as planned. ICANN defended its gTLD program, arguing that the new gTLDs will offer more protections for consumers and trademark holders than current gTLDs; that new gTLDs will provide needed competition, choice, and innovation to the domain name system; and that critics have already had ample opportunity to contribute input during a seven-year deliberative policy development process.[35] Ultimately, ICANN did not delay the initiation of the new gTLD program, and the application window was opened on January 12, 2012, as planned.

Much of the pressure on ICANN to delay the new gTLD program was directed at NTIA, given NTIA's unique relationship with ICANN. At both the December 2011 Senate and House hearings, NTIA expressed support for ICANN's planned rollout of the TLD expansion program, arguing that national governments have been able to address intellectual property, law enforcement, and consumer concerns through the Governmental Advisory Committee (GAC):

> NTIA believes that ICANN improved the new gTLD program by incorporating a significant number of proposals from the GAC. ICANN's new gTLD program also now provides law enforcement and consumer protection authorities with significantly more tools than those available in existing gTLDs to address malicious conduct. The fact that not all of the GAC's proposals were adopted as originally offered does not represent a failure of the process or a setback to governments; rather, it reflects the reality of a multistakeholder model.[36]

While NTIA stated that it would continue to monitor progress and push for necessary changes to ICANN's TLD expansion program, a key aspect of NTIA's argument for supporting ICANN's planned rollout was to preserve the integrity of the multistakeholder Internet governance process:

> NTIA is dedicated to maintaining an open, global Internet that remains a valuable tool for economic growth, innovation, and the free flow of information, goods, and services online. We believe the best way to achieve this goal is to continue to actively support and participate in multi-stakeholder Internet governance processes such as

ICANN. This is in stark contrast to some countries that are actively seeking to move Internet policy to the United Nations. If we are to combat the proposals put forward by others, we need to ensure that our multi-stakeholder institutions have provided a meaningful role for governments as stakeholders. NTIA believes that the strength of the multi-stakeholder approach to Internet policy-making is that it allows for speed, flexibility, and decentralized problem-solving and stands in stark contrast to a more traditional, top-down regulatory model characterized by rigid processes, political capture by incumbents, and in so many cases, impasse or stalemate.[37]

On January 3, 2012, NTIA sent ICANN a letter concerning implementation of the new gTLD program.[38] While NTIA recognized that the program "is the product of a six-year international multistakeholder process" and that NTIA does "not seek to interfere with the decisions and compromises reached during that process," NTIA urged ICANN to consider implementing measures to address many of the criticisms raised. Such measures would address concerns of trademark holders, law enforcement, and consumer protection. NTIA also asked ICANN to assess (after the initial application window closes and the list of prospective new gTLDs is known) whether there is a need to phase in the introduction of new gTLDs, and whether additional trademark protection measures need to be taken.

NTIA concluded its letter as follows:

> How ICANN handles the new gTLD program will, for many, be a litmus test of the viability of this approach. For its part, NTIA is committed to continuing to be an active member of the GAC and working with stakeholders to mitigate any unintended consequences of the new gTLD program.[39]

On June 13, 2012, ICANN announced it had received 1,930 applications for new gTLDs,[40] and ICANN has now moved into the evaluation phase; ICANN will decide whether or not to accept each of the 1,930 new gTLD applications. With the first round application period concluded, there remain significant issues in play as the new gTLD program goes forward. First, ICANN has stated that a second and subsequent round will take place, and that changes to the application and evaluation process will be made such that a "systemized manner of applying for gTLDs be developed in the long term."[41] ICANN's goal is to begin the second application round "within one year of the close of the application submission period for the initial round."[42] Thus, many observers are eager to see what changes may be made in the second round.

Second, when the new gTLDs go "live," many stakeholders are concerned that various forms of domain name abuse (e.g., trademark infringement, consumer fraud, malicious behavior, etc.) could manifest itself within the hundreds of new gTLD domain spaces. Thus, the effectiveness of ICANN's approach to addressing such issues as intellectual property protection of second level domain names and mitigating unlawful behavior in the domain name space will be of interest as the new gTLD program goes forward.

With respect to the new gTLD program, the GAC provides advice to the ICANN Board on any first round applications the GAC considers problematic. GAC advice can take three forms:

i.   The GAC advises ICANN that it is the consensus of the GAC that a particular application should not proceed. This will create a strong presumption for the ICANN Board that the application should not be approved.

ii.   The GAC advises ICANN that there are concerns about a particular application "dot-example." The ICANN Board is expected to enter into dialogue with the GAC to understand the scope of concerns. The ICANN Board is also expected to provide a rationale for its decision.

iii. The GAC advises ICANN that an application should not proceed unless remediated. This will raise a strong presumption for the Board that the application should not proceed unless there is a remediation method available in the Guidebook (such as securing the approval of one or more governments), that is implemented by the applicant.[43]

The GAC also issues Early Warnings to the ICANN Board in the event that any GAC member finds an application problematic for any reason. An Early Warning is an indication that a formal GAC objection is possible (either through the GAC advice process or through the formal objection process). Applicants are notified of an Early Warning against their application and given the opportunity to address the concerns or to withdraw the application (thereby qualifying for a partial refund of the application fee).

Proposed Models for Internet Governance

As discussed above, ICANN is a working example of a multistakeholder model of Internet governance, whereby a bottom-up collaborative process is used to provide Internet stakeholders with access to the policymaking process. Support for the multistakeholder model of Internet governance is reflected in international organizations such as the Organisation for Economic Cooperation and Development (OECD) and the Group of Eight (G8). For example, the OECD's *Communiqué on Principles for Internet Policy-Making* cites multistakeholderism as a central tenet of Internet governance:

> In particular, continued support is needed for the multi-stakeholder environment, which has underpinned the process of Internet governance and the management of critical Internet resources (such as naming and numbering resources) and these various stakeholders should continue to fully play a role in this framework. Governments should also work in multi-stakeholder environments to achieve international public policy goals and strengthen international co-operation in Internet governance.[44]

Similarly, at the G8 Summit of Deauville on May 26-27, 2011, the G8 issued a declaration on its renewed commitment for freedom and democracy that contained a new section on the Internet. Support for a multistakeholder model for Internet governance with a significant national government role was made explicit:

> As we support the multi-stakeholder model of Internet governance, we call upon all stakeholders to contribute to enhanced cooperation within and between all international fora dealing with the governance of the Internet. In this regard, flexibility and

transparency have to be maintained in order to adapt to the fast pace of technological and business developments and uses. Governments have a key role to play in this model.[45]

As discussed above, in 2005, the World Summit on the Information Society (WSIS) considered four models of Internet governance, of which three would have involved an intergovernmental body to oversee the Internet and the domain name system. While the WSIS ultimately decided not to pursue an intergovernmental model in 2005, some nations have again advocated an intergovernmental approach for Internet governance. For example:

- India, Brazil, and South Africa (referred to as IBSA) proposed that "an appropriate body is urgently required in the U.N. system to coordinate and evolve coherent and integrated global public policies pertaining to the Internet." The IBSA proposed body would "integrate and oversee the bodies responsible for technical and operational functioning of the Internet, including global standards setting."[46]

- In order to implement the major aspects of the IBSA proposal, the government of India proposed (in the U.N. General Assembly) the establishment of a new institutional mechanism in the United Nations for global internet-related policies, to be called the United Nations Committee for Internet-Related Policies (CIRP). CIRP would be comprised of 50 member states chosen on the basis of equitable geographical representation. The Internet Governance Forum (IGF) and four advisory stakeholder groups would provide input to CIRP, which would report directly to the General Assembly and present recommendations for consideration, adoption, and dissemination among all relevant intergovernmental bodies and international organizations.[47]

- Another group of nations, including China and the Russian Federation, proposed a voluntary "International Code of Conduct for Information Security," for further discussion in the U.N. General Assembly. The Code includes language that promotes the establishment of a multilateral, transparent, and democratic international management system to ensure an equitable distribution of resources, facilitate access for all, and ensure a stable and secure functioning of the Internet.[48]

Thus, governments such as the United States and the European Union support ICANN's multistakeholder model, while at the same time advocating increased governmental influence within that model.[49] By contrast, other nations support an expanded role for an intergovernmental model of Internet governance. The debate has been summarized by NTIA as follows:

By engaging all interested parties, multistakeholder processes encourage broader and more creative problem solving, which is essential when markets and technology are changing as rapidly as they are. They promote speedier, more flexible decision making than is common under traditional, top-down regulatory models which can too easily fall prey to rigid procedures, bureaucracy, and stalemate. But there is a challenge emerging to this model in parts of the world.... Some nations appear to prefer an Internet managed and controlled by nation-states. In December 2012, the U.S. will participate in the ITU's World Conference on International Telecommunications (WCIT). This treaty negotiation will conduct a review of the International Telecommunication Regulations (ITRs), the general principles which relate to traditional international voice telecommunication services. We expect that some states will attempt to rewrite the regulation in a manner that would exclude the contributions of multi-stakeholder organizations and instead provide for heavy-handed governmental control of the Internet, including provisions for cybersecurity and granular operational and technical requirements for private industry. We do not support any of these elements. It is critical that we work with the private sector on outreach to countries to promote the multi-stakeholder model as a credible alternative.[50]

World Conference on International Telecommunications (WCIT)

The World Conference on International Telecommunications (WCIT) was held in Dubai on December 3-14, 2012. Convened by the International Telecommunications Union (the ITU, an agency within the United Nations), the WCIT was a formal meeting of the world's national governments held in order to revise the International Telecommunications Regulations (ITRs). The ITRs, previously revised in 1988, serve as a global treaty outlining the principles which govern the way international telecommunications traffic is handled.

Because the existing 24-year-old ITRs predated the Internet, one of the key policy questions in the WCIT was how and to what extent the updated ITRs should address Internet traffic and Internet governance. The Administration and Congress took the position that the new ITRs should continue to address only traditional international telecommunications traffic, that a multistakeholder model of Internet governance (such as ICANN) should continue, and that the ITU should not take any action that could extend its jurisdiction or authority over the Internet.

As the WCIT approached, concerns heightened in the 112th Congress that the WCIT might potentially provide a forum leading to an increased level of intergovernmental control over the Internet. On May 31, 2012, the House Committee on Energy and Commerce, Subcommittee on Communications and Technology, held a hearing entitled, "International Proposals to Regulate the Internet."[51] To accompany the hearing, H.Con.Res. 127 was introduced by Representative Bono Mack expressing the sense of Congress regarding actions to preserve and advance the multistakeholder governance model. Specifically, H.Con.Res. 127 expressed the sense of Congress that the Administration "should continue working to implement the position of the United States on Internet governance that clearly articulates the consistent and unequivocal policy of the United States to promote a global Internet free from government control and preserve and advance the successful multistakeholder model that governs the Internet today." H.Con.Res. 127 was passed unanimously by the House (414-0) on August 2, 2012.

A similar resolution, S.Con.Res. 50, was introduced into the Senate by Senator Rubio on June 27, 2012, and referred to the Committee on Foreign Relations. The Senate resolution expressed the sense of Congress "that the Secretary of State, in consultation with the Secretary of Commerce, should continue working to implement the position of the United States on Internet governance that clearly articulates the consistent and unequivocal policy of the United States to promote a global Internet free from government control and preserve and advance the successful multistakeholder model that governs the Internet today." S.Con.Res. 50 was passed by the Senate by unanimous consent on September 22, 2012. On December 5, 2012—shortly after the WCIT had begun in Dubai—the House unanimously passed S.Con.Res. 50 by a vote of 397-0.

During the WCIT, a revision to the ITRs was proposed and supported by Russia, China, Saudi Arabia, Algeria, and Sudan that sought to explicitly extend ITR jurisdiction over Internet traffic, infrastructure, and governance. Specifically, the proposal stated that "Member States shall have the sovereign right to establish and implement public policy, including international policy, on matters of Internet governance."[52] The proposal also included an article establishing the right of Member States to manage Internet numbering, naming, addressing, and identification resources.

The proposal was subsequently withdrawn. However, as an intended compromise, the ITU adopted a nonbinding resolution (Resolution 3, attached to the final ITR text) entitled, "To Foster an enabling environment for the greater growth of the Internet." Resolution 3 includes language stating "all governments should have an equal role and responsibility for international Internet governance" and invites Member States to "elaborate on their respective positions on international Internet-related technical, development and public policy issues within the mandate of ITU at various ITU forums."[53]

Because of the inclusion of Resolution 3, along with other features of the final ITR text (such as new ITU articles related to spam and cybersecurity), the United States declined to sign the treaty. The leader of the U.S. delegation stated the following:

> The Internet has given the world unimaginable economic and social benefits during these past 24 years—all without UN regulation. We candidly cannot support an ITU treaty that is inconsistent with a multi-stakeholder model of Internet governance. As the ITU has stated, this conference was never meant to focus on internet issues; however, today we are in a situation where we still have text and resolutions that cover issues on spam and also provisions on internet governance. These past two weeks, we have of course made good progress and shown a willingness to negotiate on a variety of telecommunications policy issues, such as roaming and settlement rates, but the United States continues to believe that internet policy must be multi-stakeholder driven. Internet policy should not be determined by member states but by citizens, communities, and broader society, and such consultation from the private sector and civil society is paramount. This has not happened here.[54]

Of the 144 eligible members of the ITU, 89 nations signed the treaty, while 55 either chose not to sign (such as the United States) or remain undecided.[55]

While the WCIT in Dubai is concluded, the international debate over Internet governance is expected to continue in future intergovernmental telecommunications meetings and conferences. The 113th Congress is overseeing and supporting the U.S. government's continuing efforts to resist international attempts to exert control over Internet governance. On February 5, 2013, the House Committee on Energy and Commerce, Subcommittee on Communications and Technology, held a hearing entitled "Fighting for Internet Freedom: Dubai and Beyond." The hearing was held jointly with the House Committee on Foreign Affairs, Subcommittee on Terrorism, Nonproliferation, and Trade and the Subcommittee on Africa, Global Health, Global Human Rights, and International Organizations.

On April 16, 2013, H.R. 1580, a bill "To Affirm the Policy of the United States Regarding Internet Governance," was introduced by Representative Walden. Using language similar to the WCIT-related Congressional resolutions passed by the 112th Congress (S.Con.Res. 50 and H.Con.Res. 127), H.R. 1580 states that "It is the policy of the United States to preserve and advance the successful multistakeholder model that governs the

Internet." On April 17, 2013, H.R. 1580 was approved (by voice vote) by the House Committee on Energy and Commerce.

Issues for Congress

Congress plays an important role overseeing NTIA's stewardship of the domain name system and ICANN. The House Committee on Energy and Commerce and the Senate Committee on Commerce, Science, and Transportation have held numerous oversight hearings exploring ICANN's performance in general, as well as specific DNS issues that arise (e.g., the proposed gTLD expansion). Additionally, other committees, such as the House and Senate Judiciary Committees, maintain an interest in the DNS as it affects Internet policy issues such as intellectual property, privacy, and cybercrime. Since 1997, congressional committees have held 31 hearings on the DNS and ICANN.[56]

Congress has an impact on the issue of Internet governance, both via its oversight of NTIA and the DNS, and through its actions in other and more specific areas of Internet policymaking. For example, Congress continues to oversee and evaluate NTIA's strategy of supporting ICANN's multistakeholder model while opposing arguments for increased intergovernmental control. At the same time, NTIA is seeking to maximize government influence within the ICANN process (primarily through the GAC), especially in instances where Internet policy intersects with national laws addressing such issues as intellectual property, privacy, law enforcement, and cybersecurity.

One of NTIA's arguments for increasing government influence over ICANN policymaking (via the GAC) is that if governments feel their interests are not adequately addressed within the ICANN process, this perception will give support to the argument that the DNS and the Internet should be governed through a more formal intergovernmental mechanism. Congress may wish to examine where an appropriate balance exists between a sufficient level of governmental influence within the ICANN system, and an inappropriately excessive level of governmental control through the GAC that might threaten the multistakeholder model that ICANN represents.

To the extent that ICANN is successful in its endeavors and its credibility remains strong with Internet stakeholders, the argument for a multistakeholder model of Internet governance will be bolstered. By contrast, to the extent that ICANN falls short, the arguments for a growing role for some sort of formal intergovernmental body could become stronger. The following are some important issues that the 113th Congress may wish to consider as part of its oversight of NTIA's relationship with ICANN:

- How transparent and accountable is the ICANN governance structure, and to what extent do all Internet stakeholders have equal access to and influence over the ICANN policymaking process?
- How effectively does ICANN balance the interests and positions of differing stakeholders on particularly controversial issues, such as the new gTLD program? How successful will be the rollout of the gTLD program and other high-profile initiatives in the future?
- Regarding the Board of Directors and the ICANN staff, to what extent are sufficient ethics safeguards in place to prevent special interests (who may, for example, have

financial interests at stake) from exerting undue influence over ICANN policy decisions?[57]

- Should the U.S. government maintain its current legacy authority over ICANN and the DNS, and if so, how can NTIA best use this authority judiciously in order to advance U.S. government interests, while at the same time minimizing the perception by other nations (as well as the international community of Internet stakeholders) that the United States has an inappropriate level of control or influence over the Internet and the DNS?

- To what extent will ongoing and future intergovernmental telecommunications conferences (such as the December 2012 WCIT) constitute an opportunity for some nations to increase intergovernmental control over the Internet, and how effectively are NTIA and other government agencies (such as the State Department) working to counteract that threat?

Congress may also have a collateral impact on the debate over Internet governance through legislative activity related to specific areas of national Internet policy. For example, in the 112th Congress, provisions intended to protect intellectual property in the Preventing Real Online Threats to Economic Creativity and Theft of Intellectual Property Act (PROTECT IP or PIPA, S. 968) and the Stop Online Piracy Act (SOPA, H.R. 3261) sought to prohibit Internet service providers from directing Internet traffic to domain names with infringing content.[58] One of the arguments against the legislation was that any imposition of U.S. restrictions on the functioning of the DNS will, in the long run, undermine the integrity of the current multistakeholder model of Internet governance and give ammunition to those arguing for a formal intergovernmental body overseeing the Internet. For SOPA/PIPA and other Internet-related legislation, Congress may weigh arguable Internet governance impacts within the context of other arguments for and against the legislation. But the impact of domestic Internet laws and regulations on the overall Internet governance debate is an issue that may increasingly be considered by Congress.

Finally, the debate over how the Internet's domain name system is governed may have a significant impact on future debates on how other Internet policy areas are governed on a worldwide basis.[59] The ultimate success or failure of ICANN, and the multistakeholder model of Internet governance it represents, could help determine how other Internet policy issues—such as cybersecurity and privacy—are addressed.

APPENDIX. ICANN BASICS

ICANN is a not-for-profit public benefit corporation headquartered in Marina del Rey, CA, and incorporated under the laws of the state of California. ICANN is organized under the California Nonprofit Public Benefit Law for charitable and public purposes, and as such, is subject to legal oversight by the California attorney general. ICANN has been granted tax-exempt status by the federal government and the state of California.[60]

ICANN's organizational structure consists of a Board of Directors (BOD) advised by a network of supporting organizations and advisory committees that represent various Internet constituencies and interests (see **Figure A-1**). Policies are developed and issues are researched by these subgroups, who in turn advise the Board of Directors, which is responsible for making all final policy and operational decisions. The Board of Directors consists of 16 international and geographically diverse members, composed of one president, eight members selected by a Nominating Committee, two selected by the Generic Names Supporting Organization, two selected by the Address Supporting Organization, two selected by the Country-Code Names Supporting Organization, and one selected by the At-Large Advisory Committee. Additionally, there are five non-voting liaisons representing other advisory committees.

The explosive growth of the Internet and domain name registration, along with increasing responsibilities in managing and operating the DNS, has led to marked growth of the ICANN budget, from revenues of about $6 million and a staff of 14 in 2000, to revenues of $90 million and a staff of 149 forecasted for 2012.[61] ICANN is funded primarily through fees paid to ICANN by registrars and registry operators. Registrars are companies (e.g., GoDaddy, Google, Network Solutions) with which consumers register domain names.[62] Registry operators are companies and organizations that operate and administer the master database of all domain names registered in each top level domain (for example VeriSign, Inc. operates .com and .net, Public Interest Registry operates .org, and Neustar, Inc. operates .biz).[63] In 2011, ICANN received 94% of its total revenues from registry and registrar fees (49% from registry fees, 45% from registrar fees).[64]

Additionally, the collection of fees from the new generic top level domain (gTLD) program could contribute to an unprecedented level of revenue for ICANN in the years to come. For the first round of the new gTLD program, ICANN estimates revenues of $337 million from the new gTLD application fees, which is twice the amount of traditional revenues from all other sources over the next two years. After operating expenses (processing and evaluating the applications), ICANN estimates a surplus of $27.8 million from the new gTLD program.[65]

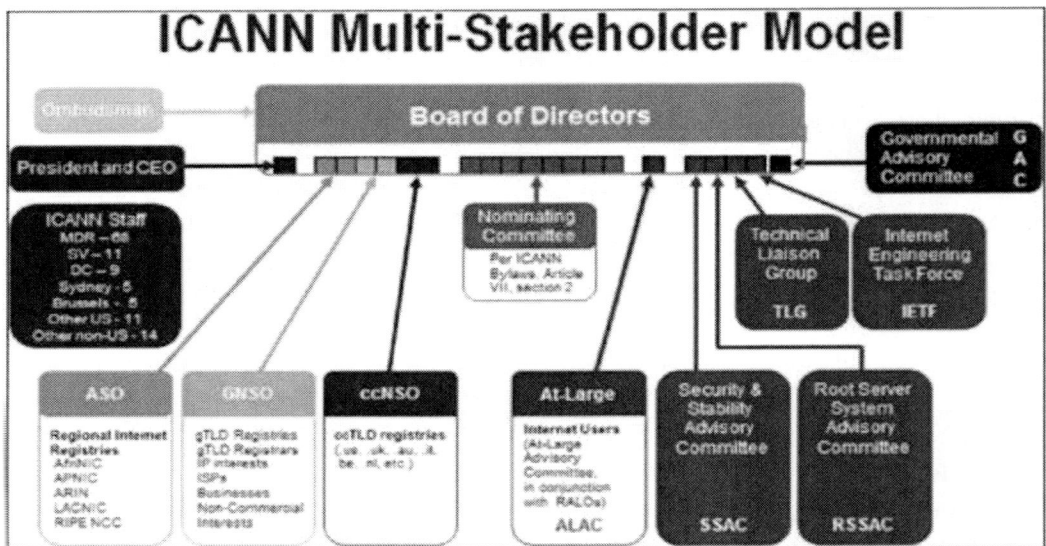

Source: ICANN; http://www.icann.org/en/groups/chart.

Figure A-1. Organizational Structure of ICANN.

End Notes

[1] Tunis Agenda for the Information Society, November 18, 2005, WSIS-05/TUNIS/DOC6(Rev.1)-E, p. 6, available at http://www.itu.int/wsis/docs2/tunis/off/6rev1.pdf.

[2] The IGP describes itself as "an alliance of academics that puts expertise into practical action in the fields of global governance, Internet policy, and information and communication technology." See http://www.internet governance.org.

[3] Milton Mueller, John Mathiason, and Hans Klein, "The Internet and Global Governance: Principles and Norms for a New Regime," *Global Governance*, vol. 13 (2007), p. 245.

[4] For more information on the Affirmation of Commitments, including the precursor agreements between DOC and ICANN such as the Joint Project Agreement and the Memorandum of Understanding, see CRS Report 97-868, *Internet Domain Names: Background and Policy Issues*, by Lennard G. Kruger.

[5] Affirmation of Commitments by the United States Department of Commerce and the Internet Corporation for Assigned Names and Numbers, September 30, 2009, available at http://www.ntia.doc.gov/ntiahome /domainname/ Affirmation_of_Commitments_2009.pdf.

[6] WHOIS is a publically available online database that provides information on domain name registrants. WHOIS is used to identify domain name holders. WHOIS policy is controversial because it encompasses two competing considerations: protecting the privacy of domain name holders versus enabling law enforcement and trademark holders to identify owners of domain names and websites engaging in criminal activities or infringing on intellectual property.

[7] The ATRT final report is available at http://www.icann.org/en/reviews/affirmation/atrt-final-recommendations-31dec10-en.pdf.

[8] NTIA, *Press Release*, "NTIA Commends ICANN Board on Adopting the Recommendations of the Accountability and Transparency Review Team," June 24, 2011, available at http://www.ntia.doc.gov/press/2011/ NTIA_Statement_06242011.html.

[9] According to the National Research Council, "The root zone file defines the DNS. For all practical purposes, a top level domain (and, therefore, all of its lower-level domains) is in the DNS if and only if it is listed in the root zone file. Therefore, presence in the root determines which DNS domains are available on the Internet." See National Research Council, Committee on Internet Navigation and the Domain Name System, *Technical Alternatives and Policy Implications, Signposts on Cyberspace: The Domain Name System and Internet Navigation*, National Academy Press, Washington, DC, 2005, p. 97.

[10] The June 30, 2005, "U.S. Principles on the Internet's Domain Name and Addressing System" stated the intention to "preserve the security and stability" of the DNS, and asserted that "the United States is committed to taking no action that would have the potential to adversely impact the effective and efficient operation of the DNS and will therefore maintain its historic role in authorizing changes or modifications to the authoritative root zone file." See http://www.ntia.doc.gov/ntiahome/domainname/USDNSprinciples_06302005.pdf.

[11] Department of Commerce, National Telecommunications and Information Administration, "Request for Comments on the Internet Assigned Numbers Authority (IANA) Functions," 76 *Federal Register* 10570, February 25, 2011.

[12] Ibid., p. 34662.

[13] See ICANN comments at http://www.ntia.doc.gov/files/ntia/icann_fnoi_comments_20110722.pdf, p. 7.

[14] Available at https://www.fbo.gov/index?s=opportunity&mode=form&id=c564af28581edb2a7b9441eccfd6391d& tab=core&_cview=0.

[15] NTIA, Press Release, "Commerce Department Awards Contract for Management of Key Internet Functions to ICANN," July 2, 2012, available at http://www.ntia.doc.gov/press-release/2012/commerce-department-awardscontract-management-key-internet-functions-icann.

[16] European Commission, "Digital Agenda: Commission welcomes improvements in new IANA contract," *Press Release*, November 14, 2011, available at http://europa.eu/rapid/pressReleasesAction.do? reference=IP /11/1345& format=HTML&aged=0&language=EN&guiLanguage=en.

[17] Working Group on Internet Governance, Report from the Working Group on Internet Governance, World Summit on the Information Society, Document WSIS-II/PC-3/DOC/5-E, August 3, 2005, available at http://www.itu.int/wsis/ docs2/pc3/html/off5/index.html.

[18] In the 109th Congress, H.Con.Res. 268 was passed unanimously by the House on November 16, 2005. S.Res. 323 was passed in the Senate by Unanimous Consent on November 18, 2005.

[19] See McCarthy, Kieren, .*nxt*, "Global Internet Governance Fight Looms," September 22, 2011, available at http://news.dot-nxt.com/2011/09/22/internet-governance-fight-looms.

[20] ICANN, *Adopted Board Resolutions, Cartegena*, December 10, 2010, available at http://www.icann.org/en/minutes/ resolutions-10dec10-en.htm#4.

[21] Letter from Chair, Governmental Advisory Committee to ICANN Chairman of the Board, March 16, 2011, available at https://gacweb.icann.org/download/attachments/1540116/20110316+GAC+Advice+on+.xxx.pdf?version=2& modificationDate=1312469527000.

[22] Letter from ICANN to Chair of GAC, February 10, 2011, available at http://icann.org/en/correspondence/jeffrey-toto-dryden-10feb11-en.pdf.

[23] ICANN must receive formal approval from NTIA for any additions of new gTLDs to the DNS. See Kevin Murphy, "US upset with ICANN over .xxx," *Domain Incite*, March 20, 2011, available at http://domainincite.com/us-upsetwith-icann-over-xxx/. India and Saudi Arabia have stated their intention to block the .xxx domain. See "xxx addresses open for business," *The Times of India*, April 19, 2011, available at http://articles.timesofindia.indiatimes.com/2011-04-19/computing/ 29446429_1_icann-suffix-websites.

[24] Kevin Murphy, "Europe asked US to delay .xxx," Domain Incite, May 5, 2011, available at http://domainincite.com/ europe-did-ask-the-us-to-delay-xxx/.

[25] Letter from Lawrence Strickling to Neelie Kroes, "Strickling letter to Kroes re: dot-xxx," *.nxt*, April 20, 2011, available at http://news.dot-nxt.com/2011/04/20/strickling-letter-kroes-xxx.

[26] Ibid.

[27] The Association of National Advertisers (ANA) has been a leading voice against ICANN's current rollout of the new gTLD program. See ANA webpage, "Say No to ICANN: Generic Top Level Domain Developments," available at http://www.ana.net/content/show/id/icann.

[28] See "Rockefeller Says Internet Domain Expansion Will Hurt Consumers, Businesses, and Non-Profits—Urges Delay," *Press Release*, Senate Committee on Commerce, Science and Transportation, December 28, 2011, available at http://commerce.senate.gov/public/index.cfm?p=PressReleases.

[29] Letter from Senator Amy Klobuchar and Senator Kelly Ayotte to ICANN, December 22, 2011, available at http://www.icann.org/en/correspondence/klobuchar-ayotte-to-beckstrom-crocker-22dec11-en.pdf.

[30] House Committee on Energy and Commerce, "Committee Urges ICANN to Delay Expansion of Generic Top-Level Domain Program," *Press Release*, December 21, 2011, available at http://energycommerce.house.gov/news/PRArticle.aspx?NewsID=9176.

[31] Ibid.

[32] Letter from Representative Goodlatte and Representative Berman to the Secretary of Commerce, December 16, 2011, available at http://www.icann.org/en/correspondence/goodlatte-berman-to-bryson-16dec11-en.pdf.

[33] Letter from FTC to ICANN, December 16, 2011, available at http://www.ftc.gov/os/closings/publicltrs /111216letterto-icann.pdf.

[34] Letter from the Chairmen and Ranking Members of the Senate and House Judiciary Committees to Rod Beckstrom, CEO, ICANN, December 27, 2011, available at http://www.icann.org/en/correspondence/leahy-to-beckstrom-27dec11- en.pdf.

[35] Testimony of Kurt Pritz, Senior Vice President, ICANN, before the House Committee on Energy and Commerce, Subcommittee on Communications and Technology, December 14, 2011, available at http://republicans.energycommerce.house.gov/Media/file/Hearings/Telecom/121411/Pritz.pdf. The gTLD expansion is also strongly supported by many in the Internet and domain name industry, see letter to Senator Rockefeller and Senator Hutchison at http://news.dot-nxt.com/sites/news.dot-nxt.com/files/gtld-industry-to-congress-gtlds-8dec11.pdf.

[36] Testimony of Fiona M. Alexander, Associate Administrator, NTIA, before the House Committee on Energy and Commerce, Subcommittee on Communications and Technology, December 14, 2011, available at http://www.ntia.doc.gov/speechtestimony/2011/testimony-associate-administrator-alexander-icann-s-top-level-domainname-progr.

[37] Ibid.

[38] Letter from Lawrence Strickling, Assistant Secretary for Communications and Information, U.S. Department of Commerce, to ICANN, January 3, 2012, available at http://www.ntia.doc.gov/other-publication/2012/ntia-letterregarding-gtld-program.

[39] Ibid.

[40] A complete list of new gTLD applications is provided at http://newgtlds.icann.org/en/program-status/application-results/strings-1200utc-13jun12-en.

[4] ICANN, *New gTLD Applicant Guidebook*, June 4, 2012, Module 1, p. 1-21, available at http://newgtlds.icann.org/en/ applicants/agb.

[42] Ibid.

[43] Ibid., Module 3, p. 3-3.

[44] Organisation for Economic Co-operation and Development, OECD High Level Meeting, The Internet Economy: Generating Innovation and Growth, *Communique on Principles for Internet Policy-Making*, June 28-29, 2011, p. 4, available at http://www.oecd.org/dataoecd/33/12/48387430.pdf.

[45] G8 Declaration, Renewed Commitment for Freedom and Democracy, G8 Summit of Deauville, May 26-27, 2011, available at http://www.g20-g8.com/g8-g20/g8/english/live/news/renewed-commitment-for-freedom-and-democracy.1314.html.

[46] IBSA Multistakeholder meeting on Global Internet Governance, *Recommendations*, September 1-2, 2011 at Rio de Janeiro, Brazil, available at http://www.culturalivre.org.br/artigos/IBSA_recommendations_ Internet_ Governance.pdf.

[47] The CIRP proposal is available at http://igfwatch.org/discussion-board/indias-proposal-for-a-un-committee-for-internet-related-policies-cirp.

[48] United Nations General Assembly, Sixty-sixth session, Item 93 of the provisional agenda, Developments in the field of information and telecommunications in the context of international security, "Letter dated 12 September 2011 from the Permanent Representatives of China, the Russian Federation, Tajikistan, and Uzbekistan to the United Nations addressed to the Secretary-General," September 14, 2011, A/66/359, available at http://blog.internetgovernance.org/ pdf/UN-infosec-code.pdf.

[49] The European Commission has been a particularly strong voice in favor of significantly increasing GAC influence on the ICANN policy process. See Kieren McCarthy, "European Commission calls for greater government control over Internet," *.nxt*, August 31, 2011, available at http://news.dot-nxt.com/2011/08/31/ec-greater-government-control.

[50] Remarks by Lawrence Strickling, Assistant Secretary of Commerce for Communications and Information, National Telecommunications and Information Administration, Department of Commerce, before the PLI/FCBA Telecommunications Policy & Regulation Institute, Washington, DC, December 8, 2011, available at http://www.ntia.doc.gov/speechtestimony/2011/remarks-assistant-secretary-strickling-practising-law-institutes-29thannual-te.

[51] Available at http://energycommerce.house.gov/hearings/hearingdetail.aspx?NewsID=9543.

[52] See Article 3A , "Proposals for the Work of the Conference," available at http://files.wcitleaks.org/public/ Merged%20UAE%20081212.pdf.

[53] International Telecommunications Union, *Final Acts*, World Conference on International Telecommunications, Dubai, 2012, Resolution 3, p. 20, available at http://www.itu.int/en/wcit-12/Documents/final-acts-wcit-12.pdf.

[54] Statement delivered by Ambassador Terry Kramer from the floor of the WCIT, December 13, 2012. U.S. Department of State, *Press Release*, "U.S. Intervention at the World Conference on International Telecommunications," December 13, 2012, available at http://www.state.gov/r/pa/prs/ps/2012/12/202037.htm.

[55] The official ITU list of signatories and non-signatories is at http://www.itu.int/osg/wcit-12/highlights/ signatories.html.

[56] For a complete list, see the Appendix in CRS Report 97-868, *Internet Domain Names: Background and Policy Issues*, by Lennard G. Kruger.

[57] See for example: *Press Release of Senator Ron Wyden*, "Wyden Calls for Ethics Rules to Prevent Revolving Door for Internet Domain Name Regulators," September 14, 2011, available at http://wyden.senate.gov /newsroom/press/ release/?id=2e414e69-1250-4ca3-ae6b-2b6091ed52cc.

[58] See CRS Report R42112, *Online Copyright Infringement and Counterfeiting: Legislation in the 112th Congress*, by Brian T. Yeh.

[59] See for example: The White House, *International Strategy for Cyberspace: Prosperity, Security, and Openness in a Networked World*, May 2011, p. 21-22, available at http://www.whitehouse.gov/sites/default /files/rss_viewer/ international_strategy_for_cyberspace.pdf.

[60] ICANN, *2008 Annual Report*, December 31, 2008, p. 24, available at http://www.icann.org/en /annualreport/annualreport-2008-en.pdf.

[61] ICANN, FY13 Operating Plan and Budget, June 24, 2012, available at http://www.icann.org/en/about/financials.

[62] A list of ICANN-accredited registrars is available at http://www.icann.org/en/registries/agreements.htm.

[63] A list of current agreements between ICANN and registry operators is available at http://www.icann.org/en/ registries/agreements.htm.

[64] ICANN Financials Dashboard, updated June 15, 2011, available at https://charts.icann.org/public/index-finance-fy11.html.

[65] ICANN, FY13 Operating Plan and Budget, June 24, 2012, p. 61, available at http://www.icann.org/en/news/ announcements/announcement-13jul12-en.htm.

ISBN: 978-1-62948-413-6
© 2013 Nova Science Publishers, Inc.

Chapter 9

INTERNET DOMAIN NAMES: BACKGROUND AND POLICY ISSUES[*]

Lennard G. Kruger

SUMMARY

Navigating the Internet requires using addresses and corresponding names that identify the location of individual computers. The Domain Name System (DNS) is the distributed set of databases residing in computers around the world that contain address numbers mapped to corresponding domain names, making it possible to send and receive messages and to access information from computers anywhere on the Internet. Many of the technical, operational, and management decisions regarding the DNS can have significant impacts on Internet-related policy issues such as intellectual property, privacy, Internet freedom, e-commerce, and cybersecurity.

The DNS is managed and operated by a not-for-profit public benefit corporation called the Internet Corporation for Assigned Names and Numbers (ICANN). Because the Internet evolved from a network infrastructure created by the Department of Defense, the U.S. government originally owned and operated (primarily through private contractors) the key components of network architecture that enable the domain name system to function. A 1998 Memorandum of Understanding (MOU) between ICANN and the Department of Commerce (DOC) initiated a process intended to transition technical DNS coordination and management functions to a private-sector not-for-profit entity. While the DOC played no role in the internal governance or day-to-day operations of the DNS, ICANN remained accountable to the U.S. government through the MOU, which was superseded in 2006 by a Joint Project Agreement (JPA). On September 30, 2009, the JPA between ICANN and DOC expired and was replaced by an Affirmation of Commitments (AoC), which provides for review panels to periodically assess ICANN processes and activities.

Additionally, a contract between DOC and ICANN authorizes the Internet Assigned Numbers Authority (IANA) to perform various technical functions such as allocating IP address blocks, editing the root zone file, and coordinating the assignment of unique

---

[*] This is an edited, reformatted and augmented version of Congressional Research Service, Publication No. 97-868, dated January 3, 2013.

protocol numbers. With the current contract due to expire on September 30, 2012, NTIA announced on July 2, 2012, the award of the new IANA contract to ICANN for up to seven years.

With the expiration of the ICANN-DOC Joint Project Agreement on September 30, 2009, the announcement of the new AoC, the renewal of the IANA contract, and the rollout of the new generic top level domain (gTLD) program, the 113th Congress and the Administration are likely to continue assessing the appropriate federal role with respect to ICANN and the DNS, and examine to what extent ICANN is positioned to ensure Internet stability and security, competition, private and bottom-up policymaking and coordination, and fair representation of the global Internet community. Controversies over the new gTLDs and the addition of the .xxx domain have led some governments to criticize the ICANN policymaking process and to suggest various ways to increase governmental influence over that process. How these and other issues are ultimately addressed and resolved could have profound impacts on the continuing evolution of ICANN, the DNS, and the Internet.

BACKGROUND AND HISTORY

The Internet is often described as a "network of networks" because it is not a single physical entity but, in fact, hundreds of thousands of interconnected networks linking hundreds of millions of computers around the world. Computers connected to the Internet are identified by a unique Internet Protocol (IP) number that designates their specific location, thereby making it possible to send and receive messages and to access information from computers anywhere on the Internet. Domain names were created to provide users with a simple location name, rather than requiring them to use a long list of numbers. For example, the IP number for the location of the THOMAS legislative system at the Library of Congress is 140.147.248.9; the corresponding domain name is thomas.loc.gov. Top Level Domains (TLDs) appear at the end of an address and are either a given country code, such as .jp or .uk, or are generic designations (gTLDs), such as .com, .org, .net, .edu, or .gov. The Domain Name System (DNS) is the distributed set of databases residing in computers around the world that contain the address numbers, mapped to corresponding domain names. Those computers, called root servers, must be coordinated to ensure connectivity across the Internet.

The Internet originated with research funding provided by the Department of Defense Advanced Research Projects Agency (DARPA) to establish a military network. As its use expanded, a civilian segment evolved with support from the National Science Foundation (NSF) and other science agencies. While there were (and are) no formal statutory authorities or international agreements governing the management and operation of the Internet and the DNS, several entities played key roles in the DNS. For example, the Internet Assigned Numbers Authority (IANA), which was operated at the Information Sciences Institute/University of Southern California under contract with the Department of Defense, made technical decisions concerning root servers, determined qualifications for applicants to manage country code TLDs, assigned unique protocol parameters, and managed the IP address space, including delegating blocks of addresses to registries around the world to assign to users in their geographic area.

NSF was responsible for registration of nonmilitary domain names, and in 1992 put out a solicitation for managing network services, including domain name registration. In 1993,

NSF signed a five-year cooperative agreement with a consortium of companies called InterNic. Under this agreement, Network Solutions Inc. (NSI), a Herndon, VA, engineering and management consulting firm, became the sole Internet domain name registration service for registering the .com, .net., and .org. gTLDs.

After the imposition of registration fees in 1995, criticism of NSI's sole control over registration of the gTLDs grew. In addition, there was an increase in trademark disputes arising out of the enormous growth of registrations in the .com domain. There also was concern that the role played by IANA lacked a legal foundation and required more permanence to ensure the stability of the Internet and the domain name system. These concerns prompted actions both in the United States and internationally.

An International Ad Hoc Committee (IAHC), a coalition of individuals representing various constituencies, released a proposal for the administration and management of gTLDs on February 4, 1997. The proposal recommended that seven new gTLDs be created and that additional registrars be selected to compete with each other in the granting of registration services for all new second level domain names. To assess whether the IAHC proposal should be supported by the U.S. government, the executive branch created an interagency group to address the domain name issue and assigned lead responsibility to the National Telecommunications and Information Administration (NTIA) of the Department of Commerce (DOC). On June 5, 1998, DOC issued a final statement of policy, "Management of Internet Names and Addresses." Called the White Paper, the statement indicated that the U.S. government was prepared to recognize and enter into agreement with "a new not-for-profit corporation formed by private sector Internet stakeholders to administer policy for the Internet name and address system."[1] In deciding upon an entity with which to enter such an agreement, the U.S. government would assess whether the new system ensured stability, competition, private and bottom-up coordination, and fair representation of the Internet community as a whole.

The White Paper endorsed a process whereby the divergent interests of the Internet community would come together and decide how Internet names and addresses would be managed and administered. Accordingly, Internet constituencies from around the world held a series of meetings during the summer of 1998 to discuss how the New Corporation might be constituted and structured. Meanwhile, IANA, in collaboration with NSI, released a proposed set of bylaws and articles of incorporation. The proposed new corporation was called the Internet Corporation for Assigned Names and Numbers (ICANN). After five iterations, the final version of ICANN's bylaws and articles of incorporation were submitted to the Department of Commerce on October 2, 1998. On November 25, 1998, DOC and ICANN signed an official Memorandum of Understanding (MOU), whereby DOC and ICANN agreed to jointly design, develop, and test the mechanisms, methods, and procedures necessary to transition management responsibility for DNS functions—including IANA—to a private-sector not-for-profit entity.

On September 17, 2003, ICANN and the Department of Commerce agreed to extend their MOU until September 30, 2006. The MOU specified transition tasks which ICANN agreed to address. On June 30, 2005, Michael Gallagher, then-Assistant Secretary of Commerce for Communications and Information and Administrator of NTIA, stated the U.S. government's principles on the Internet's domain name system. Specifically, NTIA stated that the U.S. government intends to preserve the security and stability of the DNS, that the United States would continue to authorize changes or modifications to the root zone, that governments have legitimate interests in the management of their country code top level domains, that ICANN is the appropriate technical manager of the DNS, and that dialogue related to Internet governance should continue in relevant multiple fora.[2]

On September 29, 2006, DOC announced a new Joint Project Agreement (JPA) with ICANN which was intended to continue the transition to the private sector of the coordination of technical functions relating to management of the DNS. The JPA extended through September 30, 2009, and focused on institutionalizing transparency and accountability mechanisms within ICANN. On September 30, 2009, DOC and ICANN announced agreement on an Affirmation of Commitments (AoC) to "institutionalize and memorialize" the technical coordination of the DNS globally and by a private-sector-led organization.[3] The AoC affirms commitments made by DOC and ICANN to ensure accountability and transparency; preserve the security, stability, and resiliency of the DNS; promote competition, consumer trust, and consumer choice; and promote international participation.

ICANN BASICS

ICANN is a not-for-profit public benefit corporation headquartered in Marina del Rey, CA, and incorporated under the laws of the state of California. ICANN is organized under the California Nonprofit Public Benefit Law for charitable and public purposes, and as such, is subject to legal oversight by the California attorney general. ICANN has been granted tax-exempt status by the federal government and the state of California.[4]

ICANN's organizational structure consists of a Board of Directors (BOD) advised by a network of supporting organizations and advisory committees that represent various Internet constituencies and interests (see **Figure 1**). Policies are developed and issues are researched by these subgroups, who in turn advise the Board of Directors, which is responsible for making all final policy and operational decisions. The Board of Directors consists of 15 international and geographically diverse members, composed of one president, eight members selected by a Nominating Committee, two selected by the Generic Names Supporting Organization, two selected by the Address Supporting Organization, and two selected by the Country-Code Names Supporting Organization. Additionally, there are six non-voting liaisons representing other advisory committees.

The explosive growth of the Internet and domain name registration, along with increasing responsibilities in managing and operating the DNS, has led to marked growth of the ICANN budget, from revenues of about $6 million and a staff of 14 in 2000, to revenues of $90 million and a staff of 149 forecasted for 2012.[5] ICANN is funded primarily through fees paid to ICANN by registrars and registry operators. Registrars are companies (e.g., GoDaddy, Google, Network Solutions) with which consumers register domain names.[6] Registry operators are companies and organizations who operate and administer the master database of all domain names registered in each top level domain (for example VeriSign, Inc. operates .com and .net, Public Interest Registry operates .org, and Neustar, Inc. operates .biz).[7] In 2011, ICANN received 94% of its total revenues from registry and registrar fees (49% from registry fees, 45% from registrar fees).[8]

The collection of fees from the new generic top level domain (gTLD) program could contribute to an unprecedented level of revenue for ICANN in the years to come. At the 44th

Board Meeting in Prague on June 23, 2012, the ICANN Board adopted a 2013 budget and operating plan. The plan splits the budget into two separate pots—one for the new gTLD program, the other for all other ICANN operations and activities. For the first round of the new gTLD program, ICANN estimates revenues of $337 million from the new gTLD application fees, which is twice the amount of traditional revenues from all other sources over the next two years. After operating expenses (processing and evaluating the applications), ICANN estimates a surplus of $27.8 million from the new gTLD program.[9]

ISSUES IN THE 113TH CONGRESS

Congressional committees (primarily the Senate Committee on Commerce, Science and Transportation and the House Committee on Energy and Commerce) maintain oversight on how the Department of Commerce manages and oversees ICANN's activities and policies. Other committees, such as the House and Senate Judiciary Committees, maintain an interest in other issues affected by ICANN, such as intellectual property and privacy. The **Appendix** shows a listing of congressional committee hearings on ICANN and the domain name system dating back to 1997.

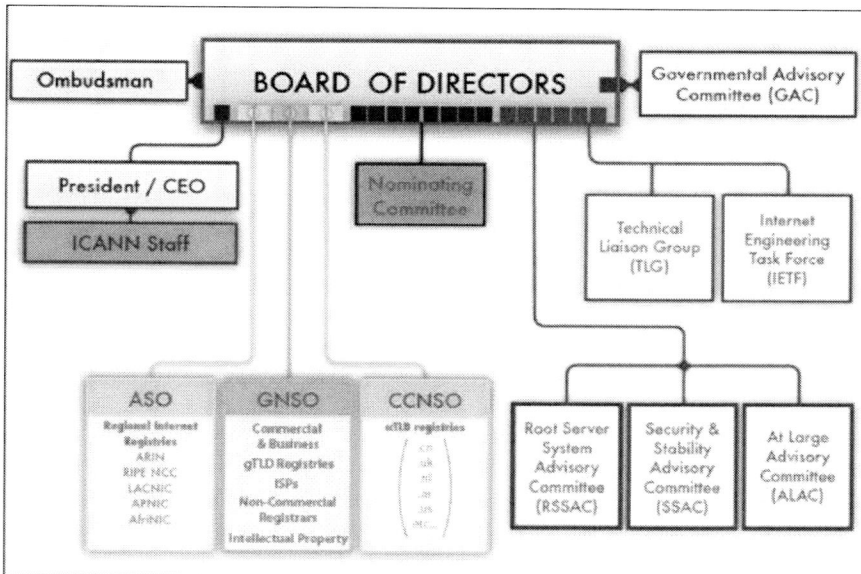

Source: ICANN (http://www.icann.org/en/structure/).

Figure 1. Organizational Structure of ICANN.

ICANN's Relationship with the U.S. Government

The Department of Commerce (DOC) has no statutory authority over ICANN or the DNS. However, because the Internet evolved from a network infrastructure created by the Department of Defense, the U.S. government originally owned and operated (primarily through private contractors such as the University of Southern California, SRI International, and Network Solutions Inc.) the key components of network architecture that enable the domain name system to function. The 1998 Memorandum of Understanding between ICANN and the Department of Commerce initiated a process intended to transition technical DNS coordination and management functions to a private-sector not-for-profit entity. While the DOC plays no role in the internal governance or day-to-day operations of ICANN, the U.S. government, through the DOC, retains a role with respect to the DNS via three separate contractual agreements. These are

- the Affirmation of Commitments (AoC) between DOC and ICANN, which was signed on September 30, 2009;
- the contract between IANA/ICANN and DOC to perform various technical functions such as allocating IP address blocks, editing the root zone file, and coordinating the assignment of unique protocol numbers; and
- the cooperative agreement between DOC and VeriSign to manage and maintain the official DNS root zone file.

Affirmation of Commitments

On September 30, 2009, DOC and ICANN announced agreement on an Affirmation of Commitments (AoC) to "institutionalize and memorialize" the technical coordination of the DNS globally and by a private-sector-led organization.[10] The AoC succeeds the concluded Joint Project Agreement (which in turn succeeded the Memorandum of Understanding between DOC and ICANN). The AoC has no expiration date and would conclude only if one of the two parties decided to terminate the agreement.

Buildup to the AoC

Various Internet stakeholders disagreed as to whether DOC should maintain control over ICANN after the impending JPA expiration on September 30, 2009. Many U.S. industry and public interest groups argued that ICANN was not yet sufficiently transparent and accountable, that U.S. government oversight and authority (e.g., DOC acting as a "steward" or "backstop" to ICANN) was necessary to prevent undue control of the DNS by international or foreign governmental bodies, and that continued DOC oversight was needed until full privatization is warranted. On the other hand, many international entities and groups from countries outside the United States argued that ICANN had sufficiently met conditions for privatization, and that continued U.S. government control over an international organization was not appropriate. In the 110th Congress, Senator Snowe introduced S.Res. 564, which stated the sense of the Senate that although ICANN had made progress in achieving the goals of accountability and transparency as directed by the JPA, more progress was needed.[11]

On April 24, 2009, NTIA issued a Notice of Inquiry (NOI) seeking public comment on the upcoming expiration of the JPA between DOC and ICANN.[12] According to NTIA, a mid-term review showed that while some progress had been made, there remained key areas where further work was required to increase institutional confidence in ICANN. These areas included long-term stability, accountability, responsiveness, continued private-sector leadership, stakeholder participation, increased contract compliance, and enhanced competition. NTIA asked for public comments regarding the progress of transition of the technical coordination and management of the DNS to the private sector, as well as the model of private-sector leadership and bottom-up policy development which ICANN represents. Specifically, the NOI asked whether sufficient progress had been achieved for the transition to take place by September 30, 2009, and if not, what should be done.

On June 4, 2009, the House Committee on Energy and Commerce, Subcommittee on Communications, Technology, and the Internet, held a hearing examining the expiration of the JPA and other issues. Most members of the committee expressed the view that the JPA (or a similar agreement between DOC and ICANN) should be extended. Subsequently, on August 4, 2009, majority leadership and majority Members of the House Committee on Energy and Commerce sent a letter to the Secretary of Commerce urging that rather than replacing the JPA with additional JPAs, the DOC and ICANN should agree on a "permanent instrument" to "ensure that ICANN remains perpetually accountable to the public and to all of its global stakeholders." According to the committee letter, the instrument should ensure the permanent continuance of the present DOC-ICANN relationship; provide for periodic reviews of ICANN performance; outline steps ICANN will take to maintain and improve its accountability; create a mechanism for implementation of the addition of new gTLDs and internationalized domain names; ensure that ICANN will adopt measures to maintain timely and public access to accurate and complete WHOIS[13] information; and include commitments that ICANN will remain a not-for-profit corporation headquartered in the United States.

Critical Elements of the AoC

Under the AoC, ICANN commits to remain a not-for-profit corporation "headquartered in the United States of America with offices around the world to meet the needs of a global community." According to the AoC, "ICANN is a private organization and nothing in this Affirmation should be construed as control by any one entity."

Specifically, the AoC calls for the establishment of review panels which will periodically make recommendations to the ICANN Board in four areas:

- *Ensuring accountability, transparency and the interests of global Internet users*—the panel will evaluate ICANN governance and assess transparency, accountability, and responsiveness with respect to the public and the global Internet community. The panel will be composed of the chair of ICANN's Governmental Advisory Committee (GAC), the chair of the Board of ICANN, the Assistant Secretary for Communications and Information of the Department of Commerce (i.e., the head of NTIA), representatives of the relevant ICANN Advisory Committees and Supporting Organizations, and independent experts. Composition of the panel will be agreed to jointly by the chair of the GAC and the chair of ICANN.
- *Preserving security, stability, and resiliency*—the panel will review ICANN's plan to enhance the operational stability, reliability, resiliency, security, and global interoperability of the DNS. The panel will be composed of the chair of the GAC, the CEO of ICANN, representatives of the relevant Advisory Committees and Supporting Organizations, and independent experts. Composition of the panel will be agreed to jointly by the chair of the GAC and the CEO of ICANN.
- *Impact of new gTLDs*—starting one year after the introduction of new gTLDs, the panel will periodically examine the extent to which the introduction or expansion of gTLDs promotes competition, consumer trust, and consumer choice. The panel will be composed of the chair of the GAC, the CEO of ICANN, representatives of the

relevant Advisory Committees and Supporting Organizations, and independent experts. Composition of the panel will be agreed to jointly by the chair of the GAC and the CEO of ICANN.

- *WHOIS policy*—the panel will review existing WHOIS policy and assess the extent to which that policy is effective and its implementation meets the legitimate needs of law enforcement and promotes consumer trust. The panel will be composed of the chair of the GAC, the CEO of ICANN, representatives of the relevant Advisory Committees and Supporting Organizations, independent experts, representatives of the global law enforcement community, and global privacy experts. Composition of the panel will be agreed to jointly by the chair of the GAC and the CEO of ICANN.

On December 31, 2010, the Accountability and Transparency Review Team (ATRT) released its recommendations to the Board for improving ICANN's transparency and accountability with respect to: Board governance and performance, the role and effectiveness of the GAC and its interaction with the Board, public input and policy development processes, and review mechanisms for Board decisions.[14] At the June 2011 meeting in Singapore, the Board adopted all 27 ATRT recommendations. According to NTIA, "the focus turns to ICANN management and staff, who must take up the challenge of implementing these recommendations as rapidly as possible and in a manner that leads to meaningful and lasting reform."[15]

DOC Contracts: IANA and VeriSign

A contract between DOC and ICANN authorizes the Internet Assigned Numbers Authority (IANA) to perform various technical functions such as allocating IP address blocks, editing the root zone file, and coordinating the assignment of unique protocol numbers. Additionally, a cooperative agreement between DOC and VeriSign (operator of the .com and .net registries) authorizes VeriSign to manage and maintain the official root zone file that is contained in the Internet's root servers that underlie the functioning of the DNS.[16]

By virtue of these legal agreements, the DOC has policy authority over the root zone file,[17] meaning that the U.S. government can approve or deny changes or modifications made to the root zone file (changes, for example, such as adding a new top level domain). The June 30, 2005, U.S. government principles on the Internet's domain name system stated the intention to "preserve the security and stability" of the DNS, and asserted that "the United States is committed to taking no action that would have the potential to adversely impact the effective and efficient operation of the DNS and will therefore maintain its historic role in authorizing changes or modifications to the authoritative root zone file."[18]

The JPA was separate and distinct from the DOC legal agreements with ICANN and VeriSign. As such, the expiration of the JPA and the establishment of the AoC did not directly affect U.S. government authority over the DNS root zone file. While ICANN has not advocated ending U.S. government authority over the root zone file, foreign governmental bodies have argued that it is inappropriate for the U.S. government to maintain exclusive authority over the DNS.

Debate was ongoing regarding negotiations over the renewal of the IANA contract between DOC and ICANN, which was due to expire on September 30, 2012. On February 25, 2011, NTIA issued a Notice of Inquiry seeking public comment on the upcoming award of a new IANA functions contract. Specific questions included whether the various IANA functions should continue to be administered by a single entity, whether changes should be made to how root zone management requests for ccTLDs are processed, and whether the contract should explicitly make reference to other entities within the Internet technical community.[19]

On June 14, 2011, NTIA released a Further Notice of Inquiry (FNOI) in which a draft Statement of Work (SOW) detailing work requirements for the IANA contract was offered for public comment.[20] Under the draft SOW, NTIA states that the separate IANA functions should continue to be operated by a single entity. The SOW would also require that requests to IANA for new gTLDs be accompanied by documentation demonstrating how the proposed new gTLD "reflects consensus among relevant stakeholders and is supportive of the global public interest."[21] On July 22, 2011, ICANN submitted comments to NTIA on the FNOI, expressing strong opposition to the proposal that requests to IANA for new gTLDs be accompanied by documentation demonstrating global public support and consensus. According to ICANN, such a step would undermine ICANN's multistakeholder model by revising the gTLD implementation and policy processes already adopted through the bottom-up decision-making process.[22]

NTIA's final contract solicitation, released as a Request for Proposal (RFP) on November 10, 2011, lessened the IANA contractor requirements for adding new gTLDs, stating that when adding new gTLDs to the root zone, the contractor must provide "specific documentation demonstrating how the process provided the opportunity for input from relevant stakeholders and was supportive of the global public interest."[23] The IANA contract solicitation specified that the contractor must be a wholly U.S. owned and operated firm or a U.S. university or college; that all primary operations and systems shall remain within the United States; and that the U.S. government reserves the right to inspect the premises, systems, and processes of all facilities and components used for the performance of the contract.

On July 2, 2012, NTIA announced the award of the new IANA contract to ICANN for up to seven years (through September 2019). The new contract includes a separation between the policy development of IANA services and the implementation by the IANA functions contractor. The contract also features a "a robust company-wide conflict of interest policy; a heightened respect for local national law; and a series of consultation and reporting requirements to increase transparency and accountability."[24]

ICANN and the International Community

Because cyberspace and the Internet transcend national boundaries, and because the successful functioning of the DNS relies on participating entities worldwide, ICANN is by definition an international organization. Both the ICANN Board of Directors and the various constituency groups who influence and shape ICANN policy decisions are composed of members from all over the world. Additionally, ICANN's Governmental Advisory Committee (GAC), which is composed of government representatives of nations worldwide, provides advice to the ICANN Board on public policy matters and issues of government concern. Although the ICANN Board is required to consider GAC advice and recommendations, it is not obligated to follow those recommendations.

Many in the international community, including foreign governments, have argued that it is inappropriate for the U.S. government to maintain its legacy authority over ICANN and the DNS, and have suggested that management of the DNS should be accountable to a higher intergovernmental body. The United Nations, at the December 2003 World Summit on the Information Society (WSIS), debated and agreed to study the issue of how to achieve greater international involvement in the governance of the Internet and the domain name system in particular. The study was conducted by the U.N.'s Working Group on Internet Governance (WGIG). On July 14, 2005, the WGIG released its report, stating that no single government should have a preeminent role in relation to international Internet governance. The report called for further internationalization of Internet governance, and proposed the creation of a new global forum for Internet stakeholders. Four possible models were put forth, including two involving the creation of new Internet governance bodies linked to the U.N. Under three of the four models, ICANN would either be supplanted or made accountable to a higher intergovernmental body. The report's conclusions were scheduled to be considered during the second phase of the WSIS held in Tunis in November 2005. U.S. officials stated their opposition to transferring control and administration of the domain name system from ICANN to any international body. Similarly, the 109th Congress expressed its support for maintaining U.S. control over ICANN (H.Con.Res. 268 and S.Res. 323).[25]

The European Union (EU) initially supported the U.S. position. However, during September 2005 preparatory meetings, the EU seemingly shifted its support towards an approach which favored an enhanced international role in governing the Internet. Conflict at the WSIS Tunis Summit over control of the domain name system was averted by the announcement, on November 15, 2005, of an Internet governance agreement between the United States, the EU, and over 100 other nations. Under this agreement, ICANN and the United States maintained their roles with respect to the domain name system. A new international group under the auspices of the U.N. was formed—the Internet Governance Forum (IGF)—which provides an ongoing forum for all stakeholders (both governments and nongovernmental groups) to discuss and debate Internet policy issues. The IGF does not have binding authority and was slated to run through 2010. In December 2010, the U.N. General Assembly renewed the IGF for another five years and tasked the U.N.'s Commission on Science and Technology for Development (CSTD) to develop a report and recommendations on how the IGF might be improved. A Working Group on Improvements to the Internet Governance Forum was formed, which includes 22 governments (including the United States) and the participation of Internet stakeholder groups.

Starting in 2010 and 2011, controversies surrounding the roll-out of new generic top level domains (gTLDs) and the addition of the .xxx TLD led some governments to argue for increased government influence on the ICANN policy development process.[26] Governments such as the United States, Canada, and the European Union, while favoring the current ICANN multistakeholder model of DNS governance, have advocated an enhanced role for the Governmental Advisory Committee (GAC) on ICANN policy decisions. Other nations—such as Brazil, South Africa, and India (referred to as IBSA)—favored the creation of an Internet policy development entity within the U.N. system, whose purview would include integrating and overseeing existing bodies (such as ICANN) that are responsible for the technical and operational functioning of the Internet. A third group of nations, including Russia and China, proposed a voluntary "International Code of Conduct for Information Security," for further discussion in the General Assembly of the U.N. The Code included language that promotes the establishment of a multilateral, transparent, and democratic international management of the Internet.

World Conference on International Telecommunications (WCIT)

The World Conference on International Telecommunications (WCIT) was held in Dubai on December 3-14, 2012. Convened by the International Telecommunications Union (the ITU, an agency within the United Nations), the WCIT was a formal meeting of the world's national governments held in order to revise the International Telecommunications Regulations (ITRs). The ITRs, previously revised in 1988, serve as a global treaty outlining the principles which govern the way international telecommunications traffic is handled.

Because the existing 24-year-old ITRs predated the Internet, one of the key policy questions in the WCIT was how and to what extent the updated ITRs should address Internet traffic and Internet governance. The Administration and Congress took the position that the new ITRs should continue to address only traditional international telecommunications traffic, that a multistakeholder model of Internet governance (such as ICANN) should continue, and that the ITU should not take any action that could extend its jurisdiction or authority over the Internet.

As the WCIT approached, concerns heightened in the 112th Congress that the WCIT might potentially provide a forum leading to an increased level of intergovernmental control over the Internet. On May 31, 2012, the House Committee on Energy and Commerce, Subcommittee on Communications and Technology, held a hearing entitled, "International Proposals to Regulate the Internet." To accompany the hearing, H.Con.Res. 127 was introduced by Representative Bono Mack expressing the sense of Congress regarding actions to preserve and advance the multistakeholder governance model. Specifically, H.Con.Res. 127 expressed the sense of Congress that the Administration "should continue working to implement the position of the United States on Internet governance that clearly articulates the consistent and unequivocal policy of the United States to promote a global Internet free from government control and preserve and advance the successful multistakeholder model that governs the Internet today." H.Con.Res. 127 was passed unanimously by the House (414-0) on August 2, 2012.

A similar resolution, S.Con.Res. 50, was introduced into the Senate by Senator Rubio on June 27, 2012, and referred to the Committee on Foreign Relations. The Senate resolution expressed the sense of Congress "that the Secretary of State, in consultation with the Secretary of Commerce, should continue working to implement the position of the United States on Internet governance that clearly articulates the consistent and unequivocal policy of the United States to promote a global Internet free from government control and preserve and advance the successful multistakeholder model that governs the Internet today." S.Con.Res. 50 was passed by the Senate by unanimous consent on September 22, 2012. On December 5, 2012—shortly after the WCIT had begun in Dubai—the House unanimously passed S.Con.Res. 50 by a vote of 397-0.

During the WCIT, a revision to the ITRs was proposed and supported by Russia, China, Saudi Arabia, Algeria, and Sudan that sought to explicitly extend ITR jurisdiction over Internet traffic, infrastructure, and governance. Specifically, the proposal stated that "Member States shall have the sovereign right to establish and implement public policy, including international policy, on matters of Internet governance." The proposal also included an article establishing the right of Member States to manage Internet numbering, naming, addressing, and identification resources.

The proposal was subsequently withdrawn. However, as an intended compromise, the ITU adopted a nonbinding resolution (Resolution 3, attached to the final ITR text) entitled, "To Foster an enabling environment for the greater growth of the Internet." Resolution 3 includes language stating "all governments should have an equal role and responsibility for international Internet governance" and invites Member States to "elaborate on their respective positions on international Internet-related technical, development and public policy issues within the mandate of ITU at various ITU forums.... "

Because of the inclusion of Resolution 3, along with other features of the final ITR text (such as new ITR articles related to spam and cybersecurity), the United States declined to sign the treaty. While the WCIT in Dubai is concluded, the international debate over Internet governance is expected to continue in future intergovernmental telecommunications meetings and conferences. The 113th Congress will likely monitor this ongoing debate and oversee the U.S. government's efforts to oppose any future proposals for intergovernmental control over the Internet and the domain name system.

Adding New Generic Top Level Domains (gTLDs)

Top Level Domains (TLDs) are the suffixes that appear at the end of an address (after the "dot"). TLDs can be either a country code such as .us, .uk, or .jp, or a generic TLD (gTLD) such as .com, .org, or .gov. Prior to ICANN's establishment, there were eight gTLDs (.com, .org, .net, .gov, .mil, .edu, .int, and .arpa). In 2000 and 2004, ICANN held application rounds for a limited number of new gTLDs; there are currently 22 gTLDs in operation. Some are reserved or restricted to particular types of organizations (e.g., .museum, .gov, .travel) and others are open for registration by anyone (.com, .org, .info).[27] Applicants for new gTLDs are typically commercial and non-profit organizations who seek to become ICANN-recognized registries that will establish and operate name servers for their TLD registry, as well as implement a domain name registration process for that particular TLD.

With the growth of the Internet and the accompanying growth in demand for domain names, debate focused on whether and how to further expand the number of gTLDs. Beginning in 2005, ICANN embarked on a long consultative process to develop rules and procedures for introducing and adopting an indefinite number of new gTLDs into the domain name system. A new gTLD can be any word or string of characters that is applied for and approved by ICANN. Between 2008 and 2011, ICANN released seven iterations of its gTLD Applicant Guidebook (essentially the rulebook for how the new gTLD program will be implemented).

On June 20, 2011, the ICANN Board of Directors voted to approve the launch of the new gTLD program, under which potentially hundreds of new gTLDs could ultimately be approved by ICANN and introduced into the DNS. Applications for new gTLDs were to be accepted from January 12 through April 12, 2012, and an application or evaluation fee of $185,000 is required.[28]

ICANN's approval of the new gTLD program has been controversial, with many trademark holders pointing to possible higher costs and greater difficulties in protecting their trademarks across hundreds of new gTLDs. Similarly, governments expressed concern over intellectual property protections, and, along with law enforcement entities, also cited concerns over the added burden of combating various cybercrimes (such as phishing and identity theft) across hundreds of new gTLDs. Throughout ICANN's policy development process, governments, through the Governmental Advisory Committee, advocated for additional intellectual property protections in the new gTLD process. The GAC also argued for more stringent rules that would allow for better law enforcement in the new domain space to better protect consumers. While changes were made, strong opposition from many trademark holders[29] led to opposition from some parts of the U.S. government towards the end of 2011, including the Senate Committee on Commerce, Science and Transportation,[30] the House Committee on Energy and Commerce,[31] the House Judiciary Committee,[32] and the Federal Trade Commission.[33]

At December 2011 House and Senate hearings, ICANN stated its intention to proceed with the gTLD expansion as planned. ICANN defended its gTLD program, arguing that the new gTLDs will offer more protections for consumers and trademark holders than current gTLDs; that new gTLDs will provide needed competition, choice, and innovation to the domain name system; and that critics have already had ample opportunity to contribute input during a seven-year deliberative policy development process.[34] Ultimately, ICANN did not delay the initiation of the new gTLD program, and the application window was opened on January 12, 2012.

On June 13, 2012, ICANN announced it had received 1,930 applications for new gTLDs,[35] including 66 geographic name applications and 116 Internationalized Domain Names (IDNs) in scripts such as Chinese, Arabic, and Cyrillic.[36] With the applications received, ICANN moved into the evaluation phase. ICANN will decide whether or not to accept each of the 1,930 new gTLD applications. The process is multi-tiered and complex. Depending on whether an extended evaluation is required, whether there are objections filed requiring dispute resolution, and whether there is string contention (where one or more qualified applicants are applying for the same gTLD), it could take anywhere from 9 to 20 months (from the time the application window closed on May 30) for a new gTLD to be approved and delegated into the domain name system (DNS). All of the rules, procedures, and policies related to the evaluation of the new gTLDs are provided in ICANN's *gTLD Applicant Guidebook, Version 2012-06-04*.[37]

With the first round application period concluded, there remain significant issues in play as the new gTLD program goes forward. First, ICANN has stated that a second and subsequent round will take place, and that changes to the application and evaluation process will be made such that a "systemized manner of applying for gTLDs be developed in the long term."[38] ICANN's goal is to begin the second application round "within one year of the close of the application submission period for the initial round."[39] Thus, many observers are eager to see what changes may be made in the second round.

Second, when the new gTLDs go "live" sometime in 2013, many stakeholders are concerned that various forms of domain name abuse (e.g., trademark infringement, consumer fraud, malicious behavior, etc.) could manifest themselves within the hundreds of new gTLD domain spaces. Thus, the effectiveness of ICANN's approach to addressing such issues as intellectual property protection of second level domain names and mitigating unlawful behavior in the domain name space will be of interest as the new gTLD program goes forward.

.xxx and Protecting Children on the Internet

Domain names have been viewed by some policymakers as a tool that could be used to protect children from obscene or indecent material on the Internet. In the 107th Congress, legislation was enacted to create a "kids-friendly top level domain name" that would contain only age-appropriate content. The Dot Kids Implementation and Efficiency Act of 2002 was signed into law on December 4, 2002 (P.L. 107-317), and authorized NTIA to require the .us registry operator (currently NeuStar) to establish, operate, and maintain a second level domain within the .us TLD (kids.us) that is restricted to material suitable for minors.

An opposite approach—establishing an adult content top level domain name that could be filtered by parents—has also been considered. In past Congresses, two bills were introduced to require the Department of Commerce to compel ICANN to establish a mandatory top level domain name (such as .xxx) for material that is deemed "harmful to minors." The bills were S. 2426 (109th Congress), which was introduced by Senator Baucus, and S. 2137 (107th Congress), which was introduced by Senator Landrieu. Neither of those bills advanced beyond introduction.

Meanwhile, as part of its process to add new generic top-level domains (gTLDs), ICANN repeatedly considered (since 2000) whether to allow the establishment of a gTLD for adult content. On June 1, 2005, ICANN announced that it had entered into commercial and technical negotiations with a registry company (ICM Registry) to operate a new ".xxx" domain, which would be designated for use by adult websites. Registration by adult websites into the .xxx domain would be purely voluntary, and those sites would not be required to give up their existing (for the most part, .com) sites.

Announcement of a possible .xxx domain proved highly controversial. With the ICANN Board scheduled to consider final approval of the .xxx domain on August 16, 2005, the Department of Commerce sent a letter to ICANN requesting that adequate additional time be provided to allow ICANN to address the objections of individuals expressing concerns about the impact of pornography on families and children and opposing the creation of a new top level domain devoted to adult content. ICANN's Governmental Advisory Committee (GAC) also requested more time before the final decision. At the March 2006 Board meeting in New Zealand, the ICANN Board authorized ICANN staff to continue negotiations with ICM Registry to address concerns raised by the DOC and the GAC. However, on May 10, 2006, the Board voted 9-5 against accepting the proposed agreement, but did not rule out accepting a revised agreement. Subsequently, on January 5, 2007, ICANN published for public comment a proposed revised agreement with ICM Registry to establish a .xxx domain. However, on March 30, 2007, the ICANN Board voted 9-5 to deny the .xxx domain, citing its reluctance to possibly assume an ongoing management and oversight role with respect to Internet content.[40]

ICM Registry subsequently challenged ICANN's decision before an Independent Review Panel (IRP), claiming that ICANN's rejection of ICM's application for a .xxx gTLD was not consistent with ICANN's Articles of Incorporation and Bylaws. On February 19, 2010, the three-person Independent Review Panel (from the International Centre for Dispute Resolution) ruled primarily in favor of ICM Registry, finding that its application for the .xxx TLD had met the required criteria, and that the ICANN Board's reversal of its initial approval "was not consistent with the application of neutral, objective and fair documented policy."[41]

The IRP decision was not binding; it was the ICANN Board of Directors' decision to determine how to proceed and whether ICM's application to operate a .xxx TLD should ultimately be approved. At ICANN's March 2010 meeting in Nairobi, the Board voted to postpone any decision about the .xxx TLD, and directed ICANN's CEO and general counsel to write a report examining possible options.[42]

On June 25, 2010, at the ICANN meeting in Brussels, the Board voted to allow ICM's .xxx application to move forward. The Board approved next steps for the application, including expedited due diligence by ICANN staff, negotiations between ICANN and ICM on a draft registry agreement, and consultation with ICANN's Governmental Advisory Committee (GAC).

At the December ICANN meeting in Cartegena, Colombia, the ICANN Board passed a resolution stating that while "it intends to enter into a registry agreement with ICM Registry for the .xxx TLD," the Board will enter into a formal consultation with the Governmental Advisory Committee on areas where the Board's decision is in conflict with GAC advice relating to the ICM application.[43]

A February 2011 letter from ICANN to the GAC acknowledged and responded to areas where approving the .xxx registry agreement with ICM would conflict with GAC advice received by ICANN.[44] With the GAC not offering approval of .xxx (and continuing to raise specific objections), the ICANN Board acknowledged that the Board and the GAC were not able to reach a mutually acceptable solution. Ultimately, on March 18, 2011, at the ICANN meeting in San Francisco, the ICANN Board approved a resolution giving the CEO or General Counsel of ICANN the authority to execute the registry agreement with ICM to establish a .xxx TLD. The vote was nine in favor, three opposed, and four abstentions. The .xxx top level domain became available to all registrants starting in December 2011.

ICANN and Cybersecurity

The security and stability of the Internet has always been a preeminent goal of DNS operation and management. One issue of recent concern is an intrinsic vulnerability in the DNS which allows malicious parties to distribute false DNS information. Under this scenario, Internet users could be unknowingly redirected to fraudulent and deceptive websites established to collect passwords and sensitive account information.

A technology called DNS Security Extensions (DNSSEC) has been developed to mitigate those vulnerabilities. DNSSEC assures the validity of transmitted DNS addresses by digitally "signing" DNS data via electronic signature. "Signing the root" (deploying DNSSEC on the root zone) is a necessary first and critical step towards protecting against malicious attacks on the DNS.[45] On October 9, 2009, NTIA issued a Notice of Inquiry (NOI) seeking public comment on the deployment of DNSSEC into the Internet's DNS infrastructure, including the authoritative root zone.[46] On June 3, 2009, NTIA and the National Institute of Standards and Technology (NIST) announced plans to work with ICANN and VeriSign to develop an interim approach for deploying DNSSEC in the root zone.[47] On June 9, 2010, NTIA filed a notice in the *Federal Register* seeking public comments on its testing and evaluation report and its intention to proceed with the final stages of domain name system security extensions implementation in the authoritative root zone.[48] On July 15, 2010, ICANN published the root zone trust anchor and root operators began to serve the signed root zone with actual keys, thereby making the signed root zone available. Ultimately, DNSSEC must be voluntarily adopted by registries, registrars, and the thousands of DNS server operators around the world in order to effectively deploy DNSSEC at all levels to maximize protection against fraudulent DNS redirection of Internet traffic.

Privacy and the WHOIS Database

Any person or entity who registers a domain name is required to provide contact information (phone number, address, email) which is entered into a public online database (the "WHOIS" database). The scope and accessibility of WHOIS database information has been an issue of contention. Privacy advocates have argued that access to such information should be limited, while many businesses, intellectual property interests, law enforcement agencies, and the U.S. government have argued that complete and accurate WHOIS information should continue to be publicly accessible. Over the past several years, ICANN has debated this issue through its Generic Names Supporting Organization (GNSO), which is developing policy recommendations on what data should be publicly available through the WHOIS database. On April 12, 2006, the GNSO approved an official "working definition" for the purpose of the public display of WHOIS information. The GNSO supported a narrow technical definition favored by privacy advocates, registries, registrars, and non-commercial user constituencies, rather than a more expansive definition favored by intellectual property interests, business constituencies, Internet service providers, law enforcement agencies, and the Department of Commerce (through its participation in ICANN's Governmental Advisory Committee). At ICANN's June 2006 meeting, opponents of limiting access to WHOIS data continued urging ICANN to reconsider the working definition. On October 31, 2007, the GNSO voted to defer a decision on WHOIS database privacy and recommended more studies. The GNSO also rejected a proposal to allow Internet users the option of listing third party contact information rather than their own private data. Currently, the GNSO is exploring several extensive studies of WHOIS.[49] On June 22, 2011, the ICANN announced the initiation of four separate studies of WHOIS, which were recommended by the Governmental Advisory Committee (GAC) in 2008. The studies will examine WHOIS "misuse," WHOIS registrant identification, WHOIS proxy and privacy "abuse," and the feasibility of a WHOIS proxy and privacy reveal study.

Meanwhile, a WHOIS policy review team, established by the Affirmation of Commitments, began its first review of WHOIS policy on October 1, 2010.[50] The team issued its final report on May 11, 2012. The report issued 16 recommendations for strengthening WHOIS, including those related to registrar compliance and improving WHOIS data accuracy and access.[51] On November 8, 2012, the ICANN Board approved a resolution directing the ICANN CEO to launch a new effort to redefine the purpose of collecting, maintaining, and providing access to gTLD registration data, and to consider safeguards for protecting that data.[52]

At present, there is no overall unified WHOIS policy. Rather WHOIS rules and requirements are primarily governed by the contractual agreements between ICANN and the hundreds of ICANNaccredited registrars. Currently, ICANN is in the process of negotiating an amended Registrar Accreditation Agreement (RAA), which is intended to address law enforcement agency recommendations on WHOIS verification, as well as other issues such as requiring registrars to maintain points of contact for reporting abuse, reseller obligations, heightened obligations relating to privacy/proxy service, and increased compliance mechanisms.[53] ICANN has released a draft RAA which reflects many of the law enforcement agency concerns with WHOIS. However, for these reforms to go into effect an agreement must be reached between ICANN and the registrars.[54]

Domain Names and Intellectual Property

Ever since the domain name system has been opened to commercial users, the ownership and registration of domain names has raised intellectual property concerns. The White Paper called upon the World Intellectual Property Organization (WIPO) to develop a set of recommendations for trademark/domain name dispute resolutions, and to submit those recommendations to ICANN. At ICANN's August 1999 meeting in Santiago, the board of directors adopted a dispute resolution policy to be applied uniformly by all ICANN-accredited registrars. Under this policy, registrars receiving complaints will take no action until receiving instructions from the domain-name holder or an order of a court or arbitrator. An exception is made for "abusive registrations" (i.e., cybersquatting and cyberpiracy), whereby a special administrative procedure (conducted largely online by a neutral panel, lasting 45 days or less, and costing about $1,000) will resolve the dispute. Implementation of ICANN's Domain Name Dispute Resolution Policy commenced on December 9, 1999. Meanwhile, the 106th Congress passed the Anticybersquatting Consumer Protection Act (incorporated into P.L. 106-113, the FY2000 Consolidated Appropriations Act). The act gives courts the authority to order the forfeiture, cancellation, and/or transfer of domain names registered in "bad faith" that are identical or similar to trademarks, and provides for statutory civil damages of at least $1,000, but not more than $100,000, per domain name identifier.

Currently, intellectual property is one of the key issues driving the debate over ICANN's addition of new generic top level domain names, with many trademark holders, industry groups, and governments arguing that a proliferation of new gTLDs could compromise intellectual property and increase the costs of protecting trademarks. Domain names have also recently been viewed as a possible way to address piracy of online content. In the 112th Congress, S. 968, the Protecting Real Online Threats to Economic Creativity and Theft of Intellectual Property Act (PROTECT IP), and H.R. 3261, the Stop Online Piracy Act (SOPA), were introduced to prohibit Internet service providers from directing Internet traffic to domain names with infringing content.[55]

CONCLUSION

Many of the technical, operational, and management decisions regarding the DNS can have significant impacts on Internet-related policy issues such as intellectual property, privacy, Internet freedom, e-commerce, and cybersecurity. As such, decisions made by ICANN affect Internet stakeholders around the world. In transferring management of the DNS to the private sector, the key policy question has always been how to best ensure achievement of the White Paper principles: Internet stability and security, competition, private and bottom-up policymaking and coordination, and fair representation of the global Internet community. What is the best process to ensure these goals, and how should various stakeholders—companies, institutions, individuals, governments—fit into this process?

Controversies over new gTLDs and .xxx have led some governments to criticize the ICANN policymaking process, and to suggest various ways to increase governmental influence over that process, whether it be an enhanced role for the GAC or a greater role for a U.N.-based or multilateral entity. With the increasing impact of the Internet on virtually all aspects of modern society, governments argue that they should have an enhanced role in developing Internet policies that will affect their citizens. On the other hand, defenders of the multistakeholder model argue that the phenomenal growth of the Internet has been and will continue to be fostered by a bottom-up, consensus approach, which serves to protect policy decisions from the political and bureaucratic control of national governments and international and multilateral institutions.

An ongoing factor in this debate is the performance of ICANN, which is seen by many as emblematic of the multistakeholder model for Internet governance. The U.S. government—through NTIA—maintains two separate instruments or agreements that provide a level of control or oversight over ICANN functions. One is the IANA contract, which was renewed through 2019. The other is the Affirmation of Commitments, which established a mechanism to review ICANN activities and policies regarding transparency and accountability, new gTLDs, DNS security and stability, and the WHOIS database. Evaluation of the progress ICANN makes in these areas could have an impact on whether the current multistakeholder model of DNS governance is maintained or altered.

The 113th Congress and the Administration are likely to continue monitoring the progress and status of ICANN under the Affirmation of Commitments. The rollout and evolution of the new gTLD program is also likely to be of particular interest. Ultimately, how these issues are addressed could have profound impacts on the continuing evolution of ICANN, the DNS, and Internet governance.

APPENDIX. CONGRESSIONAL HEARINGS ON THE DOMAIN NAME SYSTEM

Date	Congressional Committee	Topic
May 31, 2012	House Energy and Commerce	"International Proposals to Regulate the Internet"
December 14, 2011	House Energy and Commerce	"ICANN"'s Top-Level Domain Name Program"
December 8, 2011	Senate Commerce, Science and Transportation	"ICANN's Expansion of Top Level Domains"
May 4, 2011	House Judiciary	"ICANN Generic Top-Level Domains (gTLD) Oversight Hearing"
September 23, 2009	House Judiciary	"Expansion of Top Level Domains and its Effects on Competition"
June 4, 2009	House Energy and Commerce	"Oversight of the Internet Corporation for Assigned Names and Numbers (ICANN)"
September 21, 2006	House Energy and Commerce	"ICANN Internet Governance: Is It Working?"

Date	Congressional Committee	Topic
September 20, 2006	Senate Commerce, Science and Transportation	"Internet Governance: the Future of ICANN"
July 18, 2006	House Financial Services	"ICANN and the WHOIS Database: Providing Access to Protect Consumers from Phishing"
June 7, 2006	House Small Business	"Contracting the Internet: Does ICANN Create a Barrier to Small Business?"
September 30, 2004	Senate Commerce, Science and Transportation	"ICANN Oversight and Security of Internet Root Servers and the Domain Name System (DNS)"
May 6, 2004	House Energy and Commerce	"The 'Dot Kids' Internet Domain: Protecting Children Online"
July 31, 2003	Senate Commerce, Science and Transportation	"Internet Corporation for Assigned Names and Numbers (ICANN)"
September 4, 2003	House Judiciary	"Internet Domain Name Fraud – the U.S. Government's Role in Ensuring Public Access to Accurate WHOIS Data"
September 12, 2002	Senate Commerce, Science and Transportation	"Dot Kids Implementation and Efficiency Act of 2002"
June 12, 2002	Senate Commerce, Science and Transportation	"Hearing on ICANN Governance"
May 22, 2002	House Judiciary	"The Accuracy and Integrity of the WHOIS Database"
November 1, 2001	House Energy and Commerce	"Dot Kids Name Act of 2001"
July 12, 2001	House Judiciary	"The Whois Database: Privacy and Intellectual Property Issues"
March 22, 2001	House Judiciary	"ICANN, New gTLDs, and the Protection of Intellectual Property"
February 14, 2001	Senate Commerce, Science and Transportation	"Hearing on ICANN Governance"
February 8, 2001	House Energy and Commerce	"Is ICANN's New Generation of Internet Domain Name Selection Process Thwarting Competition?"
July 28, 1999	House Judiciary	"Internet Domain Names and Intellectual Property Rights"
July 22, 1999	Senate Judiciary	"Cybersquatting and Internet Consumer Protection"
July 22, 1999	House Energy and Commerce	"Domain Name System Privatization: Is ICANN Out of Control?"
October 7, 1998	House Science	"Transferring the Domain Name System to the Private Sector: Private Sector
		Implementation of the Administration's Internet 'White Paper'"
June 10, 1998	House Commerce	"Electronic Commerce: The Future of the Domain Name System"
March 31, 1998	House Science	"Domain Name System: Where Do We Go From Here?"
February 21, 1998	House Judiciary	"Internet Domain Name Trademark Protection"

(Continued)

Date	Congressional Committee	Topic
November 5, 1997	House Judiciary	"Internet Domain Name Trademark Protection"
September 30, 1997	House Science	"Domain Name System (Part 2)"
September 25, 1997	House Science	"Domain Name System (Part 1)"

End Notes

[1] Management of Internet Names and Addresses, National Telecommunications and Information Administration, Department of Commerce, *Federal Register*, Vol. 63, No. 111, June 10, 1998, 31741.

[2] See http://www.ntia.doc.gov/ntiahome/domainname/USDNSprinciples_06302005.pdf.

[3] Affirmation of Commitments by the United States Department of Commerce and the Internet Corporation for Assigned Names and Numbers, September 30, 2009, available at http://www.ntia.doc.gov/ntiahome/domainname/ Affirmation_of_Commitments_2009.pdf.

[4] ICANN, *2008 Annual Report*, December 31, 2008, p. 24, available at http://www.icann.org/en/annualreport/annualreport-2008-en.pdf.

[5] ICANN, FY2013 Operating Plan and Budget, June 24, 2012, available at http://www.icann.org/en/about/financials.

[6] A list of ICANN-accredited registrars is available at http://www.icann.org/en/registries/agreements.htm.

[7] A list of current agreements between ICANN and registry operators is available at http://www.icann.org/en/registries/ agreements.htm.

[8] ICANN Financials Dashboard, updated June 15, 2011, available at https://charts.icann.org/public/index-finance-fy11.html.

[9] ICANN, FY2013 Operating Plan and Budget, June 24, 2012, p. 61, available at http://www.icann.org/en/news/announcements/announcement-13jul12-en.htm.

[10] Affirmation of Commitments by the United States Department of Commerce and the Internet Corporation for Assigned Names and Numbers, September 30, 2009, available at http://www.ntia.doc.gov /ntiahome/domainname/Affirmation_of_Commitments_2009.pdf.

[11] In the 110th Congress, S.Res. 564 was referred to the Committee on Commerce, Science, and Transportation. It did not advance to the Senate floor.

[12] Department of Commerce, National Telecommunications and Information Administration, "Assessment of the Transition of the Technical Coordination and Management of the Internet's Domain Name and Addressing System," 74 *Federal Register* 18688, April 24, 2009.

[13] Any person or entity who registers a domain name is required to provide contact information (phone number, address, email) which is entered into a public online database (the "WHOIS" database).

[14] The ATRT final report is available at http://www.icann.org/en/reviews/affirmation/atrt-final-recommendations-31dec10-en.pdf.

[15] NTIA, *Press Release*, "NTIA Commends ICANN Board on Adopting the Recommendations of the Accountability and Transparency Review Team," June 24, 2011, available at http://www.ntia.doc.gov/press/2011/ NTIA_Statement_06242011.html.

[6] "The root zone file defines the DNS. For all practical purposes, a top level domain (and, therefore, all of its lower-level domains) is in the DNS if and only if it is listed in the root zone file. Therefore, presence in the root determines which DNS domains are available on the Internet." National Research Council, Committee on Internet Navigation and the Domain Name System: Technical Alternatives and Policy Implications, Signposts on Cyberspace: The Domain Name System and Internet Navigation, National Academy Press, Washington DC, 2005, p. 97.

[17] Milton Mueller, *Political Oversight of ICANN: A Briefing for the WSIS Summit*, Internet Governance Project, November 1, 2005, p. 4.

[1] See http://www.ntia.doc.gov/ntiahome/domainname/USDNSprinciples_06302005.pdf.

[19] Department of Commerce, National Telecommunications and Information Administration, "Request for Comments on the Internet Assigned Numbers Authority (IANA) Functions," 76 *Federal Register* 10570, February 25, 2011.

[20] National Telecommunications and Information Administration, "The Internet Assigned Numbers Authority (IANA) Functions," 76 *Federal Register* 34658-34667, June 14, 2011.

[21] Ibid., p. 34662.

[22] See ICANN comments at http://www.ntia.doc.gov/files/ntia/icann_fnoi_comments_20110722.pdf, p. 7.

[23] Available at https://www.fbo.gov/index?s=opportunity&mode=form&id=c564af28581edb2a7b9441eccfd6391d&tab=core&_cview=0.

[24] NTIA, Press Release, "Commerce Department Awards Contract for Management of Key Internet Functions to ICANN," July 2, 2012, available at http://www.ntia.doc.gov/press-release/2012/commerce-department-awardscontract-management-key-internet-functions-icann.

[25] In the 109th Congress, H.Con.Res. 268 was passed unanimously by the House on November 16, 2005. S.Res. 323 was passed in the Senate by Unanimous Consent on November 18, 2005.

[26] For more information on this issue, see CRS Report R42351, *Internet Governance and the Domain Name System: Issues for Congress*, by Lennard G. Kruger.

[27] The 21 current gTLDs are listed at http://www.iana.org/domains/root/db/#.

[28] A FAQ for the new gTLD process is available at http://newgtlds.icann.org/applicants/faqs/faqs-en.

[29] The Association of National Advertisers (ANA) has been a leading voice against ICANN's current rollout of the new gTLD program. See ANA webpage, "Say No to ICANN: Generic Top Level Domain Developments," available at http://www.ana.net/content/show/id/icann.

[30] See "Rockefeller Says Internet Domain Expansion Will Hurt Consumers, Businesses, and Non-Profits—Urges Delay," *Press Release*, Senate Committee on Commerce, Science and Transportation, December 28, 2011, available at http://commerce.senate.gov/public/index.cfm?p=PressReleases.

[31] House Committee on Energy and Commerce, "Committee Urges ICANN to Delay Expansion of Generic Top-Level Domain Program," *Press Release*, December 21, 2011, available at http://energycommerce.house.gov/news/ PRArticle.aspx?NewsID=9176.

[32] Letter from Representative Goodlatte and Representative Berman to the Secretary of Commerce, December 16, 2011, available at http://www.icann.org/en/correspondence/goodlatte-berman-to-bryson-16dec11-en.pdf.

[33] Letter from FTC to ICANN, December 16, 2011, available at http://www.ftc.gov/os/closings /publicltrs/111216letterto-icann.pdf.

[34] Testimony of Kurt Pritz, Senior Vice President, ICANN, before the House Committee on Energy and Commerce, Subcommittee on Communications and Technology, December 14, 2011, available at http://republicans.energycommerce.house.gov/Media/file/Hearings/Telecom/121411/Pritz.pdf. The gTLD expansion is also strongly supported by many in the Internet and domain name industry, see letter to Senator Rockefeller and Senator Hutchison at http://news.dot-nxt.com/sites/news.dot-nxt.com/files/gtld-industry-to-congress-gtlds-8dec11.pdf.

[35] A complete list of new gTLD applications is provided at http://newgtlds.icann.org/en/program-status/application-results/strings-1200utc-13jun12-en.

[36] Application statistics are available at http://newgtlds.icann.org/en/program-status/statistics.

[37] Available at http://newgtlds.icann.org/en/applicants/agb.

[38] *gTLD Applicant Guidebook*, Module 1, p. 1-21.

[39] Ibid.

[40] For a discussion of the constitutionality of a .xxx top level domain name, see CRS Report RL33224, *Constitutionality of Requiring Sexually Explicit Material on the Internet to Be Under a Separate Domain Name*, by Henry Cohen.

[41] International Centre for Dispute Resolution, In the Matter of an Independent Review Process: ICM Registry, LLC, Claimant, v. Internet Corporation for Assigned Names and Numbers, Respondent, Declaration of the Independent Review Panel, ICDR Case No. 50 117 T 00224 08, February 19, 2010, p. 70, available at http://safekids.com/ documents/irp-panel-declaration-19feb10-en.pdf.

[42] See possible options and public comments at http://icann.org/en/announcements/announcement-2-26mar10-en.htm.

[43] ICANN, *Adopted Board Resolutions, Cartegena*, December 10, 2010, available at http://www.icann.org/en/minutes/ resolutions-10dec10-en.htm#4.

[44] Letter from ICANN to Chair of GAC, February 10, 2011, available at http://icann.org/en/correspondence/jeffrey-to-to-dryden-10feb11-en.pdf.

[45] Internet Corporation for Assigned Names and Numbers, "DNSSEC—What Is It and Why Is It Important?" October 9, 2008, available at http://icann.org/en/announcements/dnssec-qaa-09oct08-en.htm.

[46] Department of Commerce, National Telecommunications and Information Administration, "Enhancing the Security and Stability of the Internet's Domain Name and Addressing System," 73 *Federal Register* 59608, October 9, 2008.

[47] Department of Commerce, National Institute of Standards and Technology, *NIST News Release*, "Commerce Department to Work With ICANN and VeriSign to Enhance the Security and Stability of the Internet's Domain Name and Addressing System," June 3, 2009.

[48] Department of Commerce, National Telecommunications and Information Administration, "Availability of Testing and Evaluation Report and Intent To Proceed With the Final Stages of Domain Name System Security Extensions Implementation in the Authoritative Root Zone," 74 *Federal Register* 32748, June 9, 2010.

[49] See ICANN "Whois Services" page, available at http://www.icann.org/topics/whois-services/.

[50] See ICANN "WHOIS Policy Review" page, available at http://www.icann.org/en/reviews/affirmation/review-4-en.htm.

[51] WHOIS Policy Review Team, *Final Report*, May 11, 2012, p. 7-18, available at https://community.icann.org/pages/ viewpage.action?pageId=33456480.

[52] ICANN, *Approved Board Resolutions*, "WHOIS Policy Team Report," November 8, 2012, available at http://www.icann.org/en/groups/board/documents/resolutions-08nov12-en.htm.

[53] For more information, see ICANN, "Negotiations Between ICANN and Registrars to Amend the Registrar Accreditation Agreement," available at https://community.icann.org/display/RAA/ Negotiations+Between+ ICANN+and+Registrars+to+Amend+the+Registrar+Accreditation+Agreement.

[54] ICANN, "RAA Negotiations Update – December 2012," December 14, 2012, available at http://www.icann.org/en/ news/announcements/announcement-3-14dec12-en.htm.

[55] See CRS Report R42112, *Online Copyright Infringement and Counterfeiting: Legislation in the 112th Congress*, by Brian T. Yeh.

In: Transformations in Telecommunications and Media
Editor: Irwin Cavazos

ISBN: 978-1-62948-413-6
© 2013 Nova Science Publishers, Inc.

Chapter 10

THE FIRST RESPONDER NETWORK AND NEXT-GENERATION COMMUNICATIONS FOR PUBLIC SAFETY: ISSUES FOR CONGRESS[*]

Linda K. Moore

SUMMARY

Since September 11, 2001, when communications failures contributed to the tragedies of the day, Congress has passed several laws intended to create a nationwide emergency communications capability. Yet the United States has continued to strive for a solution that assures seamless communications among first responders and emergency personnel at the scene of a major disaster. To address this problem, Congress included provisions in the Middle Class Tax Relief and Job Creation Act of 2012 (P.L. 112-96) for planning, building, and managing a new, nationwide, broadband network for public safety communications, and assigned additional spectrum to accommodate the new network. In addition, the act has designated federal appropriations of over $7 billion for the network and other public safety needs. These funds will be provided through new revenue from the auction of spectrum licenses. The cost of construction of a nationwide network for public safety is estimated by experts to be in the tens of billions of dollars over the long term, with similarly large sums needed for maintenance and operation. In expectation that public-private partnerships to build the new network will reduce costs to the public sector, the law has provided requirements and guidelines for shared use.

The act has mandated that technical standards developed for the new network incorporate commercial standards for Long Term Evolution (LTE). LTE is a fourth-generation wireless technology that bases its operating standards on the Internet Protocol (IP). IP-enabled networks and wireless devices provide higher capacity and transmission speeds than earlier generations of technology. LTE represents the convergence of wireless technology with the Internet, bringing the capacity and resiliency of packet-switched networks to emergency

---

[*] This is an edited, reformatted and augmented version of the Congressional Research Service publication, CRS Report for Congress R42543, dated May 28, 2013.

communications. It is generally believed that the use of LTE and IP standards will greatly enhance communications for emergency response and recovery.

There are many challenges for public safety leaders and policy makers in establishing IP-enabled technologies as the baseline for the development of future solutions for response and recovery. One of the immediate challenges in developing standards is the need for a clear policy on the use of spectrum for commercial and public safety LTE. Because public safety planning has lagged behind commercial efforts to build LTE networks, the work on design and development of technical requirements is incomplete. Many experts are concerned that these delays may place public safety officials at a disadvantage in negotiating with potential partners, increase costs, and add further delays in moving forward to build a nationwide broadband network. Requirements in the act for standards development may be insufficient to overcome current technical obstacles for desired network features such as roaming between public safety and commercial networks.

In addition to monitoring progress in building the new broadband network for public safety, Congress may want to consider reviewing the role of commercial networks in emergency response and recovery. Once commercial communications lines are compromised because of infrastructure failures, interdependent public safety networks are threatened and the ability to communicate vital information to the public is diminished. New policy initiatives may be needed to identify critical gaps in communications infrastructure and the means to fund the investments needed to close these gaps.

INTRODUCTION

The importance of wireless communications in emergency response has expanded in parallel with increasing reliance on mobile communications across all sectors of the American economy. The consequences of failure in emergency communications networks have also grown, as the nation witnessed on September 11, 2001, and in the days that followed, as first responders and other emergency workers struggled to communicate with each other. The need for robust emergency communications was again underlined by network failures in the wakes of Hurricanes Katrina and Rita, in 2005. Fixing the problems of communications interoperability and operability that hampered response and recovery in these and other catastrophic events has been and remains a long-term goal of policy makers.

After September 11, many experts recognized that a first responder communications network with national coverage would provide the standards and connectivity needed for interoperability and survivability. The National Commission on Terrorist Attacks Upon the United States (9/11 Commission) also recognized the role of networks in providing interoperability, citing the Army Signal Corps as a possible model in recommendations to Congress.[1]

From 2002 through 2007 Congress passed several laws intended to provide the Department of Homeland Security with the tools to plan for a national network. Efforts fell short of congressional expectations, however, in part because federal resources were directed to maintaining local jurisdiction in decision-making at the expense of coordinating a nationwide network.[2]

With the passage of the Middle Class Tax Relief and Job Creation Act of 2012 (P.L. 112-96) on February 22, 2012, the Administration, Congress, the public safety sector, and many other stakeholders have come together to begin the process of developing, constructing, and operating a nationwide network designed to meet public safety communications needs. The act has given government agencies and public safety officials new tools for providing nationwide availability of state-of-the art communications capability for emergency response and recovery. A new network is to be built to provide broadband communications using Internet Protocol (IP) standards to support high-speed delivery (broadband) of data-rich content and video. The IP-based technology mandated for the nationwide public safety broadband network is Long Term Evolution (LTE). Mission critical voice communications using standards designed for Land Mobile Radio (LMR) will be carried on separate networks. In time, many anticipate that IP standards for radios will replace LMR, bringing new economies of scale and higher levels of performance. The development of a unified effort to provide a national network places the nation on the path to achieve the long-sought goal of robust, interoperable communications for first responders.

Vulnerabilities from Commercial Network Failures

One of the many considerations in planning and building FirstNet is how it may be interoperable with commercial infrastructure. Part of the impetus for building a new network dedicated primarily to public safety is to assure that additional investments are made to harden the network. Hardening refers to protective measures such as redundant network connectivity, auxiliary power generators, and reinforcement of cell tower structures. The design for the new network, still in the early stages of fact-finding, will almost certainly rely in part on existing commercial installations such as towers, switches, network connections, and data and operations centers.

The devastation caused by Superstorm Sandy to parts of the Atlantic Seaboard in October 2012 dramatically underscored the value of hardening communications networks. Whereas approximately 25% of commercial wireless towers were out of service immediately after the storm, emergency communications links in New York City remained fully operational.[3] Since September 11, 2001, New York City has made substantial investments in network survivability and interoperability. Some surrounding communities that have not upgraded their networks reported problems, however.

The deployment of FirstNet is intended to overcome existing shortcomings and assure communications for first responders everywhere in the country. The kinds of investments in emergency communications undertaken in localities such as New York City and envisioned for the nationwide public safety network have for the most part not been made by the commercial wireless industry. Measures proposed by the FCC, after Hurricanes Katrina and Rita in 2005, to require back-up power for towers and operations centers, were successfully challenged by the wireless industry.

Key New Legislative Provisions to Improve Public Safety Communications

A program to provide nationwide coverage for public safety communications is to be developed and managed by a new authority created in Title VI (known as the Public Safety Spectrum Act or Spectrum Act) of the Middle Class Tax Relief and Job Creation Act of 2012 (P.L. 112-96). The First Responder Network Authority, or FirstNet, has been established by the act and given broad powers to ensure that the nationwide public safety broadband network is built, maintained, and kept up-to-date as technology evolves.[4] In consultation with federal and state authorities, FirstNet will develop proposals to construct and manage the network with partners from the private sector, among others. Following is a discussion of major provisions in the act that pertain to public safety communications, including provisions to improve the nation's 911 emergency call system.

Among federal agencies designated by the act to provide consultation and support are the Federal Communications Commission (FCC), the National Telecommunications and Information Administration (NTIA), the National Institute of Standards and Technology (NIST), and the Office of Emergency Communications (OEC). The FCC manages commercial and non-federal spectrum use, including spectrum allocated to public safety. The NTIA manages federal spectrum resources and, along with NIST, is an agency within the Department of Commerce. OEC is part of the Office of Cybersecurity and Communications, Department of Homeland Security.

Spectrum Assignment

Radio frequency spectrum is an essential resource for wireless communications. The energy in electronic telecommunications transmissions converts airwaves into signals to deliver voice, text, and images. These signal frequencies are allocated for specific purposes, such as television broadcasting or WiFi,[5] and assigned to specific users through licenses. Allocating sufficient spectrum for wireless emergency communications has long been a concern for Congress. The Balanced Budget Act of 1997 (P.L. 105-33), for example, directed the FCC to allocate 24 MHz[6] of spectrum in the 700 MHz band for public safety use.[7]

With the passage of the Middle Class Tax Relief and Job Creation Act of 2012, some existing public safety licenses in the 700 MHz band[8] and an additional license (known as the D Block),[9] together totaling 22 MHz, have been designated by Congress to support a broadband communications network for public safety. As required by the act, the initial, 10-year license was assigned by the FCC to FirstNet. It is renewable for an additional 10 years, on condition that FirstNet has met its duties and obligations under the act.[10]

A total of 34 MHz of spectrum capacity will therefore be available for public safety networks within the 700 MHz band: the 22 MHz designated for broadband and 12 MHz allocated for narrowband communications, primarily voice.[11] Additionally, there are public safety networks on adjacent frequencies within the 800 MHz band. Time and technological advances may someday bring these spectrum assets together, but at present there are three distinct public safety network technologies in use or planned within the 700 MHz and 800 MHz bands. These are: broadband communications at 700 MHz; interoperable narrowband communications at 700 MHz; and narrowband communications at 800 MHz. Some of the narrowband networks at 700 MHz and 800 MHz can share infrastructure and radios but older narrowband networks at 800 MHz are often not easily integrated with narrowband networks being built on 700 MHz frequencies.

All of the 700 MHz band spectrum assigned for public safety use can support broadband networks. At present, however, there is no tested technology to deliver voice communications over LTE broadband that meets first responder needs. The act gives the FCC the authority to "... allow the narrowband spectrum to be used in a flexible manner, including usage for public safety broadband communications.... "subject to technical and interference protection measures.[12] This provision might open an opportunity for early broadband network build-outs by public safety agencies that want to be in the vanguard of using LTE voice communications technology.

The act requires that public safety users return frequencies known as the T-Band.[13] These are frequencies between 470 and 512 MHz allocated for television that have been made available for public safety use in 11 urban areas.[14] Since the transition to digital television, radio transmissions on some of these frequency assignments have experienced interference and the public safety agencies that use them are considering moving to new networks at 700 MHz. Other areas have recently invested to upgrade networks built on the T-Band frequencies and are concerned about the loss of this communications capacity. The act requires that the FCC act by February 2021 to establish a relocation plan that would free up the T-Band for reassignment through competitive bidding. Proceeds from the auctions of T-Band frequencies are to be available for grants to cover relocation costs.[15] There are no requirements in the law as to how the NTIA, the designated grants administrator, is to structure the grant program or determine eligible costs, although the agency might decide to follow procedures for reallocating federal spectrum.

Some of the earliest spectrum assignments for public safety are in channels below 512 MHz. Public safety and other license-holders in designated channels below 512 MHz are required to reband their holdings to conform to an FCC mandate to improve spectrum efficiency.[16] This narrowbanding requirement, as it is called, requires that assigned channels be reduced from a width of 25 khz to 12.5 khz, thereby freeing up new spectrum capacity for public safety and other uses. The deadline to meet the narrowbanding requirement was January 1, 2013. To accommodate public safety license holders in the T-Band that now fall under requirements established in the act, the FCC has ruled to exempt them from the narrowbanding requirements.[17]

Other spectrum assets available for public safety communications include 50 MHz of spectrum at 4940-4990 MHz (4.9 GHz).[18] Current technology limits these frequencies to local area networks covering a small area. Many experts believe that short-range applications can be incorporated with broadband networks to provide additional resources and better coverage, for example in responding to emergencies in high-rise buildings. Spectrum at 4.9 GHz is also suitable for communicating with unmanned aerial vehicles (UAVs). Benefits to public safety uses of UAVs include search and rescue and fire spotting. UAVs can also be used as airborne antennas to expand the reach of wireless communications. Current deployment of UAVs is the subject of debate among policy makers, however, because of unresolved questions over privacy protection, air space safety, and other concerns.

Although not specifically required by the act, several federal agencies have broad powers to undertake research and development that might further goals for improved performance of emergency communications systems, and more efficient and effective use of all spectrum resources allocated for public safety use. Many policy makers believe that additional technological development and planning should be undertaken, although FirstNet's mandate appears to limit it to the public safety broadband network to be operated on the spectrum licensed to it.

Expenditures and Revenue Sources

The cost of building a new wireless communications network is likely to be in the tens of billions of dollars.[19] To meet these costs, the expectation is that FirstNet will have access to existing infrastructure for some of the network's components and that it will be able to invest through partnerships—with commercial wireless carriers or other secondary users of its spectrum and infrastructure—that generate revenue.

The Middle Class Tax Relief and Job Creation Act of 2012 provides over $7 billion in funding directed either to FirstNet and states participating in the nationwide network, or as grants to states that have opted out of participating in the FirstNet nationwide network program, but have qualified to build their state's portion of the nationwide network. There is an initial loan of $2 billion (repayable from spectrum-license auction proceeds) to set up FirstNet and begin its operation.[20] The remaining $5 billion will become available as auctions for spectrum licenses are concluded and the revenues deposited in the Public Safety Trust Fund.

Public Safety Trust Fund

The law provides for transfers from a Public Safety Trust Fund that is created by the act to receive revenues from designated auctions of spectrum licenses.[21] The designated amounts are to remain available through FY2022, after which any remaining funds are to revert to the Treasury, to be used for deficit reduction. Auction proceeds are to be distributed in the following priority:

- To the NTIA, to reimburse the Treasury for funds advanced to cover the initial costs of establishing FirstNet: not to exceed $2 billion.
- To the State and Local Implementation Fund for a grant program: $135 million.
- To the Network Construction Fund for costs associated with building the nationwide network and for grants to states that qualify to build their own networks: $7 billion, reduced by the amount advanced to establish FirstNet.
- To NIST for public safety research: $100 million.
- To the Treasury for deficit reduction: $20.4 billion.
- To the NTIA and the National Highway Traffic Safety Administration for a grant program to improve 911 services: $115 million.
- To NIST for public safety research: $200 million.
- To the Treasury for deficit reduction: any remaining amounts from designated auction revenues.

Network Construction Fund

The Network Construction Fund is established in the Treasury to be used by FirstNet for expenditures on construction, maintenance, and related expenses to build the nationwide network required in the act, and by the NTIA for grants to those states that qualify to build their own radio access network links to the FirstNet core infrastructure.[22]

FirstNet: Limit on Expenditures

The act caps FirstNet's administrative expenses at $100 million in total over the first 10 years of operation. Costs attributed to oversight and audits are not included in the expense cap.[23]

FirstNet: Fee Income and Other Revenue

FirstNet has the authority to obtain grants and to receive payment for the use of network capacity licensed to FirstNet and of network infrastructure "constructed, owned, or operated" by FirstNet.[24] Specifically, FirstNet is authorized to collect network user fees from public safety and secondary users[25] and to receive payments under leasing agreements in public-private partnerships.[26] These partnerships may be formed between FirstNet and a secondary user for the purpose of constructing, managing, and operating the network. The agreements may allow access to the network on a secondary basis for services other than public safety. FirstNet and its partners may also receive payments for leasing access to infrastructure, such as towers.[27] The act requires that these fees be sufficient each year to cover annual expenses of FirstNet to carry out required activities,[28] with any remaining revenue going to network construction, operation, maintenance, and improvement.[29] There is a prohibition on providing service directly to consumers; this does not impact the right to collect fees from a secondary user or enter into leasing agreements.[30]

State and Local Implementation Fund

The State and Local Implementation Fund was allocated $135 million from the Public Safety Trust Fund. The NTIA, which administers the grant program for this fund, may borrow up to the full amount.[31] The grants are to be made available to all 56 states and territories to develop a plan on how to use a nationwide public safety broadband network to meet their emergency communications needs. The program is to be established as a matching grant program. Federal grants from the fund are not to exceed 80% of the projected cost to the state, however, the NTIA may make the decision to waive the matching funds requirement.[32] The distribution of available funds among the states will be established by the NTIA in consultation with FirstNet.[33]

Funding is planned for distribution in two phases. The first phase will provide funding for initial planning and related activities. The deadline for completed applications for phase one was March 19, 2013. The second phase will address states' needs in preparing for additional consulting with FirstNet, and for planning to undertake data collection activities.[34] Expenditures by the NTIA from the State and Local Implementation Fund were reported at $300,000 for FY2012 for administrative costs. Disbursements for administrative costs and grants funding are estimated at $124,958,000 (base) for FY2013 and $9,700,000 for FY2014.[35] The announced amount available for the first phase of grants from the fund is $121.5 million.[36]

Other Sources of Funds

The construction of this new network represents a significant investment for all participants. State public safety agencies have multiple obligations to build or upgrade, and equip, other networks and may not be in a position to contribute to building and maintaining the new broadband network. The ability of FirstNet to procure funding from the private sector may be crucial to its success.

Planning Authority

The Middle Class Tax Relief and Job Creation Act of 2012 creates FirstNet as an independent entity within the NTIA and empowers it to oversee the establishment of an interoperable broadband network for public safety. The act requires that state and local agencies have a consulting role in the development, deployment, and operation of the nationwide network. The act further provides an opportunity for states to build their own radio access networks within the framework of the nationwide broadband network.

FirstNet

FirstNet is to be headed by a board of 15 members of which 12 are appointed by the Secretary of Commerce according to criteria established by the law, which are intended to provide both representation from key stakeholders and expertise. The other three members of the board are the Secretary of the Department of Homeland Security, the Attorney General of the United States, and the Director of the Office of Management and Budget. The Secretary of Commerce is required to appoint a chairman of the board for an initial term of two years.[37] FirstNet has the statutory authority to exercise all powers specifically granted by the act and "such incidental powers as shall be necessary."[38] Appointments to the board were announced on August 20, 2012.[39]

FirstNet is required to create a public safety advisory committee to assist in carrying out its mandate.[40] The committee is to take "all actions necessary to ensure the building, deployment, and operation" of the network in consultation with federal, state, tribal, and local public safety entities, the Director of NIST, the FCC, and the public safety advisory committee.[41] There are no requirements in the statute as to the composition of the committee. By-laws adopted at the organizing meeting of the First Net Board of Directors on September 25, 2012 created a Public Safety Advisory Committee.[42] It was further agreed that the members of the committee would be chosen from the Advisory Committee to SAFECOM, within the Department of Homeland Security, to be chosen in consultation with the Secretary of Homeland Security. The organizations chosen to be represented on the committee were announced on February 20, 2013.[43] State and local government interests are represented through a subcommittee of PSAC.

Other board committees established in the By-Laws are the Executive Committee; the Planning and Technology Committee; the Audit, Budget and Finance Committee; and the Governance and Personnel Committee.

FirstNet appears to be an autonomous organization, with broad powers to carry out its mandate, within the requirements established by the law. It has for example sole power to select the program's manager and its agents, consultants, and other experts subject to the requirement that they be chosen "in a fair, transparent, and objective manner."[44] In managing proposals and contracts, it is to "take such other actions as may be necessary" to accomplish the network buildout.[45] The selection of General Manager for FirstNet was announced on April 23, 2013.[46]

As part of its management of the network, FirstNet is required at a minimum:

- To establish network policies, including development of detailed requests for proposals to build the network, and operational matters such as terms of service and billing practices.[47]
- To consult with states on expenditures, as part of the preparation of policies and requests for proposals.[48]
- To enter into agreements to use existing communications infrastructure, including commercial and federal infrastructure, "to the maximum extent economically desirable."[49]
- To ensure the construction, maintenance, operation, and improvement of the broadband network, taking into account new and evolving technologies.[50]
- To enter into agreements with commercial networks to allow public safety roaming on their networks.[51]
- To represent the interests of the network's users before standards-setting boards, in consultation with NIST, the FCC, and its own Public Safety Advisory Committee.[52]

As a first step in developing plans for the FirstNet infrastructure, a network proposal was submitted at the September 25, 2012 board meeting.[53] The NTIA subsequently issued a Notice of Inquiry requesting comments by November 1, 2012.[54] Additional input will be gathered during regional workshops and consultations with states, tribes, territories, and localities—scheduled to occur throughout the summer of 2013—beginning with a regional workshop held May 15-16, 2013.[55]

State and Local Participation

Every state has one or more agencies that plan for public safety, homeland security, and emergency communications. Most states have a Statewide Interoperability Coordinator (SWIC) to administer its Statewide Communication Interoperability Plan (SCIP).[56] SCIPs are written to conform with federal guidelines and requirements, such as the National Emergency Communications Plan. FirstNet is required to consult with regional, state, tribal, and local authorities regarding decisions such as those concerning the costs of the policies it formulates, as required in the law, including expenditures for the core network, placement of towers, coverage areas, security, and priority access for local users. Consultation will be through a state-selected coordinator as specified in the act.[57] Appointment of an individual or governmental body as the point-of-contact is also required as a condition of state participation and eligibility to receive grants established by the act.[58] States may decide to use the existing SWIC as the required single point-of-contact or may choose to appoint a separate coordinator.

The governor of each state is to be notified by FirstNet when it has completed its requests for proposals regarding construction, operation, maintenance, and improvement of a nationwide network. The governor or his designee will receive the details of the proposed plans and notification of the amount of funding available to the state if it participates in the FirstNet program.[59]

A state that does not want to participate in FirstNet must submit an alternative plan for construction, operation, maintenance, and improvement of the radio access network within the state. The state must demonstrate to the FCC, which the law requires to review the plan, that its planned network would comply with minimum technical requirements and be interoperable with FirstNet. The state has 90 days to agree to participate or to notify FirstNet, the NTIA, and the FCC of its intent to deploy its own part of the radio access network, and an additional 180 days to provide its plan to the FCC.[60]

If the FCC does not approve the plan, the state might be obliged to participate in FirstNet.[61] If a state's plan is approved it will be eligible to apply for a grant, administered by the NTIA, that will be funded from the Network Construction Fund created by the act. The amount available will be less than what would have been provided if the state had opted in to the FirstNet program, because the grant will be applied only toward building the radio access network and may be subject to matching grant requirements. Approval of the grant is contingent on meeting additional requirements established by the NTIA, including sustainability, timeliness, cost-effectiveness, security, coverage, and services that are comparable to FirstNet.[62] The state would be required to pay a user fee for access to FirstNet.[63] It would not be permitted to enter commercial markets or lease access to its network except through a public-private partnership. Any revenue to the state from a partnership must be used only for costs associated with its broadband network.[64]

Some industry observers have expressed concern about the impact on the success of the nationwide broadband network if many states choose to build their own radio access networks. The cost to FirstNet of building the nationwide network may go up, for example, if anticipated economies of scale are diminished. It may be more difficult for FirstNet to negotiate the partnerships that are expected to provide much of the needed funding for the network. A state that has its plans approved by the FCC may not be able to meet stipulated requirements when its network is built; absent any action by the FCC to enforce technical requirements, the goal of seamless interoperability across all broadband systems may be jeopardized. States may also have difficulty in finding the funds to complete radio access network build-outs, leaving significant gaps in what is intended to be nationwide coverage. The law only identifies two options for a state: join FirstNet or build a statewide radio access network subject to the provisions of the act. The act does not include specific provisions for a state that chooses to build its own network without opting out of FirstNet, although providing such an option is likely within FirstNet's charter. The act also is silent on whether states may choose to opt-out of the broadband network entirely, choosing neither to join FirstNet nor to build a broadband network on the frequencies assigned to FirstNet. Some states may prefer to concentrate their resources on improving mission-critical voice networks and acquire broadband access from a commercial provider or through other means.

One advantage for states building their own radio access networks on FirstNet spectrum is that they will have greater control over any partnerships formed and on expenditures within their states. Although the act requires states to use any revenue from partnerships only to cover costs associated with the state's network, the states will be able to make their own decisions about priorities, with more confidence that revenues will be available when needed. Although there are many potential benefits for states to participate in a nationwide network, such as economies of scale, more secure and robust communications, and a unified base for collaborative efforts, there are also a number of risks, especially if FirstNet fails to deliver promised benefits. The success of FirstNet as an accepted planning authority and leader may depend on whether it makes a compelling business case in the requests for proposals required by the act.

FirstNet's plans for partnerships with the private sector and the nature of the network development plans proposed to each state may be of particular interest to Congress as an early indicator of the viability of FirstNet in meeting the goals required by the act.

Federal Governance

Federal governance of the nationwide public safety broadband network, as required by the Middle Class Tax Relief and Job Creation Act of 2012, is primarily through consultation and oversight. Planning, investment, operating, and other related decisions are to be made by the FirstNet board and the experts it is to hire on a permanent or consultative basis. The designated appropriate congressional committees are, in the Senate, the Committee on Commerce, Science, and Transportation; in the House, the Committee on Energy and Commerce.[65] These committees and other committees with jurisdiction are likely to take an active role in oversight, many believe.

Examples of statutory obligations for Congress and the Administration in the direction of FirstNet include the following.

Membership on FirstNet board. The members of the FirstNet board are to be chosen by the Secretary of Commerce, within the parameters established in the act. The Department of Homeland Security, the Attorney General, and the Office of Management and Budget each have one member on the board in permanence. The Secretary of Commerce is required to appoint a chairman of the board for an initial term of two years.[66]

Grant programs for planning. The NTIA is to establish and administer the State and Local Implementation Fund. Grant provisions are to be in accordance with decisions made by FirstNet.[67]

Grant programs for state networks. The NTIA is to administer grants from the Network Construction Fund to states that qualify to build their own radio access networks and choose to apply for a grant.[68]

Spectrum leases for state networks. The NTIA sets the terms and is responsible for enforcing the requirement that states qualifying to build their radio access networks must sublease spectrum through FirstNet, the assigned license-holder.[69]

License review. The FCC is required to review the initial 10-year license assigned to FirstNet and consider its renewal based on performance criteria.[70]

Performance review. The Government Accountability Office (GAO), within 10 years, is to prepare a report assessing the effectiveness of FirstNet with recommendations on "what action Congress should take" regarding the mandated termination of authority.[71]

Fee schedule. The NTIA is to review and approve the annual schedule of fees charged to public safety agencies and other users for access to FirstNet's resources.[72]

Annual audit. The Secretary of Commerce is to contract for an annual audit of FirstNet's finances and activities. The reports are to be submitted to Congress, the President, and FirstNet.[73]

Report to Congress. FirstNet is required to submit annual reports to Congress on its "operations, activities, financial conditions, and accomplishments."[74]

Although there are several platforms for oversight and guidance provided in the act, it seems likely that the primary responsibility for monitoring progress will fall to the NTIA. The agency may choose to seek assistance from other agencies beyond what is specified in the act, possibly through memoranda of understanding.

Public-Private Partnerships

Partnerships are expected to play a critical role in building and operating the network. Electric utility companies, for example, are upgrading their networks to meet Smart Grid requirements,[75] and some companies have expressed an interest in partnering with FirstNet or state authorities. Some commercial wireless service providers have also expressed an interest in working in partnership with public safety entities to develop and operate new broadband networks.

The Middle Class Tax Relief and Job Creation Act of 2012 requires FirstNet to issue "open, transparent, and competitive" requests for proposals to private sector entities for building, operating, and maintaining the network[76] that leverage to the extent "economically desirable" existing commercial wireless infrastructure, in order to expedite network deployment.[77] It is charged with managing and overseeing the resulting contracts or agreements. As part of a separate requirement to assure substantial rural coverage during all phases of deployment, the act requires that industry proposals and contracts include, if possible, partnerships with existing commercial mobile providers.[78]

Decisions by FirstNet about the network's design, construction, and operation are likely to have a significant impact on commercial participation in a public safety broadband network or networks. These decisions may also influence decision-making by states as to whether or not to pursue radio area network construction independently or through their own partnerships.

Congress may be interested in the composition of private sector partnerships formed by FirstNet and individual states, not only for their business plans but also for the inclusion of a wide variety of stakeholders. For example, are rural and tribal wireless carriers included as business partners? Do secondary access agreements support services that meet social goals, such as for telemedicine, or are they exclusively for commercial purposes? Is competition in providing wireless services being enhanced or hindered?

Infrastructure

Infrastructure for the new network includes operations centers, towers, antennae, and other communications equipment, as well as radios and the software that links them to the network. For wireless communications, an important infrastructure component is the network that links radio towers to communications backbones. These networks, which usually operate over fiber-optic cable or microwave connection, are typically referred to as backhaul.

The Middle Class Tax Relief and Job Creation Act of 2012 requires FirstNet to establish a nationwide, interoperable public safety network,[79] with a "single, national network architecture that evolves with technological advancement...."[80] Network infrastructure components that are specifically required include

- Core network of national and regional data centers and other elements, all based on commercial standards.
- Connectivity between the radio access network and the public Internet or the Public Switched Telephone Network, or both.
- Network cell site equipment, antennas, and backhaul equipment, based on commercial standards, to support wireless devices operating on frequencies designated for public safety broadband.

FirstNet is required to leverage existing infrastructure by entering into agreements to use commercial or other communications infrastructure including federal, state, tribal, or local infrastructure.[81] Planned phases for infrastructure deployment are to include "substantial rural coverage."[82]

FirstNet's ability to build the required network may depend on the timeliness, scope, and outcome of its negotiations to share infrastructure with other parties in order to focus resources on providing elements deemed essential for public safety use of broadband communications.

Timeframe

The requirements of the Middle Class Tax Relief and Job Creation Act of 2012 must be substantially met and the viability of the project demonstrated no later than the end of FY2022, if not sooner. The State and Local Implementation Fund and the Network Construction Fund expire in 2022, with any balances reverting to the Treasury. By 2022, the GAO must have begun an assessment of the performance of FirstNet and the FCC must decide whether or not to renew the licenses for the public safety broadband network. Within this 10-year timeframe, there are few deadlines beyond requirements for the initial establishment of the planning and implementation framework.

Many of the important steps for building the network have no required deadline. Some milestones, such as rural coverage, are mandated in the act, but the deadlines are not specified. There are, for example, no deadlines in provisions that require FirstNet to:

- Develop requests for proposals that include a requirement for timetables; and to consult with states on establishing state and local planning processes.[83]
- Complete the request for proposal process that is to be given to each state governor regarding the request for proposal and its details, and the funding level for each state as determined by the NTIA.[84]

Mandated deadlines for states include

- Within 90 days of receipt of notice from FirstNet, the governor shall choose either to participate in deployment of FirstNet or to conduct its own deployment within the state.[85]
- Within 180 days of giving notice to opt out of FirstNet, the governor shall complete requests for proposals for a state network.[86]

No deadline is established in the statute for the FCC to approve or disapprove state proposals for their own portion of the nationwide broadband network.[87] There are also no specified deadlines for a state to apply to the NTIA for a grant to construct the radio access network and to lease spectrum capacity from First Net, if FCC approval is received for a state network.[88] However, one condition of eligibility for a grant to a state to build its own radio access network is that the state's plan must demonstrate "the ability to complete the project within specified comparable deadlines.... "[89]

FirstNet and the FCC may need to be expeditious in completing all steps for the preparation, review, and acceptance of requests for proposals so that construction of the required core network begins in a timely manner. Too many delays in administrative processes may erode the feasibility of the project.

Next Generation 9-1-1

Today's 911 system is built on an infrastructure of analog technology that does not support many of the features that most Americans expect to be part of an emergency response. Efforts to splice newer, digital technologies onto this aging infrastructure have created points of failure where a call can be dropped or misdirected, sometimes with tragic consequences. Callers to 911, however, generally assume that the newer technologies they are using to place a call are matched by the same level of technology at the 911 call centers, known as Public Safety Answering Points (PSAPs). However, this is not always the case. To modernize the system to provide the quality of service that approaches the expectations of its users will require that the PSAPs and state, local, and possibly federal emergency communications authorities invest in new technologies. As envisioned by most stakeholders, these new technologies—collectively referred to as Next Generation 911 or NG9-1-1—should incorporate Internet Protocol standards. An IP-enabled emergency communications network that supports 911 will facilitate interoperability and system resilience; improve connections between 911 call centers; provide more robust capacity; and offer flexibility in receiving and managing calls. The same network can also serve wireless broadband communications for public safety and other emergency personnel, as well as other purposes.

Recognizing the importance of providing effective 911 service, Congress has passed three major bills supporting improvements in the handling of 911 emergency calls. The Wireless Communications and Public Safety Act of 1999 (P.L. 106-81) established 911 as the number to call for emergencies and gave the Federal Communications Commission (FCC) authority to regulate many aspects of the service. The most recent of these laws, the NET 911 Improvement Act of 2008 (P.L. 110-283), required the preparation of a National Plan for migrating to an IP- enabled emergency network. Responsibility for the plan was assigned to the E-911 Implementation Coordination Office (ICO), created to meet requirements of an earlier law, the ENHANCE 911 Act of 2004 (P.L. 108-494). Authorization for the ICO terminated on September 30, 2009. ICO was jointly administered by the National Telecommunications and Information Administration and the National Highway Traffic Safety Administration.

The Middle Class Tax Relief and Job Creation Act of 2012 re-establishes the federal 9-1-1 Implementation Coordination Office (ICO) to advance planning for next-generation systems and to administer a grant program.[90] ICO is to provide matching grants to eligible state or local governments or tribal organizations for the implementation, operation, and migration of various types of 911 and IP-enabled emergency services, and for public safety personnel training.[91] States that have diverted fees collected for 911 services are not eligible for grants under the program.[92] Based on the act's prioritized plan for funding programs with spectrum license auction revenue, the funds for the grant program will be made available only after $27.635 billion of available auction revenue has been applied to other purposes.

Provisions in the act regarding 911 programs include

- The GAO is required to study how states assess fees on 911 services and how those fees are used.[93]
- The General Services Administration is required to prepare a report on 911 capabilities of multi-line telephone systems in federal facilities and the FCC is to seek comment on the feasibility of improving 911 identification for calls placed through multi-line telephone systems.[94]
- The FCC is to assess the legal and regulatory environment for development of NG9-1-1 and barriers to that development, including state regulatory roadblocks.[95] The FCC is also to (1) initiate a proceeding to create a specialized Do-Not-Call registry for public safety answering points, and (2) to establish penalties and fines for autodialing (robocalls) and related violations.[96]
- ICO, in consultation with NHTSA and DHS is to report on costs for requirements and specifications of NG9-1-1 services, including an analysis of costs, and assessments and analyses of technical uses.[97]
- Immunity and liability protections are provided—to the extent consistent with specified provisions of the Wireless Communications and Public Safety Act of 1999—for various users and providers of Next Generation 911 and related services, including for the release of subscriber information.[98]

The act also requires FirstNet to promote integration of the nationwide public safety broadband network with PSAPs.[99] Since the NTIA has responsibilities for both ICO and FirstNet, the agency is in a position to improve interoperability between PSAPs and First Responders as they move to common IP-based platforms.

Technology and Standards

Standardization of network components, including radios, is generally considered essential to achieving interoperability, improving service, and reducing operating costs. The mandated standard for the new public safety network is Long Term Evolution (LTE), with technical requirements based on commercial standards for LTE.[100] LTE is a fourth-generation (4G) technology based on the Internet Protocol. The commercial sector has begun the transition to operating on IP-enabled networks such as LTE. Wireless carriers around the world are installing LTE networks for consumers and planning for the next generation of LTE: LTE Advanced.[101] LTE Advanced technologies will be able to operate across noncontiguous spectrum bands, thereby increasing channel widths for greater capacity and performance. Most experts agree that LTE Advanced will facilitate the transition to new technologies by making it easier and less expensive to phase out older infrastructure.

FirstNet

The Middle Class Tax Relief and Job Creation Act of 2012 requires FirstNet to assure nationwide standards for use of and access to the network it is tasked with developing. The act specifies the use of commercial standards for some of the network components.[102]

To promote competition, devices for public safety network radios and other wireless devices are required to be built to open, non-proprietary, commercially available standards, "capable of being used by any public safety entity and by multiple vendors across all broadband networks operating in the 700 MHz band" and backward compatible with existing commercial networks where necessary and feasible.[103]

FCC

The act required the FCC to establish a Technical Advisory Board for First Responder Interoperability, and set out criteria for the selection and participation of board members.[104] The primary purpose of the board was to agree on minimum technical requirements for nationwide interoperability on the public safety broadband network. The Interoperability Board was required to develop these technical recommendations in consultation with the NTIA, NIST, and the OEC.[105] The board's technical recommendations are required to be based on commercial standards for LTE.[106] The establishment of minimum technical requirements has a two-fold purpose. One, the requirements are to be presented to the Board of Directors of FirstNet as recommended requirements for interoperability. Second, the minimum technical requirements are to be used by the FCC as a standard of interoperability for evaluating state plans in cases where states have asked to build their own radio access networks.

In the report it submitted,[107] the Interoperability Board, in addition to minimum technical standards, also provided additional considerations that it judged to be important for achieving interoperability.

NIST

The Director of NIST, in consultation with the FCC, DHS, and the National Institute of Justice, Department of Justice, is to "conduct research and assist with the development of standards, technologies and applications to advance wireless public safety communications."[108] More specifically, in consultation with FirstNet and the Public Safety Advisory Committee, NIST is to

- Document technical requirements for public safety wireless communications.
- Accelerate the development of interoperability between currently deployed systems and the public safety broadband network.
- Establish a research plan and direct research for next-generation wireless public safety needs.
- Accelerate the development of broadband network features such as mission- critical voice, prioritization, and authentication.

- Accelerate the development of communications equipment and technology to facilitate the eventual migration of public safety narrowband communications to the public safety broadband network.[109]

Furthermore, the Director of NIST, in consultation with FirstNet and the FCC, "shall ensure the development of a list of certified devices and components meeting appropriate protocols and standards for public safety and commercial vendors" for those seeking to have the use of the public safety broadband network.[110]

Need for Standards Development

Narrowband and broadband networks for public safety will by most accounts be incompatible with each other and with other networks for the foreseeable future.[111] Only a small part of the existing public safety infrastructure is expected to be usable in the development of new networks at 700 MHz. To maximize the utility of new investments in infrastructure and radios, many believe that standards that support public safety applications for IP-enabled technologies must be completed in the early stages of planning and building. Just as access to the Internet has revolutionized business and social cultures worldwide, the transition to IP-enabled networks is likely to expand the capability and scope of emergency communications.

The act variously requires NIST, the FCC, and the NTIA[112] to develop standards and take steps to improve spectrum efficiency and support the development of the next generation of wireless technology. These agencies already have a number of initiatives in place, notably the Public Safety Communications Research program (PSCR). PSCR provides research, development, and testing to advance public safety communications interoperability. The program is a joint effort between NIST's Office of Law Enforcement Standards and NTIA's Institute for Telecommunication Sciences and is sponsored by the Office for Interoperability and Compatibility at DHS, and the Department of Justice Community Oriented Policing Services.[113]

The funding for the federal research and development efforts described in the act is provided from spectrum license auction revenue. The timing of the auctions and the prioritization for distributing auction revenues are such that the funds designated for research and development may not be available for several years, if at all. Some of the act's provisions require the FCC to auction designated spectrum within three years.[114] Auction procedures require several steps that are published for comment before final rulemaking, and the process typically takes a year or more before an auction commences. The first round of funding for NIST ($100 million) would occur once the proceeds from spectrum license auctions deposited in the Public Safety Trust Fund surpass $7.135 billion. The second funding round for NIST would occur after deposits reach $27.75 billion. Although resources in existing federal programs may be shifted to give priority to the implementation of the Middle Class Tax Relief and Job Creation Act of 2012,[115] the federal government may not be able to fund all of the standards and other technological research that is required by the act or needed for public safety. Timely development of public safety applications for LTE and LTE Advanced may come primarily from the private sector, where some vendors are developing components needed for the broadband network and its devices. To meet its responsibilities under the act,

FirstNet may choose to allocate some of the funding provided to it by the act, or raise additional funds, to facilitate standards development.

If no solution is found to coordinate private and public work on standards development and new technologies for emergency communications, the development of IP-enabled technologies for public safety may continue to lag behind that of the commercial sector, perpetuating the high costs and inefficiencies that have plagued first responder communications for decades.

INTEROPERABILITY WITHIN THE 700 MHz BAND

In its *National Broadband Plan*, the FCC indicated that it wanted to make commercial networks in the 700 MHz band available for public safety use and requested that Congress confirm the FCC's authority to act.[116] The Middle Class Tax Relief and Job Creation Act of 2012 provides the FCC with statutory authority to establish rules in the public interest to improve the ability of public safety networks to roam on commercial space and to gain priority access.[117]

FirstNet and the states that build their own networks are empowered by the act to enter into agreements with commercial providers that would allow public safety network users to roam on partnering networks. Agreements might also cover rules for priority access in times of high demand for network capacity. Priority access can take several forms, such as "ruthless pre- emption," in which non-public-safety transmissions are immediately terminated to make way for emergency communications, or negotiated priority agreements that might, for example, place public safety users at the head of the line for network access as capacity becomes available. The act stipulates that the FCC's authority may not require roaming or priority access unless (1) the public safety and commercial networks are technically compatible; (2) the commercial network is reasonably compensated; and (3) access does not preempt or otherwise terminate or degrade existing traffic on the commercial network.[118] Within these limits, the FCC appears to have some leeway to use its regulatory authority to support public safety in negotiations with partners. The FCC cannot, under the act, mandate ruthless pre-emption, although the act does not preclude contractual negotiations that would allow it.

The act's provisions for roaming and priority access do not require a commercial vendor to make additional investments to insure technical compatibility, and the act's language might be interpreted as precluding an FCC mandate to that effect. Interpretation and enforcement of the compatibility provision may pose an obstacle to achieving desired levels of network interoperability and cross-network roaming because the current technical standards for the 700 MHz band preclude affordable full-spectrum roaming, that is, the ability of any network within the 700 MHz to roam on any other network within the 700 MHz band. Full-spectrum roaming is considered by many to provide advantages for public safety and also for the public at large. For example, it makes more network capacity available for shared emergency communications of all types, not just for first responders. Many believe that full-spectrum access supports competitiveness among wireless carriers—in particular assisting small wireless carriers serving rural areas to offer new broadband services—by providing access to all customers within the band.

Achieving full-spectrum roaming on the 700 MHz band requires modifications of technical requirements for LTE, the preferred technology for mobile broadband within the 700 MHz band. The standards for LTE are agreed on a global basis by the 3GPP, a standards setting group.[119] For a number of technical considerations, including maximizing spectrum efficiency and minimizing interference across spectrum channels, the 3GPP divided the 700 MHz commercial spectrum into different band classes. As documented by the FCC,[120] the 70 MHz of commercial spectrum within the 700 MHz band is the only non-interoperable commercial service band.

The band classes apply to spectrum blocks that were determined by the FCC. In preparing for an auction of spectrum licenses, the FCC follows a number of procedural steps and seeks comment on planned actions. A key first step is to develop a band plan for the spectrum and make decisions about geographical coverage and technical requirements. Allocation of spectrum blocks within the 700 MHz band occurred in several steps. To protect public safety allocations for narrowband networks in the upper end of the band from harmful interference, the FCC divided the band into an Upper 700 MHz Band and a Lower 700 MHz band. Both bands included paired spectrum licenses, that is, each license had two sets of channels, one for the uplink and the other for the downlink. The Lower 700 MHz Band conformed to industry standards for global cellular bands in the placement of the uplink and downlink; in the upper 700 MHz Band, the direction of the uplink and downlink were reversed. The auction of the majority of licenses for the 700 MHz band concluded on March 18, 2008. Standards for LTE (Release 8) were finalized in December of the same year.[121] A schematic of the auction blocks and band classes is provided in **Figure 1**.

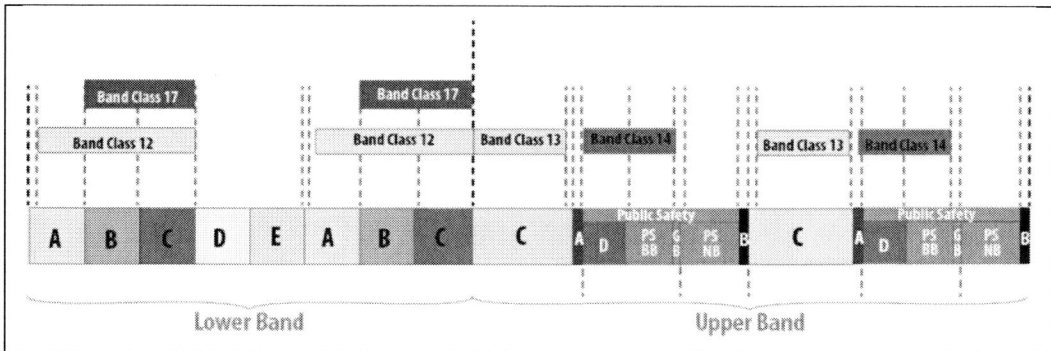

Source: FCC Notice of Proposed Rulemaking, "Interoperability of Mobile Use Equipment Across paired Commercial Spectrum Blocks in the 700 MHz Band," March 21, 2012.

Note: Within the public safety bands, spectrum allocated for broadband is BB, spectrum designated for narrowband is NB and guard bands to protect against interference are GB.

Figure 1. License Blocks and Band Classes in the 700 MHz Band.

The 700 MHz auction gave Verizon Wireless and AT&T Mobility dominant holdings in Band Class 17, corresponding to licenses in the B Block and C Block in the Lower 700 MHz Band, and in Band Class 13, corresponding to licenses in the C Block of the Upper 700 MHz Band. Many of the licenses in the A Block were acquired by smaller carriers. Because of restrictions on usage, licenses for frequencies in the A Block were less expensive than licenses in the B and C Blocks. The restrictions were largely, but not solely, to protect transmissions on these frequencies from potential interference from high-power TV broadcast signals located in TV channel 51, which is adjacent to the lower end of the 700 MHz band.

Public safety holdings include the D Block, which was originally designated for commercial use and assigned Band Class 14 by 3GPP. Band Class 14 was extended to include the part of the spectrum allocated to the nationwide public safety broadband network, as the FCC's band plan called for the two spectrum blocks to be shared. The narrowband public safety spectrum holdings at 700 MHz were not part of the 3GPP standards-setting process, as they are not commercial networks. Some licenses, such as the D and E Blocks in the Lower 700 MHz band, were auctioned before the development of LTE and have different technical requirements.

As a consequence of incompatibility among the band classes, competition in the commercial sector and interoperability for public safety communications may be at risk. The difference between the standards for Band Class 12 and the other band classes has reportedly had the effect of isolating the A Block license-holders from the mainstream of development for LTE devices. Producers of the new equipment have tailored their first offerings for LTE deployment for the markets with the greatest demand: Band Classes 13 and 17. Smaller carriers have reported to the FCC that they are having difficulty in acquiring equipment, especially handsets, for Band Class 12, the only technical standard applicable for licenses in the A Block.

To address the concerns of carriers with licenses in the A Block, the FCC has opened a proposed rulemaking to address how to mitigate interference in Band Class 12 and to increase interoperability within the lower 700 MHz. The rulemaking does not specifically address the impact on public safety roaming and interoperability.

As is the case for Band Class 12, the costs of developing and producing the chipsets, software, and other components for equipment operating on Band Class 14 are likely to be spread across a relatively smaller customer base, increasing marginal costs and the prices paid by users. Because the band classes are not interoperable across the 700 MHz band, public safety network users are likely to incur not only higher costs for equipment to operate within their assigned frequencies, but also higher costs for roaming and priority access on commercial channels.

If FirstNet and any statewide network deployed in Band Class 14 choose different commercial partners, roaming and priority access may be limited. A network agreement, for example, with C Spire Wireless (formerly Cellular South) might lead to the development of equipment to operate on Band Classes 12 and 14; an agreement with Verizon, Band Classes 13 and 14; with AT&T, Band Classes 17 and 14. Some experts have raised concerns about the limitations that will be placed on public safety roaming as a result of the lack of interoperability within the 700 MHz band. The cost of public safety radios that can operate on all band classes will be high, if they can be engineered at all. A radio that only works on Band Class 14 and one other band class may be cut off from roaming in areas where that other band class has no coverage. Many industry experts, however, believe that it is preferable for the nationwide public safety broadband network to have more than one commercial partner, to improve network redundancy and capacity, and to promote competition that might reduce costs in the long term.

Many believe that full-spectrum interoperability will, in the long term, maximize the benefits of LTE and LTE Advanced technologies deployed on the 700 MHz band. Coordinating development of 700 MHz band standards among network participants provides an opportunity to maximize the benefits inherent in IP-enabled networks for the safety of the general public. For example, it is possible to create smart phone applications that can link personal mobile devices to emergency command centers, integrating information from those devices into an action plan for response and recovery. Fully implemented within the 700 MHz band, these communications links might help emergency situation managers determine where to most effectively deploy emergency medical service personnel, firefighters, HazMat teams, utility repair crews, or other response and recovery personnel, as appropriate. As a situation stabilizes, evaluations about evacuation routes, shelters, and other post-disaster services could be expedited and information disseminated through wireless and other emergency alert systems.

Whether standards for LTE use within the 700 MHz band can be changed, how they might be changed, and when, are subjects of intense debate within the wireless industry. The debate may be fully resolved only through the transition to LTE Advanced. Development and change in wireless technology are fast-paced. A rapid transition from LTE to LTE Advanced might eliminate separate band classes, negating the need for interoperable work-arounds, some say.

THE FUTURE

One of the goals of effective spectrum management is to create opportunities for the development of innovative technologies. Wireless technology transforms air into desirable services, providing an engine for economic growth and development. The evolution of public safety communications has lagged behind the commercial sector and the military in receiving the benefits of recent innovations. By providing some of the resources needed to build IP-enabled networks for first responders, Congress has created an opportunity to expand the reach and effectiveness of emergency communications for the entire population.

Many experts in advanced communications technology believe that the transition to IP-enabled technologies is likely to bring about the convergence of commercial, military, and emergency response (federal and nonfederal) technologies on common, interoperable platforms.[122] In this view, compatible communications devices will be differentiated by applications developed by stakeholders to meet their mission needs. Infrastructure, spectrum, and mobile devices will be sharable, and it is envisaged that sharing will be encouraged.

Infrastructure and Spectrum Management

The wireless industry has sometimes compared their licensed radio frequency channels to traffic lanes.[123] These proprietary lanes (access rights assigned for specific frequencies) are used by the vehicles (wireless devices) approved by and often obtained directly from the carrier owning the access rights. Wireless customers have access to their carriers' lanes and usually to other carriers' lanes (roaming privileges) if their vehicles are of the same type used by the lane's owner. Adding more lanes to the highway, that is, providing more spectrum for carriers to expand their networks, is the policy advocated by most representatives of the wireless industry. An increasing number of wireless technology experts are arguing, however, that wireless technology has reached a stage of development where it is possible to allow for a broader variety of vehicles by building a new type of highway. This shift in technology is deemed by many to be a crucial factor in public safety use of commercial technologies. It is argued that as long as there are different classes of wireless devices built to operate on a limited number of predetermined frequencies, the military, public safety, and a number of other users with specific requirements, such as the railroad industry and utilities, will be constrained by high costs distributed among a small number of users.

The successful introduction of the iPhone in 2007 accelerated the convergence of the Internet with mobile technology. The United States is a global leader in wireless technology innovation and adoption.[124] A convergence of technology policy and spectrum policy may be needed to help the country maintain its leadership in innovation. As the nation moves to develop next-generation technologies for public safety communications, wireless industry policies and emergency response and recovery policies may need close coordination to provide a comprehensive safety net forged from sturdy communications links. FirstNet and its private-sector partners have the opportunity to lay the foundation for next-generation communications for public safety.

CONSIDERATIONS FOR CONGRESS

Congress may want to pursue oversight by, for example, monitoring the participation of states in FirstNet and how state decisions to participate or not participate are affecting network build-out. Congress may also wish to evaluate the level of collaboration in standards development among FirstNet, commercial participants in the network, and federal agencies charged with standards development and certification for network components available to public safety. A comparison of adoption rates for new wireless technologies by the private sector, the military, and public safety agencies may also be of interest to Congress in identifying additional legislative or administrative actions that might further advance the transition to needed new technologies.[125]

Congress may also evaluate the role of commercial infrastructure in supporting emergency communications, for the general public as well as among first responders. Underinvestment in commercial communications infrastructure may need greater consideration in emergency planning at all levels of government.

End Notes

[1] Discussed in Congressional Research Service General Distribution Memorandum, "Communications Support for Public Safety: The 9/11 Commission Report and Alternative Approaches," by Linda K. Moore, August 25, 2004, and in CRS Report RL31375, *Emergency Communications: Meeting Public Safety Spectrum Needs* by Linda K. Moore, 2002- 2003 (out of print; available from the author).

[2] Some of the actions by Congress and by federal agencies were summarized in testimony by Linda K. Moore, Specialist in Telecommunications Policy, Congressional Research Service, before the House Committee on Homeland Security, Subcommittee on Emergency Preparedness, Response, and Communications, "Ensuring Coordination and Cooperation: A Review of Emergency Communications Offices Within the Department of Homeland Security," November 17, 2011. The GAO has also addressed these issues in reports such as *Emergency Communications: Various Challenges Likely to Slow Implementation of a Public Safety Broadband Network*, February 2012, GAO-12-343 at http://www.gao.gov/assets/590/588795.pdf. CRS reports

on the topic include CRS Report R41842, *Funding Emergency Communications: Technology and Policy Considerations*; CRS Report R40859, *Public Safety Communications and Spectrum Resources: Policy Issues for Congress*; CRS Report RL34054, *Public-Private Partnership for a Public Safety Network: Governance and Policy*; CRS Report RL33838, *Emergency Communications: Policy Options at a Crossroads*, all by Linda K. Moore.

[3] "NY Public Safety Systems Remained at 100% During Sandy," Government Technology, November 15, 2012, http://www.govtech.com/public-safety/NY-Public-Safety-Systems-During-Sandy.html.

[4] P.L. 112-96, Section 6204 (a).

[5] WiFi, for wireless fidelity, operates on unlicensed frequencies that are not assigned to a specific owner but instead are available to support any device approved by the FCC.

[6] Spectrum is segmented into bands of radio frequencies and typically measured in cycles per second, or hertz. Standard abbreviations for measuring frequencies include kHz—kilohertz or thousands of hertz; MHz—megahertz, or millions of hertz; and GHz—gigahertz, or billions of hertz. The 700 MHz band includes radio frequencies from 698 MHz to 806 MHz.

[7] 47 U.S.C. §309 (j) (14).

[8] 763-768 MHz, 793-798 MHz, 768-769 MHz and 798-799 MHz; P.L. 112-96, Section 6201.

[9] 758-763 MHz and 788-793 MHz; P.L. 112-96, Section 6101.

[10] P.L. 112-96, Section 6201.

[11] 769-775 MHz and 799-805 MHz.

[12] P.L. 112-96, Section 6102.

[13] P.L. 112-96, Section 6103.

[14] Metropolitan areas: Boston, MA, Chicago, IL, Dallas/Fort Worth, TX, Houston, TX, Los Angeles, CA, Miami, FL, New York, NY/Newark NJ, Philadelphia, PA, Pittsburgh, PA, San Francisco/Oakland, CA, and Washington, DC.

[15] The National Public Safety Telecommunications Council (NPSTC) prepared a report that provided an overview of T- Band assignments, some of the problems created by the act's requirements, and possible alternative solutions. NPSTC, *T-Band Report*, March 15, 2013; link to PDF at http://www.npstc.org/, "NPSTC Releases T Band Report."

[16] Details at http://transition.fcc.gov/pshs/public-safety-spectrum/narrowbanding.html.

[17] FCC, "Waiver of Narrowbanding Deadlines for T-Band (470-512 MHz) Licenses," Docket No. WT 99-87, released April 26, 2012.

[18] The FCC has opened a proceeding regarding the 4.9 GHz public safety band, proposing rules and asking for comment on a number of issues to improve spectrum efficiency and encourage greater use of the 4.9 GHz band for public safety broadband communications, http://www.fcc.gov/document/comment-and-reply-comments-date-5th- fnprm-49-ghz-band.

[19] Some cost estimates for building and operating a public safety broadband network are provided in CRS Report R41842, *Funding Emergency Communications: Technology and Policy Considerations*, by Linda K. Moore.

[20] P.L. 112-96, Section 6207.

[21] P.L. 112-96, Section 6413.

[22] P.L. 112-96, Section 6206 (e).

[23] P.L. 112-96, Section 6207 (b).

[24] P.L. 112-96, Section 6206 (b) (4).

[25] P.L. 112-96, Section 6208 (a) (1).

[26] P.L. 112-96, Section 6208 (a) (2).

[27] P.L. 112-96, Section 6208 (a) (3).

[28] P.L. 112-96, Section 6208 (b).

[29] P.L. 112-96, Section 6208 (d).

[30] P.L. 112-96, Section 6212.

[31] P.L. 112-96, Section 6301.

[32] P.L. 112-96, Section 6302 (b).

[33] P.L. 112-96, Section 6302 (a).

[34] Announcement of Federal Funding Opportunity at http://www.ntia.doc.gov/files/ntia/publications/sligp_ffo_02062013.pdf.

[35] U.S. Department of Commerce, National Telecommunications and Information Administration, FY2014 Budget as Presented to Congress, April 2013; State and Local Implementation Fund, Exhibit 10.

[36] NTIA, "NTIA Announces Availability of $121.5 Million in State Grants to Assist with FirstNet Planning," February 6, 2013 (http://www.ntia.doc.gov/press-release/2013/ntia-announces-availability-1215-million-state-grants-assist-firstnet-planning) and "State and Local Implementation Grant Program Federal Funding Opportunity," February 6, 2013 (http://www.ntia.doc.gov/other-publication/2013/sligp-federal-funding-opportunity).

[37] P.L. 112-96, Section 6204.

[38] P.L. 112-96, Section 6206 (a).

[39] Announcement and background information at http://www.ntia.doc.gov/other-publication/2012/acting-secretary-rebecca-blank-announces-board-directors-first-responder-netw.

[40] P.L. 112-96, Section 6205 (a).

[41] P.L. 112-96, Section 6206 (b) (1).

[42] Board Resolution 1, By-Laws, http://www.ntia.doc.gov/files/ntia/publications/ firstnet_resolution_no._1_on_bylaws_adopted_9.25.12.pdf.

[43] NTIA, "FirstNet Names members of Public Safety Advisory Committee," February 20, 2013, http://www.ntia.doc.gov/press-release/2013/firstnet-names-members-public-safety-advisory-committee.

[44] P.L. 112-96, Section 6205 (b) (1).

[45] P.L. 112-96, Section 6206 (b) (4) (D).

[46] NTIA Press Release, "FirstNet Board Announces Hiring of Bill D'Agostino, Jr. as General Manager," April 23, 2013, at http://www.ntia.doc.gov/press-release/2013/firstnet-board-announces-hiring-bill-d-agostino-jr-general-manager.

[47] P.L. 112-96, Section 6206 (c) (1).

[48] P.L. 112-96, Section 6206 (c) (2).

[49] P.L. 112-96, Section 6206 (c) (3).

[50] P.L. 112-96, Section 6206 (c) (4).

[51] P.L. 112-96, Section 6206 (c) (5).

[52] P.L. 112-96, Section 6206 (c) (7).

[53] FirstNet Network (FNN) Proposal, http://www.ntia.doc.gov/files/ntia/publications/firstnet_fnn_presentation_09-25- 2012_final.pdf.

[54] Deadline extended to November 9, 2012. Notice at http://www.ntia.doc.gov/federal-register-notice/2012/notice-inquiry-firstnet-conceptual-network-architecture; comments posted at http://www.ntia.doc.gov/federal-register-notice/ 2012/comments-nationwide-interoperable-public-safety-broadband-network-noi.

[55] NTIA Press Release, "FirstNet to Begin Consultations with State, Tribal, Territorial and Local Entities Nationwide," May 3, 2013, at http://www.ntia.doc.gov/press-release/2013/firstnet-begin-consultations-state-tribal-territorial-and- local-entities-nationwi.

[56] See "Statewide Interoperability Coordinators" at http://www.dhs.gov/files/programs/gc_1286986920144.shtm.

[57] P.L. 112-96, Section 6206 (c) (2) (B).

[58] P.L. 112-96, Section 6302 (d).

[59] P.L. 112-96, Section 6302 (e) (1).

[60] P.L. 112-96, Section 6302 (e) (2) and (3).

[61] P.L. 112-96, Section 6302 (e) (3) (C) (iv).

[62] P.L. 112-96, Section 6302 (e) (3) (D).

[63] P.L. 112-96, Section 6302 (f).

[64] P.L. 112-96, Section 6302 (g).

[65] P.L. 112-96, Section 6001 (3).

[66] P.L. 112-96, Section 6204.

[67] P.L. 112-96, Section 6302 (a).

[68] P.L. 112-96, Section 6302 (e) (3) (C) (iii) (I).

[69] P.L. 112-96, Section 6302 (e) (3) (C) (iii) (II).

[70] P.L. 112-96, Section 6201 (b).

[71] P.L. 112-96, Section 6206 (g).

[72] P.L. 112-96, Section 6208 (c).

[73] P.L. 112-96, Section 6209.

[74] P.L. 112-96, Section 6210.

[75] "Smart Grid" is the name given to the evolving electric power network as new information technology systems and capabilities are incorporated. See also CRS Report R41886, *The Smart Grid and Cybersecurity—Regulatory Policy and Issues*, by Richard J. Campbell.

[76] P.L. 112-96, Section 6206 (b) (1) (B).

[77] P.L. 112-96, Section 6206 (b) (1) (C).

[78] P.L. 112-96, Section 6206 (b) (3).

[79] P.L. 112-96, Section 6202 (a).

[80] P.L. 112-96, Section 6202 (b).

[81] P.L. 112-96, Section 6206 (c) (3).

[82] P.L. 112-96, Section 6206 (b) (3).

[83] P.L. 112-96, Section 6206, (c) (1) and (2).

[84] P.L. 112-96, Section 6206 (c).

[85] P.L. 112-96, Section 6302, (e) (2).

[86] P.L. 112-96, Section 6302, (e) (3) (B).

[87] P.L. 112-96, Section 6302 (e) (3) (C) (i).

[88] P.L. 112-96, Section 6302, (e) (3) (C) (iii).

[89] P.L. 112-96, Section 6302, (e) (3) (D) (i) (III).

[90] P.L. 112-96, Section 6503, "Section 158 "(a).

[91] P.L. 112-96, Section 6503, "Section 158 "(b).

[92] P.L. 112-96, Section 6503, "Section 158 "(c).

[93] P.L. 112-96, Section 6505.

[94] P.L. 112-96, Section 6504.

[95] P.L. 112-96, Section 6509.

[96] P.L. 112-96, Section 6507.

[97] P.L. 112-96, Section 6508.

[98] P.L. 112-96, Section 6506.

[99] P.L. 112-96, Section 6206 (b) (2) (C).

[100] P.L. 112-96, Section 6203 (c) (2).

[101] Also known as 3GPP Release 10, see http://www.3gpp.org/LTE-Advanced.

[102] P.L. 112-96, Section 6206 (b) (1) (A).

[103] P.L. 112-96, Section 6206 (b) (2) (B).

[104] P.L. 112-96, Section 6203.

[105] P.L. 112-96, Section 6203 (c) (1).

[106] P.L. 112-96, Section 6203 (c) (2).

[107] *Recommended Minimum Technical Requirements to Ensure Nationwide Interoperability for the Nationwide Public Safety Broadband Network*, prepared by the Technical Advisory Board for First Responder Interoperability, Final Report, May 22, 2012, at http://www.fcc.gov/document/recommendations-interoperability-board.

[108] P.L. 112-96, Section 6303 (a).

[109] P.L. 112-96, Section 6303 (b) (1 – 5).

[110] P.L. 112-96, Section 6206 (c) (6).

[111] Discussed in GAO report, *Emergency Communications: Various Challenges Likely to Slow Implementation of a Public Safety Broadband Network*, February 2012, GAO-12-343.

[112] In addition to assigning NTIA responsibilities to develop public safety broadband communications, the act also specifies the NTIA's responsibility to promote efficient use of spectrum by the federal government. P.L. 112-96, Section 6410.

[113] More information is available at the PSCR website at http://www.ntia.doc.gov/category/public-safety. PSCR activities were discussed in testimony by Mary H. Saunders, Director, Standards Coordination Office, NIST before the House Committee on Homeland Security, Subcommittees on Emergency Preparedness, Response, and Communications and Cybersecurity, Infrastructure Protection, and Security Technologies, "First Responder Technologies: Ensuring a Prioritized Approach for Homeland Security Research and Development," May 9, 2012.

[114] P.L. 112-96, Section 6401 (b).

[115] The PSCR, for example, has changed its plans for testing public safety interoperability in response to provisions in the act, http://www.pscr.gov/about_pscr/press/broadband/pscr_to_focus_on_public-safety_broadband_interoperability_tests_042012-mission_critical.pdf.

[116] FCC, *Connecting America: The National Broadband Plan*, http://www.broadband.gov/download-plan/.

[117] P.L. 112-96, Section 6211.

[118] P.L. 112-96 Section 6211.

[119] The 3G Partnership Project, known in the United States as 4G Americas, is a consensus-driven, international partnership of industry-based telecommunications standards bodies.

[120] FCC, "Promoting Interoperability in the 700 MHZ Commercial Spectrum," Notice of Proposed Rulemaking, WT Docket No. 12-69, released March 21, 2012.

[121] LTE mobile broadband standard at http://www.3gpp.org/LTE.

[122] See for example reports and meeting discussions of the Visiting Committee on Advanced Technology established by NIST, http://www.nist.gov/director/vcat/.

[123] CTIA–The Wireless Association, for example, provides a "Spectrum 101" graphic that uses improvements in transportation as an analogy for changes in the wireless environment, http://files.ctia.org/pdf/Spectrum_Brochure_111111.pdf.

[124] House Committee on Homeland Security, Subcommittee on Emergency Preparedness, Response, and Communications, "Growing the Wireless Economy Through Innovation," April 18, 2012, spoken and written testimony from various witnesses at the hearing.

[125] Additional analysis in CRS Memorandum CD1289, "Establishment of the First Responder Network Authority," September 19, 2012, available upon request.

In: Transformations in Telecommunications and Media   ISBN: 978-1-62948-413-6
Editor: Irwin Cavazos   © 2013 Nova Science Publishers, Inc.

Chapter 11

THE FEDERAL COMMUNICATIONS COMMISSION: CURRENT STRUCTURE AND ITS ROLE IN THE CHANGING TELECOMMUNICATIONS LANDSCAPE[*]

Patricia Moloney Figliola

SUMMARY

The Federal Communications Commission (FCC) is an independent federal agency with its five members appointed by the President, subject to confirmation by the Senate. It was established by the Communications Act of 1934 (1934 Act) and is charged with regulating interstate and international communications by radio, television, wire, satellite, and cable. The mission of the FCC is to ensure that the American people have available—at reasonable cost and without discrimination—rapid, efficient, nation- and world-wide communication services, whether by radio, television, wire, satellite, or cable.

Although the FCC has restructured over the past few years to better reflect the industry, it is still required to adhere to the statutory requirements of its governing legislation, the Communications Act of 1934. The 1934 Act requires the FCC to regulate the various industry sectors differently. Some policymakers have been critical of the FCC and the manner in which it regulates various sectors of the telecommunications industry—telephone, cable television, radio and television broadcasting, and some aspects of the Internet. These policymakers, including some in Congress, have long called for varying degrees and types of reform to the FCC. Most proposals fall into two categories: (1) procedural changes made within the FCC or through congressional action that would affect the agency's operations or (2) substantive policy changes requiring congressional action that would affect how the agency regulates different services and industry sectors. Nine bills have been introduced during the 112th Congress that would change the operation of the FCC.

For FY2014, the FCC has requested a budget of $359,299,000. The FCC's budget is derived from regulatory fees collected by the agency rather than through a direct

---

[*] This is an edited, reformatted and augmented version of a Congressional Research Service publication, CRS Report for Congress RL32589, prepared for Members and Committees of Congress, from www.crs.gov, dated July 18, 2013.

appropriation. The fees, often referred to as "Section (9) fees," are collected from license holders and certain other entities (e.g., cable television systems) and deposited into an FCC account. The law gives the FCC authority to review the regulatory fees and to adjust the fees to reflect changes in its appropriation from year to year. It may also add, delete, or reclassify services under certain circumstances.

In the 113th Congress, three hearings have been held on FCC oversight, reform, and management and four bills have been introduced that would affect the manner in which the FCC conducts its business.

Overview of the Federal Communications Commission

The Federal Communications Commission (FCC) is an independent federal agency with its five members appointed by the President, subject to confirmation by the Senate. It was established by the Communications Act of 1934 (1934 Act or "Communications Act")[1] and is charged with regulating interstate and international communications by radio, television, wire, satellite, and cable.[2] The mission of the FCC is to ensure that the American people have available, "without discrimination on the basis of race, color, religion, national origin, or sex, a rapid, efficient, Nationwide, and worldwide wire and radio communication service with adequate facilities at reasonable charges."[3]

The 1934 Act is divided into titles and sections that describe various powers and concerns of the Commission.[4]

- Title I—FCC Administration and Powers. The 1934 Act originally called for a commission consisting of seven members, but that number was reduced to five in 1983. Commissioners are appointed by the President and approved by the Senate to serve five-year terms; the President designates one member to serve as chairman. No more than three commissioners may come from the political party of the President. Title I empowers the Commission to create divisions or bureaus responsible for specific work assigned and to structure itself as it chooses.

- Title II—Common carrier regulation, primarily telephone regulation, including circuit-switched telephone services offered by cable companies. Common carriers are communication companies that provide facilities for transmission but do not originate messages, such as telephone and microwave providers. The 1934 Act limits FCC regulation to interstate and international common carriers, although a joint federal-state board coordinates regulation between the FCC and state regulatory commissions.

- Title III—Broadcast station requirements. Much existing broadcast regulation was established prior to 1934 by the Federal Radio Commission and most provisions of the Radio Act of 1927 were subsumed into Title III of the 1934 Act. Sections 303-307 define many of the powers given to the FCC with respect to broadcasting; other sections define limitations placed upon it. For example, Section 326 of Title III prevents the FCC from exercising censorship over broadcast stations. Also, parts of the U.S. code are linked to the Communications Act. For example, 18 U.S.C. 464 makes obscene or indecent language over a broadcast station illegal.

- Title IV—Procedural and administrative provisions, such as hearings, joint boards, judicial review of the FCC's orders, petitions, and inquiries.
- Title V—Penal provisions and forfeitures, such as violations of rules and regulations.
- Title VI—Cable communications, such as the use of cable channels and cable ownership restrictions, franchising, and video programming services provided by telephone companies.
- Title VII—Miscellaneous provisions and powers, such as war powers of the President, closed captioning of public service announcements, and telecommunications development fund.

FCC Leadership

The FCC is directed by five commissioners appointed by the President and confirmed by the Senate for five-year terms (except when filling an unexpired term). The President designates one of the commissioners to serve as chairperson. Only three commissioners may be members of the same political party. None of them can have a financial interest in any Commission-related business. The commissioners are:

- Mignon Clyburn (confirmed by the Senate on July 24, 2009), Acting Chairman;
- Jessica Rosenworcel (confirmed by the Senate on May 7, 2012); and
- Ajit Pai (confirmed by the Senate on May 7, 2012).

Two seats are vacant due to the resignations of Chairman Julius Genachowski and Commissioner Robert McDowell. President Obama nominated Mr. Tom Wheeler as Chairman on May 1, 2013,[5] and the Senate Committee on Commerce, Science, and Transportation held a nomination hearing on June 18, 2013,[6] but declined to vote to confirm him. President Obama has not yet nominated anyone for the open Republican seat at the Commission.

News reports have stated that it is unlikely that the Senate will vote on Mr. Wheeler's nomination until a Republican nominee has been named. One name that has been reported is Mr. Mike O'Reilly, who is currently on the staff of Senator John Cornyn.[7]

FCC Structure

The day-to-day functions of the FCC are carried out by 7 bureaus and 10 offices. The current basic structure of the FCC was established in 2002 as part of the agency's effort to better reflect the industries it regulates. The seventh bureau, the Public Safety and Homeland Security Bureau, was established in 2006.

The bureaus process applications for licenses and other filings, analyze complaints, conduct investigations, develop and implement regulatory programs, and participate in hearings, among other things. The offices provide support services. Bureaus and offices often collaborate when addressing FCC issues.[8] The bureaus hold the following responsibilities:

- Consumer and Governmental Affairs Bureau—Addresses all types of consumer-related matters from answering questions and responding to consumer complaints to distributing consumer education materials.
- Enforcement Bureau—Enforces FCC rules, orders, and authorizations.
- International Bureau—Administers the FCC's international telecommunications policies and obligations.
- Media Bureau—Develops, recommends, and administers the policy and licensing programs relating to electronic media, including cable television, broadcast television, and radio in the United States and its territories.
- Public Safety and Homeland Security Bureau—Addresses issues such as public safety communications, alert and warning of U.S. citizens, continuity of government operations and continuity of operations planning, and disaster management coordination and outreach.
- Wireless Telecommunications Bureau—Handles all FCC domestic wireless telecommunications programs and policies.[9] Wireless communications services include cellular, paging, personal communications services, public safety, and other commercial and private radio services. This bureau also is responsible for implementing the competitive bidding authority for spectrum auctions.
- Wireline Competition Bureau—Administers the FCC's policies concerning common carriers—the companies that provide long distance and local service to consumers and businesses. These companies provide services such as voice, data, and other telecommunication transmission services.

FCC Strategic Plan

The current FCC Strategic Plan covers the five-year period FY2012–FY2016. The plan outlines eight goals:

- Connect America: Maximize Americans' access to—and the adoption of—affordable fixed and mobile broadband where they live, work, and travel.
- Maximize Benefits of Spectrum: Maximize the overall benefits of spectrum for the United States.
- Protect and Empower Consumers: Empower consumers by ensuring that they have the tools and information they need to make informed choices; protect consumers from harm in the communications market.
- Promote Innovation, Investment, and America's Global Competitiveness: Promote innovation in a manner that improves the nation's ability to compete in the global economy, creating a virtuous circle that results in more investment and in turn enables additional innovation.
- Promote Competition: Ensure a competitive market for communications and media services to foster innovation, investment, and job creation and to ensure consumers have meaningful choice in affordable services.

- Public Safety and Homeland Security: Promote the availability of reliable, interoperable, redundant, rapidly restorable critical communications infrastructures that are supportive of all required services.
- Advance Key National Purposes: Through international and national interagency efforts, advance the use of broadband for key national purposes.
- Operational Excellence: Make the FCC a model for excellence in government by effectively managing the Commission's human, information, and financial resources; by making decisions based on sound data and analyses; and by maintaining a commitment to transparent and responsive processes that encourage public involvement and best serve the public interest.

The FCC has identified performance objectives associated with each strategic goal. Commission management annually develops targets and measures related to each performance goal to provide direction toward accomplishing those goals. Targets and measures are published in the FCC's Performance Plan, submitted with the Commission's annual budget request to Congress. Results of the Commission's efforts to meet its goals, targets, and measures are found in the FCC's Annual Performance Report published each February. The FCC also issues a Summary of Performance and Financial Results every February, providing a concise, citizen-focused review of the agency's accomplishments.

FCC OPERATIONS: BUDGET, AUTHORIZATION, AND REPORTING TO CONGRESS

Since the 110th Congress, the FCC has been funded through the Financial Services (House) and Financial Services and General Government (Senate) appropriations process as a single line item. Previously, it was funded through what is now the Commerce, Justice, Science appropriations process, also as a single line item.

Most or all of the FCC's budget is derived from regulatory fees collected by the agency rather than through a direct appropriation.[10] The fees, often referred to as "Section (9) fees," are collected from license holders and certain other entities (e.g., cable television systems) and deposited into an FCC account. The law gives the FCC authority to review the regulatory fees and to adjust the fees to reflect changes in its appropriation from year to year. Most years, appropriations language prohibits the use by the Commission of any excess collections received in the current fiscal year or any prior years. These funds remain in the FCC account and are not made available to other agencies or agency programs nor redirected into the Treasury's general fund.

FCC FY2014 Budget

For FY2014, the FCC has requested a budget of $359,299,000, with no direct appropriation (i.e., the entire budget will be funded through auction proceeds).[11] It includes requests for funding to: (1) support Commission-wide information technology needs through extending the enterprise storage; (2) support for reform of the Universal Service Fund

Support Program; (3) space consolidation and facilities improvement that will reduce lease arrangements that are not cost effective and improve efficiencies; (4) create a Do-Not-Call registry for telephone numbers used by Public Safety Answering Points; (5) provide resources for mission-critical systems to ensure that they are operational during a Continuity of Operations event; and (6) provide contract funding to support mandatory audits for the Office of the Inspector General. The budget submission also includes a request to decrease the spending of Auctions funding from $98.7 million to $89.4 million to support the timely implementation of the Auctions Incentive program.

FCC Authorization

The FCC was last formally authorized in the FCC Authorization Act of 1990 (P.L. 101-396). Since that time, five bills have been introduced that would have reauthorized the FCC, but none were signed into law.

- 108th Congress, S. 1264, FCC Reauthorization Act of 2003, Senator John McCain;[12]
- 104th Congress, H.R. 1869, Federal Communications Commission Authorization Act, Representative Jack Fields;
- 103rd Congress, H.R. 4522, Federal Communications Commission Authorization Act, Representative Edward Markey, and 103rd Congress, S. 2336, Federal Communications Commission Authorization Act, Senator Daniel Inouye; and
- 102nd Congress, S. 1132, Federal Communications Commission Authorization Act, Senator Daniel Inouye.

FCC Reporting to Congress

The FCC publishes four reports for Congress.

- *Strategic Plan.* The Strategic Plan is the framework around which the FCC develops its yearly Performance Plan and Performance Budget. The FCC is to submit its next four-year Strategic Plan by February 2014, in accordance with the Government Performance and Results Modernization Act of 2010, P.L. 111-352.
- *Performance Budget.* The annual Performance Budget includes performance targets based on the FCC's strategic goals and objectives, and serves as the guide for implementing the Strategic Plan. The Performance Budget becomes part of the President's annual budget request.
- *Agency Financial Report.* The annual Agency Financial Report contains financial and other information, such as a financial discussion and analysis of the agency's status, financial statements, and audit reports.
- *Annual Performance Report.* At the end of the fiscal year, the FCC publishes an Annual Performance Report that compares the agency's actual performance with its targets.[13]

All of these reports are available on the FCC website.[14]

FCC-RELATED CONGRESSIONAL ACTION— 113TH CONGRESS15

Three hearings have been held on FCC oversight, reform, and management and four bills have been introduced that would affect the manner in which the FCC conducts its business.

Hearings

FCC Oversight. On March 12, 2013, the Senate Committee on Commerce, Science, and Transportation held a hearing on issues related to FCC oversight.[16] All five FCC commissioners testified.[17]

"Improving the FCC Process." On July 11, 2013, the House Committee on Energy and Commerce Subcommittee on Communications and Technology held a hearing on, "Improving the FCC Process." The hearing was held to review two draft bills based on two bills that were passed by the House of Representatives in the 112th Congress: H.R. 3309, the Federal Communications Commission Process Reform Act, and H.R. 3310, the Federal Communications Commission Consolidated Reporting Act. The draft bills are intended to "minimize the potential for procedural failings and abuse, and to improve agency transparency, efficiency, and accountability."[18] The committee heard from one panel: Stuart M. Benjamin, Douglas B. Maggs Chair in Law and Associate Dean for Research, Duke Law; Larry Downes, Internet industry analyst and author; Robert M. McDowell, former FCC Commissioner and Visiting Fellow, Hudson Institute; Randolph J. May, President, Free State Foundation; Richard J. Pierce Jr., Lyle T. Alverson Professor of Law, George Washington University Law School; and, James Bradford Ramsay, General Counsel, National Association of Regulatory Utility Commissioners.

Nominations. The Senate Committee on Commerce, Science, and Transportation held a hearing on the nomination of Mr. Tom Wheeler on June 18, 2013.[19] As previously noted, the Senate has not yet held a vote on this nomination.

Legislation

Commission Collaboration

Federal Communications Commission Collaboration Act (H.R. 539, S. 245)
- *Status.* H.R. 539 was introduced by Representative Anna Eshoo in the House Committee on Energy and Commerce on February 6, 2013. The bill was referred to the Subcommittee on Communications and Technology on February 8, 2013. This bill is identical to H.R. 1009, introduced by Representative Eshoo in the 112th Congress. S. 245 was introduced by Senator Amy Klobuchar in the Senate Committee on Commerce, Science, and Transportation on February 7, 2013.

- *Summary.* These bills would amend the Communications Act of 1934 to allow, notwithstanding a specified open meeting provision, three or more commissioners of the Federal Communications Commission (FCC) to hold a meeting that is closed to the public to discuss official business if (1) no agency action is taken; (2) each person present is an FCC commissioner or employee; (3) for each political party of which any commissioner is a member, at least one commissioner who is a member of the respective party is present, and, if any commissioner has no party affiliation, at least one unaffiliated commissioner is present; and (4) an attorney from the FCC's Office of General Counsel is present. It would require public disclosure of the meeting, attendees, and matters discussed.

Cost-Benefit Analysis of Proposed Rulemaking

FCC Analysis of Benefits and Costs ("ABCs") Act of 2013 (H.R. 2649)
- *Status.* H.R. 2649 was introduced by Representative Robert Latta in the House Committee on Energy and Commerce on July 10, 2013. This bill is also called the FCC Process Reform Act of 2013.
- *Summary.* This bill would amend the Communications Act of 1934 to reform the FCC by requiring an analysis of benefits and costs during the rulemaking process and creating certain presumptions regarding regulatory forbearance and biennial regulatory review determinations.

Independent Agency Regulatory Analysis Act of 2013 (S. 1173)
- *Status.* S. 1173 was introduced by Senator Rob Portman in the Senate Committee on Homeland Security and Governmental Affairs on June 18, 2013.
- *Summary.* This bill would authorize the President to require an independent regulatory agency to: (1) comply, to the extent permitted by law, with regulatory analysis requirements applicable to other federal agencies; (2) publish and provide the Administrator of the Office of Information and Regulatory Affairs with an assessment of the costs and benefits of a proposed or final economically significant rule (i.e., a rule that is likely to have an annual effect on the economy of $100 million or more and is likely to adversely affect sectors of the economy in a material way) and an assessment of costs and benefits of alternatives to the rule; and (3) submit to the Administrator for review any proposed or final economically significant rule. It would also prohibit judicial review of the compliance or noncompliance of an independent regulatory agency with the requirements of this act.

APPENDIX A. FCC-RELATED CONGRESSIONAL ACTION— 112TH CONGRESS

One hearing was held on FCC oversight and nine bills were introduced that would affect the manner in which the FCC conducts its business.

Hearings

On February 16, 2012, the House Committee on Energy and Commerce Subcommittee on Communications and Technology held a hearing on the budget and spending of the FCC.[20] FCC Chairman Julius Genachowski; Mr. David H. Hunt, FCC Inspector General; and Mr. Scott Barash, Chief Executive Officer of the Universal Service Administrative Company, testified.

Legislation

Cost-Benefit Analysis of Proposed Rulemaking

FCC Analysis of Benefits and Costs Act of 2011 (H.R. 2289)

H.R. 2289 Status. H.R. 2289, also called the "FCC ABCs Act," was introduced by Representative Robert Latta in the House Committee on Energy and Commerce on June 22, 2011. The bill was referred to the Subcommittee on Communications and Technology June 22, 2011. *H.R. 2289 Summary.* This bill would require the FCC to include in each notice of proposed rule making and in each final rule issued by the FCC an analysis of the benefits and costs of such proposed rule or final rule. It would prohibit any appropriations for the express purpose of carrying out this act.

Commission Collaboration

Federal Communications Commission Process Reform Act (H.R. 3309) Federal Communications Commission Process Reform Act (S. 1784) Telecommunications Jobs Act (S. 1817)

H.R. 3309 and S. 1784 Status. H.R. 3309 was introduced by Representative Greg Walden in the House Committee on Energy and Commerce on November 2, 2011. It was reported (H.Rept. 112- 414[21]) on March 19, 2012, and referred to the Senate the next day, where it was read and referred to the Committee on Commerce, Science, and Transportation. S. 1784 was introduced by Senator Dean Heller in the Senate Committee on Commerce, Science, and Transportation on November 2, 2011.

H.R. 3309 and S. 1784 Summary. This bill would require the FCC to (1) survey the state of the marketplace through a notice of inquiry before initiating every new rulemaking; (2) identify a market failure, consumer harm, or regulatory barrier to investment before adopting "economically significant" rules, as well as demonstrate that the benefits of the regulation outweigh the costs; (3) make the full text of a rule available to the public for 30 days of comments and 30 days of reply comments prior to voting on the proposed rule, and issue a final rule within three years; and (4) set "shot clocks" for orders, decisions, reports, or actions.

S. 1817 Status. S. 1817 was introduced by Senator Dean Heller in the Senate Committee on Commerce, Science, and Transportation on November 8, 2011. *S. 1817 Summary.* This bill is substantially similar to S. 1784.

Federal Communications Commission Collaboration Act (H.R. 1009)

H.R. 1009 Status. H.R. 1009 was introduced by Representative Anna Eshoo in the House Committee on Energy and Commerce on March 10, 2011. The bill was referred to the Subcommittee on Communications and Technology on March 15, 2011.

H.R. 1009 Summary. This bill would amend the Communications Act of 1934 to allow, notwithstanding a specified open meeting provision, three or more commissioners of the Federal Communications Commission (FCC) to hold a meeting that is closed to the public to discuss official business if (1) no agency action is taken; (2) each person present is an FCC commissioner or employee; (3) for each political party of which any commissioner is a member, at least one commissioner who is a member of the respective party is present, and, if any commissioner has no party affiliation, at least one unaffiliated commissioner is present; and (4) an attorney from the FCC's Office of General Counsel is present. It would require public disclosure of the meeting, attendees, and matters discussed.

Report Consolidation and Paperwork Reduction

Federal Communications Commission Consolidated Reporting Act (H.R. 3310)
Federal Communications Commission Consolidated Reporting Act (S. 1780)

H.R. 3310 and S. 1780 Status. H.R. 3310 was introduced by Representative Steve Scalise in the House Committee on Energy and Finance on November 2, 2011, and was reported (H.Rept. 112- 443[22]) on April 18, 2012. On May 30, 2012, the bill was passed by the House and it was referred to the Senate Committee on Science, Commerce, and Transportation, on June 4, 2012. *S. 1780* was introduced by Senator Dean Heller in the Senate Committee on Commerce, Science, and Transportation on November 2, 2011.

H.R. 3310 and S. 1780 Summary. This bill would amend the Communications Act of 1934 to consolidate the reporting obligations of the FCC to improve oversight and reduce reporting burdens.

Enhancing the Technical Expertise of the Commission

FCC Technical Expertise Capacity Heightening Act (S. 611)
FCC Commissioners' Technical Resource Enhancement Act (H.R. 2102)

S. 611 Status: S. 611 was introduced by Senator Olympia Snowe in the Senate Committee on Commerce, Science, and Transportation on March 17, 2011.

S. 611 Summary. This bill is substantially similar to its companion bill, H.R. 2102, but unlike that bill, S. 611 also includes a requirement that the FCC "enter into an arrangement with the National Academy of Sciences to complete a study of the technical policy decision making and the technical personnel at the Commission."

H.R. 2102 Status. H.R. 2102 was introduced by Representative Cliff Stearns in the House Committee on Energy and Commerce on June 2, 2011. The bill was referred to the Subcommittee on Communications and Technology on June 3, 2011.

H.R. 2102 Summary. This bill would amend the Communications Act of 1934 to permit each commissioner of the FCC to appoint an electrical engineer or computer scientist to provide technical consultation and to interface with the Office of Engineering and Technology and other FCC bureaus and technical staff. It would require such engineer or scientist to hold an undergraduate or graduate degree in his or her field of expertise.

APPENDIX B. FCC-RELATED GOVERNMENT ACCOUNTABILITY OFFICE STUDIES

The Government Accountability Office (GAO) has conducted two studies since 2008 related to the operation of the FCC.

FCC: Regulatory Fee Process Needs to Be Updated (August 2012)[23]

The FCC must, by law, assess annual regulatory fees on telecommunications entities to recover most or all of its appropriations—about $336 million in fiscal year (FY) 2011. Recently, the agency stated that it was planning to consider reforms to its regulatory fee process. GAO was asked to assess the FCC's:

- process for assessing regulatory fees among industry sectors; and
- regulatory fee collections over the past 10 years, and alternative approaches to assessing regulatory fees.

For this assessment, GAO:

- reviewed FCC data and documents;
- interviewed officials from the FCC and the telecommunications industry;
- identified alternative approaches to assessing regulatory fees; and
- met with five fee-funded U.S. and Canadian regulatory agencies.

GAO found that the FCC is currently assessing regulatory fees based on obsolete data, with limited transparency. The Communications Act requires fees to be based on the number of full-time equivalents (FTE) that perform regulatory tasks in certain bureaus, among other things. The FCC based its FY2011 regulatory fee assessments on its FY1998 FTE data. It has not updated the assessment on updated FTE data in part to avoid fluctuations in fees from year to year. FCC officials stated that the agency has complied with its statutory authority since the statute does not prescribe a specific time to update its FTE analysis. As a result, after 13 years, FCC has not validated the extent to which its fees correlate to its workload.

The GAO recommended that:

- Congress consider whether FCC's excess fees should be appropriated for FCC's use or, if not, what their disposition should be; and the FCC should:
  a) perform an updated FTE analysis and require at least biennial updates going forward;
  b) determine whether and how to revise the current fee schedule, including the number and type of fee categories;
  c) increase the transparency of its regulatory fee process; and
  d) consider the approaches of other fee-funded regulatory agencies.

The FCC agreed with GAO's recommendations.

Enforcement Program Management (February 2008)[24]

According to the GAO analysis of FCC data, between 2003 and 2006, the number of complaints received by the FCC totaled about 454,000 and grew from almost 86,000 in 2003 to a high of about 132,000 in 2005. The largest number of complaints related to violations of the do-not-call list and telemarketing during prohibited hours. The FCC processed about 95% of the complaints it received. It also opened about 46,000 investigations and closed about 39,000; approximately 9% of these investigations were closed with an enforcement action and about 83% were closed with no enforcement action. The GAO was unable to determine why these investigations were closed with no enforcement action because the FCC does not systematically collect these data. The FCC told GAO that some investigations were closed with no enforcement action because no violation occurred or the data were insufficient.

The GAO noted that the FCC assesses the impact of its enforcement program by periodically reviewing certain program outputs, such as the amount of time it takes to close an investigation, but it lacks management tools to fully measure its outcomes. Specifically, FCC has not set measurable enforcement goals, developed a well-defined enforcement strategy, or established performance measures that are linked to the enforcement goals. The GAO stated in its report that without key management tools, FCC may have difficulty assuring Congress and other stakeholders that it is meeting its enforcement mission.

The GAO found that limitations in FCC's current approach for collecting and analyzing enforcement data constitute the principal challenge the agency faces in providing complete and accurate information on its enforcement program. These limitations, according to the GAO, make it difficult to analyze trends; determine program effectiveness; allocate Commission resources; or accurately track and monitor key aspects of all complaints received, investigations conducted, and enforcement actions taken.

End Notes

[1] The Communications Act of 1934, 47 U.S.C. §151 et seq., has been amended numerous times, most significantly in recent years by the Telecommunications Act of 1996, P.L. 104-104, 110 Stat. 56 (1996). References in this report are to the 1934 Act, as amended, unless indicated. A compendium of communications-related laws is available from the House Committee on Energy and Commerce at http://energycommerce.house.gov/108/pubs/108-D.pdf. It includes selected Acts within the jurisdiction of the Committee, including the Communications Act of 1934, Telecommunications Act of 1996, Communications Satellite Act of 1962, National Telecommunications and Information Administration Organizations Act, Telephone Disclosure and Dispute Resolution Act, Communications Assistance for Law Enforcement Act, as well as additional communications statutes and selected provisions from the United States Code. The compendium was last amended on December 31, 2002.

[2] See "About the FCC," at http://www.fcc.gov/aboutus.html.

[3] 47 U.S.C. §151.

[4] When Congress established the FCC in 1934, it merged responsibilities previously assigned to the Federal Radio Commission, the Interstate Commerce Commission, and the Postmaster General into a single agency, divided into three bureaus, Broadcast, Telegraph, and Telephone. See Analysis of the Federal Communications Commission, Fritz Messere, at http://www.oswego.edu/~messere/FCC1.html and the Museum of Broadcast Communications Archive at http://www.museum.tv/archives/etv/F/htmlF/federalcommu/federalcommu.htm for additional information on the history of the FCC.

[5] http://www.whitehouse.gov/blog/2013/05/01/president-obama announces-his-nominees-fcc-chair-and-fhfa-director.

[6] The hearing was held in two sessions, online at http://www.c-spanvideo.org/program/ChairNom and http://www.cspanvideo.org/program/ChairNomi.

[7] http://www.politico.com/story/2013/07/wheeler-gop-nominee

[8] FCC Fact Sheet, http://www.fcc.gov/cgb/consumerfacts/aboutfcc.html.

[9] Except those involving satellite communications broadcasting, including licensing, enforcement, and regulatory functions. These functions are handled by the International Bureau.

[10] The Omnibus Budget Reconciliation Act of 1993 (P.L. 103-66, 47 U.S.C. §159) requires that the FCC annually collect fees and retain them for FCC use to offset certain costs incurred by the Commission. The FCC implemented the regulatory fee collection program by rulemaking on July 18, 1994.

[11] http://www.fcc.gov/document/fcc-fy-2014-budget.

[12] For more information, see S.Rept. 108-140, at http://www.gpo.gov/fdsys/pkg/CRPT-108srpt140/pdf/CRPT-108srpt140.pdf.

[13] OMB Circular A-136 allows agencies the option of producing (1) two separate reports, an Agency Financial Report and an Annual Performance Report, or (2) a consolidated Performance and Accountability Report. The same information is provided to Congress in either case. The FCC elected the first option for FY2011. Also, in addition to the reports it submits to Congress, the FCC publishes an annual Summary of Performance and Financial Information, which is a citizen-focused summary of the FCC's yearly activities.

[14] http://www.fcc.gov/encyclopedia/fcc-strategic-plan.

[15] The 110th Congress assigned responsibility for FCC appropriations process to the Subcommittee on Financial Services within the Committee on Appropriations, where has remained.

[16] Information about this hearing, including a video of the hearing, is available at http://www.commerce.senate.gov/public/index.cfm?p=Hearings&ContentRecord_id=18f83ea5-3d7d-4ef9-92ad-30a3421c11d3&ContentType_id= 14f995b9-dfa5-407a-9d35-56cc7152a7ed&Group_id=b06c39af-e033-4cba-9221-de668ca1978a.

[17] A number of organizations published summaries of the hearing, for example: Bloomberg BNA, http://dailyreport.bna.com/drpt/7010/split_display.adp?fedfid=29986801&vname=dernotallissues&wsn=499841500& searchid=19900305&doctypeid=1&type=date&mode=doc&split=0&scm=7010&pg=1; Benton Foundation, http://benton.org/node/147694; and Reuters, http://www.reuters.com/article/2013/03/12/us-usa-fcc-oversightidUSBRE92B17420130312.

[18] http://docs.house.gov/meetings/IF/IF16/20130711/101107/HHRG-113-IF16-20130711-SD002.pdf. Changes between the bills from the 112th Congress and the 113th Congress are highlighted in this document. The draft bills are available at http://docs.house.gov/meetings/IF/IF16/20130711/101107/BILLS-113pih-FCCConsolidatedReportingAct.pdf and http://docs.house.gov/meetings/IF/IF16/20130711/101107/BILLS-113pih-FCCProcessReformAct.pdf.

[19] The hearing was held in two sessions, online at http://www.c-spanvideo.org/program/ChairNom and http://www.cspanvideo.org/program/ChairNomi.

[20] Information about this hearing, including a video of the hearing, is available at http://energycommerce.house.gov/hearings/hearingdetail.aspx?NewsID=9278.

[21] http://www.gpo.gov/fdsys/pkg/CRPT-112hrpt414/pdf/CRPT-112hrpt414.pdf.

[22] http://www.gpo.gov/fdsys/pkg/CRPT-112hrpt443/pdf/CRPT-112hrpt443.pdf.

[23] GAO, Report to the Ranking Member, Committee on Energy and Commerce, and Ranking Member, Subcommittee on Telecommunications and the Internet, Committee on Energy and Commerce, House of Representatives, "FCC: Regulatory Fee Process Needs to be Updated," August 10, 2012, http://www.gao.gov/assets/600/593506.pdf.

[24] GAO, Report to the Chairman, Subcommittee on Telecommunications and the Internet, Committee on Energy and Commerce, House of Representatives, "FCC Has Made Some Progress in the Management of Its Enforcement Program but Faces Limitations, and Additional Actions Are Needed," February 15, 2008, http://www.gao.gov/new.items/ d08125.pdf.

In: Transformations in Telecommunications and Media   ISBN: 978-1-62948-413-6
Editor: Irwin Cavazos   © 2013 Nova Science Publishers, Inc.

Chapter 12

THE NATIONAL TELECOMMUNICATIONS AND INFORMATION ADMINISTRATION (NTIA): ISSUES FOR THE 113TH CONGRESS*

Linda K. Moore

SUMMARY

The National Telecommunications and Information Administration (NTIA), a bureau of the Department of Commerce, is the executive branch's principal advisory office on domestic and international telecommunications and information policies. Its mandate is to provide greater access for all Americans to telecommunications services, support U.S. efforts to open foreign markets, advise on international telecommunications negotiations, and fund research for new technologies and their applications.

NTIA also manages the distribution of funds for several key grant programs. Its role in managing radio frequency spectrum allocated for federal use includes addressing policies for sharing, and monitoring and resolving questions regarding usage, including causes of interference. It is responsible for identifying federal spectrum that can be transferred to commercial use through the auction of spectrum licenses, conducted by the Federal Communications Commission. Many of the NTIA's responsibilities are shared with other agencies.

With the passage of the Middle Class Tax Relief and Job Creation Act of 2012 (P.L. 112-96), in February 2012, Congress has given the NTIA new responsibilities in spectrum management and the support of public safety initiatives. The 113th Congress may wish to review the NTIA's performance in meeting its obligations under the act. Policy makers may also wish to consider if some of the NTIA's shared obligations might be effectively and efficiently transferred to its partners, allowing the NTIA to focus on communications policies that are considered by many to be key to future economic growth and development.

---

* This is an edited, reformatted and augmented version of a Congressional Research Service publication, CRS Report for Congress R42886, prepared for Members and Committees of Congress, from www.crs.gov, dated May 22, 2013.

INTRODUCTION

The National Telecommunications and Information Administration (NTIA) is one of 12 bureaus in the U.S. Department of Commerce (DOC). The NTIA frequently works with other executive branch agencies to develop and present the Administration's position on key policy matters. It represents the executive branch in both domestic and international telecommunications and information policy activities. Policy areas in which the NTIA acts as the representative of the Administration include international negotiations regarding global agreements on the Internet and spectrum management, and domestic use of spectrum resources by federal agencies.

In recent years, one of the responsibilities of the NTIA has been to oversee the transfer of some radio frequencies from the federal domain to the commercial domain. Many of these frequencies have subsequently been auctioned to the commercial sector and the proceeds paid into the U.S. Treasury. As part of President Obama's Wireless Initiative, the NTIA is charged with identifying electromagnetic spectrum that might be transferred from the federal sector to commercial wireless use.[1]

This spectrum might be auctioned as licenses for exclusive commercial use, made available for sharing between federal and commercial users, or repurposed in some other way that meets the stated goal of the Wireless Initiative to add 500 MHz of spectrum for wireless broadband.[2] Congress also has required the NTIA to take actions to release spectrum from federal to commercial use.[3]

The NTIA administers some grants programs created by Congress, including the Broadband Technology Opportunities Program (BTOP).[4] BTOP grant programs are in the final stages of completion. As required by the Middle Class Tax Relief and Job Creation Act of 2012 (P.L. 112- 96), the NTIA is commencing a $135 million grant program to help states plan for participation in a new, nationwide public safety broadband network. To deploy the new network, the act established the First Responder Network Authority, or FirstNet, as an independent agency within the NTIA and assigned the agency various responsibilities to support FirstNet.

FirstNet is funded through the Public Safety Trust Fund, established by Congress to receive revenues from auctions of certain spectrum licenses. FirstNet received an advance of nearly $2 billion from the U.S. Treasury against expected proceeds of sales of spectrum licenses. Another $5 billion in funding is expected from the Public Safety Trust Fund as auction revenues are deposited in the account.

FISCAL YEAR APPROPRIATIONS
AND BUDGET REQUESTS

The President's budget request for the NTIA for FY2014 is $52.1 million for salaries and expenses, a net increase of $5.2 million over the amount requested for FY2013. The Consolidated and Further Continuing Appropriations Act, 2013 provided $45.1 million for salaries and expenses for FY2013; no funding was appropriated for separate programs.

Table 1. NTIA: Fiscal Year Appropriations 2007-2013
(in millions of dollars)

Funding	FY2008	FY2009	FY2010	FY2011	FY2012	FY2013[a]
NTIA Total	$36.3	$39.2	$40.0	$41.6	$45.6	$45.1
Administration, salaries and expenses	$17.5	$19.2	$20.0	$41.6	$45.6	$45.1
PTFPC[b]	$18.8	$20.0	$20.0	0	0	0

Source: Annual Reports, Department of Commerce and Congressional Appropriations, as Enacted.
  Appropriations for ongoing grant programs are not included.
[a] Includes deduction for rescission but not sequestration.
[b] The grant program for the Public Telecommunications Facilities, Planning and Construction (PTFPC)
  program was terminated by Congress in FY2011.

In FY2010, the Public Telecommunications Facilities Program (PTFP) represented half of the NTIA's budget appropriations. In FY2011, the total enacted budget appropriations amount for the NTIA increased by 4% to $41.6 million; funding for the PTFP was transferred to administrative expenses and salaries. According to the NTIA, the initial increase of $21.6 million from FY2010 to FY2011 in funding for salaries and expenses was largely attributable to the costs of administration of a $4.7 billion program for broadband deployment, as required by the American Recovery and Reinvestment Act of 2009 (P.L. 111-5).[5] In FY2012 requests for funding to administer grant programs totaled $32.3 million, 70% of the fiscal year budget request.[6]

For FY2013, $25.8 million in funding was designated to administer the remaining broadband grant programs, primarily BTOP. The FY2014 request for broadband grant program oversight is for $24.7 million, roughly 40% of the total budget request.

The NTIA also receives funding from sources such as fees charged to federal agencies for spectrum management services. Reimbursable funding for FY2014 is estimated at $37.5 million, of which spectrum management fees from federal agencies are projected to be $28.9 million;[7] these fees were estimated at $28.7 million for FY2013.[8] The balance is attributable to reimbursable projects in telecommunications technology research.

The FY2014 budget request includes $7.5 million and staff increases of eight FTEs for a new program to develop a spectrum monitoring system and assess technologies for sharing radio frequency spectrum. The program would encompass pilot projects over a period of two years in ten major metropolitan areas.

Both Congress and the Administration have required the NTIA to take new actions in identifying and releasing additional radio frequency spectrum for wireless broadband use. To meet these obligations, funding of $1.25 million and five FTEs are requested for the Office of Spectrum Management for Wireless Broadband Access, a new line item.

Increases in staffing for some programs are offset in part by a reduction of four FTEs to administer broadband programs, and a reduction of 7 FTEs at the Institute of Telecommunication Sciences (ITS).

Total FTEs for both directly funded and reimbursable programs is estimated at 309 FTEs, compared to 302 FTEs for FY2013 (under the Continuing Resolution) and 257 FTEs in FY2012.

PROGRAMS

The NTIA fulfills many responsibilities for different constituencies. As the agency responsible for managing spectrum used by federal agencies, the NTIA often works in consultation with the Federal Communications Commission (FCC) on matters concerning spectrum access, technology, and policy. The FCC regulates private sector, state, local, and tribal spectrum use. Because many spectrum issues are international in scope and negotiated through treaty-making, the NTIA and the FCC collaborate with the Department of State in representing American interests. The NTIA also participates in interagency efforts to develop Internet policy.[9] The NTIA and the National Institute of Standards (NIST) have adjoining facilities on the Department of Commerce campus in Boulder, CO, where they collaborate on research projects with each other and with other federal agencies, such as the FCC.

The NTIA worked with the Rural Utilities Service in coordinating grants made through BTOP. The NTIA collaborates with NIST, the FCC, and the Department of Homeland Security (DHS) in providing expertise and guidance to grant recipients using BTOP funds to build new wireless networks for broadband communications.

As described by the NTIA,[10] its policies and programs are administered through

- The Office of Spectrum Management (OSM), which formulates and establishes plans and policies that ensure the effective, efficient, and equitable use of the spectrum both nationally and internationally. Through the development of long range spectrum plans, the OSM works to address future federal government spectrum requirements, including public safety operations and the coordination and registration of federal government satellite networks. The OSM also handles the frequency assignment needs of the federal agencies and provides spectrum certification for new federal agency radio communication systems.

- The Office of Policy Analysis and Development (OPAD), which is the domestic policy division of the NTIA. OPAD supports the NTIA's role as principal adviser to the Executive Branch and the Secretary of Commerce on telecommunications and information policies by conducting research and analysis and preparing policy recommendations.

- The Office of International Affairs (OIA), which develops and implements policies to enhance U.S. companies' ability to compete globally in the information technology and communications (ICT) sectors. In consultation with other U.S. agencies and the U.S. private sector, OIA participates in international and regional fora to promote policies that open ICT markets and encourage competition.

- The Institute for Telecommunication Sciences (ITS), which is the research and engineering laboratory of the NTIA. ITS provides technical support to the NTIA in advancing telecommunications and information infrastructure development, enhancing domestic competition, improving U.S. telecommunications trade opportunities, and promoting more efficient and effective use of the radio spectrum.

- The Office of Telecommunications and Information Applications (OTIA), which administers grant programs that further the deployment and use of technology in America, and the advancement of other national priorities. In the past, the OTIA has awarded grants from the Public Telecommunications Facilities Program, which was terminated by Congress in FY2011. The program supported new construction for public broadcasting stations and other organizations.

- The Office of Public Safety Communications, which was created by the NTIA at the end of 2012, to administer some provisions of the Middle Class Tax Relief and Job Creation Act of 2012, Title VI, also known as the Spectrum Act.

For budget purposes, the category of salaries and expenses is organized into four sub-activities: Domestic and International Policies; Spectrum Management; Telecommunication Sciences Research; and Broadband Programs.

Termination of the Public Telecommunications Facilities Program

Effective FY2011, Congress terminated grant funding for the Public Telecommunications Facilities Program (PTFP). In FY2010, the program received $20 million in funding to support broadcast and non-broadcast projects. Approximately half of the grant monies went to public radio and television stations to replace equipment. Another 25% of grant funds were awarded to bring radio and television services to unserved or underserved communities. Other awards included grants to 16 public television and radio stations to cover costs of converting from analog to digital broadcasting. These grants helped the Public Broadcasting Service to maintain and improve its critical role in the current Emergency Alert system (EAS) and new initiatives for Wireless Emergency Alerts (also known as commercial mobile alerts).[11] For example, the satellite communications network that supports EAS is operated by the National Public Radio, public television stations provide back-up for Wireless Emergency Alerts to mobile devices, and public television and radio stations provide emergency alerts and information to otherwise unserved communities.

SPECTRUM ACT

The most recent legislative action to provide more spectrum for commercial services was included in provisions of Title VI of the Middle Class Tax Relief and Job Creation Act of 2012 (P.L. 112-96).[12] Title VI is generally referred to as the Spectrum Act or the Public Safety and Spectrum Act.

Public Safety

The Spectrum Act has given the NTIA responsibilities to create and support FirstNet in planning, building and managing a new, nationwide, broadband network for public safety communications.[13] The NTIA will also be responsible for managing the Public Safety Trust Fund, created by the act, which remains in effect through FY2022.

The NTIA has created an Office of Public Safety Communications to oversee the State and Local Implementation Fund grant process. The new office will also help FirstNet with procurement issues. The Office will manage service-level agreements for the agency to supply administrative, technical, staffing, and other resources, as requested, to FirstNet.

The act also re-establishes the federal 9-1-1 Implementation Coordination Office (ICO) to plan for next-generation systems (NG 9-1-1) and to administer a grant program.[14] ICO is to be jointly administered by the NTIA and the National Highway Traffic Safety Administration (NHTSA). ICO is to provide matching grants for improvements in the implementation of 911 emergency services, and other purposes, from a grant program authorized at $115 million. Based on the act's prioritized plan for funding programs with spectrum license auction revenue, the funds for the grant program will be made available only after $27.635 billion of available auction revenue has been applied to other purposes. ICO, in consultation with NHTSA and DHS, is to report on costs for requirements and specifications of NG 9-1-1 services, including an analysis of costs, and assessments and analyses of technical uses.

Public Safety Trust Fund and FirstNet

The NTIA is to assure that some of the auction revenues designated for the Public Safety Trust Fund are placed in the Network Construction Fund, which is to be established as an account in the Treasury. The fund is to be used by FirstNet for expenditures on construction, maintenance, and related expenses to build the nationwide network required in the act, and by the NTIA for grants to those states that qualify to build their own radio access network links to FirstNet. The NTIA is also to facilitate payments to states that participate in the deployment of the network. The FY2014 budget estimate shows that $1.908 billion is to be available in the Public Safety Trust Fund, of which $1.902 billion is to be obligated for purchases of goods and services from government accounts.[15] For FY2014, $257 million is designated for the Network Construction Fund.[16]

The act established a State and Local Implementation Fund and required the NTIA, in consultation with FirstNet, to establish grant program requirements. Grants from this fund will be available to all 56 states and territories to support planning, consultation, data collection, education, and outreach activities. Expenditures by the NTIA from the State and Local Implementation Fund were reported at $300,000 for FY2012 for administrative costs. Disbursements for administrative costs and grants funding are estimated at $124,958,000 (base) for FY2013 and $9,700,000 for FY2014.[17] The announced amount available for the first phase of grants from the fund is $121.5 million.[18]

BTOP Grants and FirstNet

Grants under the BTOP program included seven projects to develop broadband communications for public safety.[19] After the passage of the Spectrum Act, the NTIA partly suspended funding to these projects in order to allow the FirstNet board of directors time to evaluate how the projects might be coordinated with plans for a nationwide network. Furthermore, FirstNet was assigned the sole, national license for public safety broadband; under the Spectrum Act separate lease agreements are required for spectrum access. In February 2013, the board agreed to move forward with negotiations on leasing agreements.[20]

Spectrum Reallocation

The Spectrum Act updated existing and specified new procedures for spectrum to be reallocated from federal government to commercial use. Under the act, the NTIA is required to work with the FCC to identify specific bands for release to commercial use.

The act also addressed how spectrum resources might be repurposed from federal to commercial use through auction or sharing, and how the cost of such reassignment would be defined and compensated, among other provisions. Although spectrum sharing to facilitate the transition from federal to commercial use is supported in the act's provisions, the NTIA has been required to give priority to reallocation options that assign spectrum for exclusive, non-federal uses through competitive bidding.

The act has required the establishment of a Technical Panel within the NTIA to review transition plans that each federal agency must prepare in accordance with provisions in the act. The Technical Panel is required to have three members qualified as a radio engineer or technical expert. The Director of the Office of Management and Budget, the Assistant Secretary of Commerce for Communications and Information, and the Chairman of the FCC have been required to appoint one member each. A discussion and interpretation of provisions of the act as regards the technical panel and related procedural requirements such as dispute resolution have been published by the NTIA as part of the rulemaking process.[21]

SPECTRUM POLICY

The Administration and Congress have taken steps to increase the amount of radio frequency spectrum available for mobile services such as access to the Internet. The increasingly popular smart phones and tablets require greater spectrum capacity (broadband) than the services of earlier generations of cell phones. Proposals from policy makers to use federal spectrum to provide commercial mobile broadband services include:

- Clearing federal users from designated frequencies for transfer to the commercial sector through a competitive bidding system.
- Sharing federal frequencies with specific commercial users.
- Improving the efficiency of federal spectrum use and management.[22]
- Using emerging technologies for network management to allow multiple users to share spectrum as needed.

The NTIA supports the Administration's policy goal of increasing spectrum capacity for mobile broadband by 500 MHz.[23] To this purpose, the NTIA, with input from the Policy and Plans Steering Group (PPSG),[24] has produced a 10-year plan and timetable that identifies bands of spectrum that might be available for commercial wireless broadband service. As part of its planning efforts, the NTIA prepared a "Fast Track Evaluation" of spectrum that might be made available in the near future. [25] Specific recommendations were to make available 15 MHz of spectrum from frequencies between 1695 MHz and 1710 MHz, and 100 MHz of spectrum within bands from 3550 MHz to 3650 MHz. The fast track evaluation also recommended studying two 20 MHz bands to be identified within 4200-4400 MHz for possible repurposing, and placement for consideration of this proposal on the agenda of the World Radio Conference (WRC-2015) scheduled for 2015-2016. The World Radio Conference, held approximately every four years, is the primary forum for negotiating international treaties on spectrum use.

Many decisions regarding the use of federal spectrum are made through the Interdepartmental Radio Access Committee, IRAC.[26] IRAC membership comprises representatives of all branches of the U.S. military and a number of federal department agencies affected by spectrum management decisions.[27] The NTIA is advised regarding broader spectrum policy issues by the Commerce Spectrum Advisory Committee (CSMAC), a Federal Advisory Committee. The committee was created in 2004 and is comprised of experts from outside the federal government.[28] The Office of Management and Budget also influences agency spectrum management through budget planning and recommendations.

Administration Policy on Spectrum Sharing

The President's Council of Advisors on Science and Technology (PCAST) has endorsed increasing spectrum capacity through new technology that increases efficiency and allows for shared use of spectrum resources. In a report, *Realizing the Full Potential of Government Held Spectrum to Spur Economic Growth*, the council has proposed that up to 1000 MHz of additional spectrum capacity could be provided through shared access between the federal government and commercial providers.[29] The report identified existing technologies for sharing that could be used as platforms. The report's recommendations included the development of new spectrum policies based on spectrum-sharing. The report stated that "the norm for spectrum use should be sharing," and that the White House should take actions to advance toward this goal. The NTIA has supported this policy statement[30] and its plans to test spectrum sharing are in response to the PCAST goal.

Reallocating Federal Spectrum

Working through the PPSG, the NTIA studied federal spectrum use by more than 20 agencies with over 3,100 separate frequency assignments in the 1755-1850 MHz band.[31] After evaluating the multiple steps involved in transferring current uses and users to other frequency locations, the NTIA concluded that it would cost $18 billion to clear federal users from all 95 MHz of the band.

Based on this assessment, the report included recommendations for seeking ways for federal and commercial users to share many of the frequencies, although some frequencies were identified to be cleared for auction to the private sector. At a hearing of the House Committee on Energy and Commerce, Subcommittee on Communications and Technology, [32] the GAO provided testimony regarding its preliminary findings on spectrum sharing[33] and followed up with a report.[34] Both the hearing and the report indicated that spectrum sharing technology and policies were largely undeveloped. Some of the options to encourage sharing spectrum, as identified by the GAO, include considering spectrum usage fees to provide economic incentive for more efficient use and sharing; identifying more spectrum that could be made available for unlicensed use; encouraging research and development of technologies that can better enable sharing; and improving and expediting regulatory processes related to sharing. Given the challenges for implementing spectrum sharing policies, the GAO found that further study by the NTIA and the FCC was needed.

The NTIA assumptions for the estimates of the cost of relocating federal agencies from the 1755- 1850 MHz band were challenged in a congressional hearing, leading to a request to the GAO to examine the process. In particular, the NTIA was criticized during the hearing by some committee members for not separately evaluating the 1755-1780 MHz band, which might be auctioned separately with another spectrum band already available for commercial use.

GAO Cost Estimates for Spectrum Reallocation

In a hearing before the Senate Committee on Armed Services, Subcommittee on Strategic Forces,[35] the GAO presented preliminary findings on Department of Defense (DOD) estimates of reallocation costs from some radio frequencies.[36] The GAO evaluated DOD relocation cost estimates for frequencies at 1755 MHz-1850 MHz and reported that the "preliminary cost estimate substantially or partially met GAO's identified best practices." In particular, the GAO noted the variable nature of a number of assumptions for costs and revenues, such as the characteristics of the spectrum to which services would be relocated, the availability of new technology, and market demand for spectrum.

INTERNET POLICY

Working with other stakeholders the NTIA leads and participates in interagency efforts to develop Internet policy. In addition, the NTIA works with other governments and international organizations to discuss and reach consensus on relevant Internet policy issues.

Along with the Executive Office of the President, the Office of the Secretary of Commerce, and department bureaus NIST and the International Trade Administration (ITA), the NTIA plays a role in the Internet Policy Task Force, created in 2010 by the Secretary of Commerce.[37] One of the NTIA's functions on the Task Force is to assist in the establishment of a code of conduct on mobile application transparency.[38]

The NTIA is the lead Executive Branch agency on issues relating to the Domain Name System (DNS) and supports a multi-stakeholder approach to the coordination of the DNS to ensure the long-term viability of the Internet as a force for innovation and economic growth.[39]

RESEARCH

The Institute for Telecommunication Sciences, located in Boulder, CO, is the research and engineering arm of the NTIA. ITS provides core telecommunications research and engineering services to promote: enhanced domestic competition and new technology deployment; advanced telecommunications and information services; foreign trade opportunities for American telecommunication firms; and more efficient use of spectrum.

ISSUES FOR THE 113TH CONGRESS

Principal activities for FY2014 as cited by the NTIA[40] are:

- Evaluate options for repurposing federal spectrum for commercial wireless broadband use. This includes new expenditures to develop a spectrum monitoring system.
- Oversee the activities of FirstNet.
- Lead the formation of domestic and international Internet policies such as for data privacy and the free flow of information globally.
- Monitor broadband grants awarded under the American Recovery and Reinvestment Act of 2009.

Many of the NTIA's functions are performed in conjunction with other agencies. The NTIA's role as liaison may lead to overlapping responsibilities, leading to duplication of effort across departments and agencies. At the same time, rapid advances in communications technology have changed the mission of the NTIA in areas such as spectrum policy. As an example, policy makers may wish to consider if some of the NTIA's shared obligations might be effectively and efficiently transferred to its partners, allowing the NTIA to focus on communications policies that are considered by many to be key to future economic growth and development. As it reviews communications and spectrum policy, the 113th Congress may also choose to consider if the current structure of the NTIA might be better aligned to its new responsibilities.

For purposes of oversight, Congress may—for example—choose to examine the efficacy of the NTIA's spectrum management activities, and to evaluate the agency's compliance with the Spectrum Act (P.L. 112-96, Title VI). Oversight might cover requirements of the act regarding the transfer of spectrum from federal to commercial use and the act's provisions for public safety.

End Notes

[1] The White House, Office of the Press Secretary, "President Obama Details Plan to Win the Future Through Expanded Wireless Access," Fact Sheet, February 10, 2011, http://www.whitehouse.gov/the-press-office/2011/02/10/presidentobama-details-plan-win-future-through-expanded-wireless-access.

[2] Spectrum is segmented into bands of radio frequencies and typically measured in cycles per second, or hertz. Standard abbreviations for measuring frequencies include kHz—kilohertz or thousands of hertz; MHz—megahertz, or millions of hertz; and GHz—gigahertz, or billions of hertz.

[3] P.L. 112-96, Sections 6401 and 6701

[4] For a discussion of BTOP grants, see CRS Report R41775, Background and Issues for Congressional Oversight of ARRA Broadband Awards, by Lennard G. Kruger.

[5] This amount was later reduced by Congress to $4.4 billion.

[6] U.S. Department of Commerce, National Telecommunications and Information Administration, FY2013 Budget as Presented to Congress, February 2012.

[7] U.S. Department of Commerce, National Telecommunications and Information Administration, FY2014 Budget as Presented to Congress, April 2013.

[8] U.S. Department of Commerce, National Telecommunications and Information Administration, FY2013 Budget as Presented to Congress, February 2012.

[9] For background information on NTIA's role in U.S. Internet policy, see CRS Report 97-868, Internet Domain Names: Background and Policy Issues, by Lennard G. Kruger.

[10] See http://www.ntia.doc.gov/about.

[11] Background information on FEMA and FCC websites, such as http://www.fema.gov/emergency

[12] Provisions in Title VI of the act are discussed in CRS Report R40674, Spectrum Policy in the Age of Broadband: Issues for Congress, by Linda K. Moore.

[13] Actions taken by the NTIA in establishing and assisting FirstNet are documented at http://www.ntia.doc.gov/category/public-safety.

[14] Previous legislation and for NG9-1-1 is discussed in CRS Report R41208, Emergency Communications: Broadband and the Future of 911 , by Linda K. Moore.

[15] U.S. Department of Commerce, National Telecommunications and Information Administration, FY2014 Budget as Presented to Congress, April 2013; Public Safety Trust Fund, Exhibit 16.

[16] U.S. Department of Commerce, National Telecommunications and Information Administration, FY2014 Budget as Presented to Congress, April 2013; Network Construction Fund, Exhibit 6.

[17] U.S. Department of Commerce, National Telecommunications and Information Administration, FY2014 Budget as Presented to Congress, April 2013; State and Local Implementation Fund, Exhibit 10.

[18] NTIA, "NTIA Announces Availability of $121.5 Million in State Grants to Assist with FirstNet Planning," February 6, 2013 (http://www.ntia.doc.gov/press-release/2013/ntia-announces-availability-1215-million-state-grants-assistfirstnet-planning) and "State and Local Implementation Grant Program Federal Funding Opportunity," February 6, 2013 (http://www.ntia.doc.gov/other-publication/2013/sligp-federal-funding-opportunity).

[19] Locations are Adams County, CO; Charlotte, NC; State of Mississippi; Los Angeles, CA; San Francisco Bay Area, CA; northern New Jersey; and New Mexico.

[20] NTIA Press Release, "FirstNet Board Director Sue Swenson Provides Update on Status of BTOP Negotiations," March 28, 2013, at http://www.ntia.doc.gov/press-release/2013/firstnet-board-member-sue-swenson-provides-updatestatus-btop-negotiations-0.

[21] NTIA, Notice of Proposed Rulemaking, July 17, 2012, and replies, docket no. 110627357-2209-03 at http://www.ntia.doc.gov/federal-register-notice/2012/technical-panel-and-dispute-resolution-board-nprm.

[22] The Government Accountability Office (GAO) issued a report: Spectrum Management: NTIA Planning and Processes Need Strengthening to Promote the Efficient Use of Spectrum by Federal Agencies, April 2011, GAO-11- 352.

[23] Broadband refers here to the capacity of the radio frequency channel. A broadband channel can quickly transmit live video, complex graphics, and other data-rich information as well as voice and text messages, whereas a narrowband channel might be limited to handling voice, text, and some graphics.

[24] Created in response to Department of Commerce recommendations to improve spectrum efficiency through better management, see http://www.ntia.doc.gov/legacy/reports/specpolini/factsheetspecpolini_06242004.htm.

[25] NTIA, An Assessment of Near-Term Viability of Accommodating Wireless Broadband Systems in the 1675-1710 MHZ, 1755-1780 MHz, 3500-3650 MHz, and 4200-4220 MHz, 4380-4400 MHZ Bands (President's Spectrum Plan Report), November 15, 2010, at http://www.ntia.doc.gov/report/2010/assessment-near-term-viability-accommodatingwireless-broadband-systems-1675-1710-mhz-17.

[26] See http://www.ntia.doc.gov/category/irac.

[27] Members are listed at http://www.ntia.doc.gov/page/irac-functions-and-responsibilities.

[28] See http://www.ntia.doc.gov/category/csmac.

[29] Recommendations of the President's Council of Advisors on Science and Technology, Realizing the Full Potential of Government-Held Spectrum to Spur Economic Growth, released July 20, 2012, http://www.whitehouse.gov/sites/ default/files/microsites/ostp/pcast_spectrum_report_final_july_20_2012.pdf.

[30] NTIA Blog, "Supporting Innovative Approaches to Spectrum Sharing," posted March 11, 2013 at http://www.ntia.doc.gov/blog/2013/supporting-innovative-approaches-spectrum-sharing.

[31] U.S. Department of Commerce, An Assessment of the Viability of Accommodating Wireless Broadband in the 1755- 1850 MHz Band, March 2012, at http://www.ntia.doc.gov/report/2012/assessment-viability-accommodating-wirelessbroadband-1755-1850-mhz-band.

[32] Hearing, House of Representatives, Committee on Energy and Commerce, Subcommittee on Communications and Technology, "Creating Opportunities Through Improved Government Spectrum Efficiency," September 13, 2012.

[33] GAO, Spectrum Management: Federal Government's Use of Spectrum and Preliminary Information on Spectrum Sharing, September 13, 2012, GAO-12-1018T at http://www.gao.gov/products/GAO-12-1018T .

[34] GAO, Spectrum Management: Incentives, Opportunities, and Testing Needed to Enhance Spectrum Sharing, November 14, 2012, GAO-13-7 at http://gao.gov/products/GAO-13-7.

[35] Hearing, Senate, Committee on Armed Services, Subcommittee on Strategic Forces, "Oversight: Military Space Programs and Views on DoD Usage of the Electromagnetic Spectrum," April 24, 2013.

[36] GAO, Spectrum Management: Preliminary Findings on Federal Relocation Costs and Auction Revenues, April 24, 2013, GAO-13-563T at http://www.gao.gov/products/GAO-13-563T.

[37] See The Department of Commerce, Internet Policy Task Force, Commercial Data Privacy and Innovation in the Internet Economy: A Dynamic Policy Framework, http://www.commerce.gov/sites/default/files/ documents /2010/ december/iptf-privacy-green-paper.pdf.

[38] Up-to-date details can be found at http://www.ntia.doc.gov/other-publication/2013/privacy-multistakeholder-processmobile-application-transparency.

[39] See CRS Report R42351, Internet Governance and the Domain Name System: Issues for Congress, by Lennard G. Kruger.

[40] The President's Budget for Fiscal Year 2014, Appendix.

In: Transformations in Telecommunications and Media
Editor: Irwin Cavazos

ISBN: 978-1-62948-413-6
© 2013 Nova Science Publishers, Inc.

Chapter 13

HOW THE SATELLITE TELEVISION EXTENSION AND LOCALISM ACT (STELA) UPDATED COPYRIGHT AND CARRIAGE RULES FOR THE RETRANSMISSION OF BROADCAST TELEVISION SIGNALS[*]

Charles B. Goldfarb

SUMMARY

The Satellite Television Extension and Localism Act of 2010 (STELA), P.L. 111-175, modified the copyright and carriage rules for satellite and cable retransmission of broadcast television signals. The legislation was needed to reauthorize (through December 31, 2014) certain expiring provisions in the Copyright Act and the Communications Act and to update the language in those acts to reflect the transition from analog to digital transmission of broadcast signals, as well as to address certain public policy issues. Had the expiring provisions not been reauthorized, satellite operators would have lost access to a statutory compulsory copyright license and to statutory relief from retransmission consent requirements. This would have made it difficult, if not impossible, for them to retransmit certain distant broadcast signals to their subscribers, including signals providing otherwise unavailable broadcast network programming.

The Copyright Act and Communications Act distinguish between the retransmission of local signals—the broadcast signals of stations located in the same local market as the subscriber—and distant signals. Statutory provisions block or restrict the retransmission of many distant broadcast signals in order to foster local programming. These provisions typically take the form of defining which households are "served" or "unserved" by local broadcasters, with unserved households eligible to receive distant signals. But there are many grandfather clauses and other exceptions built into the rules that allow households to receive

---

[*] This is an edited, reformatted and augmented version of the Congressional Research Service publication, CRS Report for Congress R41274, dated January 3, 2013.

otherwise proscribed distant signals. STELA generally retained, and in some cases expanded upon, these grandfathered and exceptional cases.

STELA provided broadcasters two new incentives to use their digital technology to broadcast multiple video streams (to "multicast"). It clarified that royalty fees are payable to copyright owners of the materials on non-primary digital voice streams as well as primary streams, thus encouraging broadcasters (who often hold some of those copyrights) to expand their multicasting. STELA specifically gave broadcasters the incentive to undertake such multicasting to offer otherwise unprovided network programming in so-called "short markets"—markets that do not have network affiliates for all four major networks. It did this by defining households that can receive the programming of a particular network from the non-primary multicast video stream of a local broadcaster as being served, rather than unserved, with respect to that network, thus prohibiting satellite operators from retransmitting to those households distant signals that carry that network's programming. The local broadcaster can then seek retransmission consent payments from satellite operators. Several other provisions in STELA also were intended to reduce the number of short markets or increase flow of distant network signals into short markets.

Today, satellite operators are allowed, but not required, to offer subscribers the signals of the broadcast stations in their local market. Until enactment of STELA, the satellite operators chose not to offer this "local-into-local" service in many small markets, preferring to use their satellite capacity to provide additional high definition and other programming to larger, more lucrative markets. The costs associated with providing local-into-local service in small markets may exceed the revenues. STELA provided DISH Network, which had been subject to a permanent court injunction that in effect prohibited it from retransmitting to its subscribers the signals of distant broadcast stations, the opportunity to have that injunction waived if it provided local-into- local service in all 210 local markets in the United States, which it began doing on June 3, 2010.

STELA did not address the issue of "orphan counties"—counties located in one state that are assigned to a local market, as defined by the Nielsen Media Research designated market areas, for which the principal city and most or all of the local broadcast stations are in another state.

OVERVIEW OF STELA

The Satellite Television Extension and Localism Act of 2010 (STELA), P.L. 111-175,[1] extended, updated, and modified provisions in the Copyright Act[2] and the Communications Act[3] relating to the retransmission of broadcast television signals by satellite television and cable television providers. Among other things, STELA:

- Reauthorized through December 31, 2014, expiring provisions that provide satellite carriers access to a simple statutory compulsory copyright license and free satellite carriers from retransmission consent requirements, when retransmitting to their subscribers the signals of certain broadcast stations located outside the subscribers' local markets ("distant signals"). Had these provisions expired, it would have been difficult, if not impossible, for satellite operators to provide to their subscribers broadcast network programming that the subscribers are unable to receive from their local broadcasters.

- Revised provisions in copyright and communications law to take into account the transition from analog to digital transmission of broadcast signals.

- Created an incentive for broadcasters, who often hold copyrights on of the programming they broadcast, to use their digital capabilities to offer multiple video streams ("multicasting") by requiring satellite operators to pay royalty fees for the programming on the non-primary, as well as primary, video streams.

- Provided local broadcasters in markets that currently do not have network affiliates for all four major networks (so-called "short markets") the incentive to offer the programming of the currently unavailable networks on their non- primary digital video streams. Specifically, STELA defined households that can receive the programming of a particular network from the non-primary multicast video streams of a local broadcaster as being "served" rather than "unserved" with respect to that network, thus prohibiting satellite operators from retransmitting to those households distant signals that carry that network's programming and allowing the broadcaster to seek retransmission consent payments.

- Freed DISH Network of a permanent court injunction against retransmitting the signals of distant network stations into short markets in exchange for the requirement to make available to its subscribers in *each of the 210 local markets* in the United States the signals of all the full-power broadcast stations in the local market. To meet that requirement, on June 3, 2010, DISH began providing such "local-into-local" service to the 29 local markets it had not been serving.

- Modified the rules governing which households are eligible to receive distant signals from satellite carriers, generally grandfathering those households that currently receive such signals. These rule changes, which attempt to better reflect the current market and technological environment, may increase the number of households that qualify to receive distant signals.

- Modified the copyright administrative procedures, reporting requirements, royalty fees, filing fees, and non-compliance penalties for the improper retransmission of broadcast television signals by both satellite carriers and cable operators.

- Changed the statutory licenses applicable to the copyrighted material on the retransmitted signals of "significantly viewed" broadcast stations,[4] low power broadcast stations, and other statutorily exceptional[5] stations.

- Required satellite operators to make available to their subscribers all the programming of non-commercial television stations that is in high-definition format.

- Required the Register of Copyrights to submit a report on market-based alternatives to statutory licensing and also required the Comptroller General to submit a report on changes to carriage requirements currently imposed on multichannel video programming distributors (MVPDs) and to Federal Communications Commission (FCC) regulations that might be required if Congress were to phase-out the current statutory satellite and cable licensing requirements.[6]

STELA did not address the situation in which a county has been assigned to a local market for which the principal city is in another state and the television stations located in that local market primarily address the needs of households in that other state, rather than providing news, sports, and other programming of interest to the county. There had been a number of legislative proposals intended to address this "orphan county" issue, but none was included in STELA. But STELA did require the FCC to submit a report on the in-state broadcast programming available to households that receive the signals of broadcast stations that are considered, by statute and rule, to be local but are located in a different state. The FCC submitted its report on August 29, 2011.[7]

BACKGROUND

Congress has constructed a regulatory framework for the retransmission of broadcast television signals by satellite television operators through a series of laws—the 1988 Satellite Home Viewer Act (SHVA),[8] the Satellite Home Viewer Act of 1994,[9] the 1999 Satellite Home Viewer Improvement Act (SHVIA),[10] the 2004 Satellite Home Viewer Extension and Reauthorization Act (SHVERA),[11] and most recently STELA. These laws have fostered satellite provision of MVPD service and, as satellite has become a viable competitor to cable television, have attempted to make the regulatory regimes for satellite and cable more similar. Today, the regulatory framework for satellite exists alongside an analogous, but in some significant ways different, regulatory framework for cable.[12]

The various provisions in these satellite acts created new sections or modified existing sections in the Copyright Act and the Communications Act of 1934. Under current law, in order to retransmit a broadcaster's signals to its subscribers, a satellite operator or a cable operator, with certain exceptions, must obtain a license from the copyright holders of the content contained in the broadcast for use of that *content* and also must obtain the consent of the broadcaster for retransmission of the broadcast *signal*. The statutory provisions addressing copyright are in the Copyright Act and are administered by the Copyright Office in the Library of Congress; those provisions addressing signal retransmission are in the Communications Act and are administered by the FCC. But in several cases, the provisions in one act are conditioned on meeting conditions prescribed in the other act or meeting rules adopted by the agency that administers the other act.

The satellite and cable regulatory frameworks attempt to balance a number of longstanding, but potentially conflicting, public policy goals—most notably, localism, competitive provision of video services, support for the creative process, and preservation of free over-the-air broadcast television. They also attempt to balance the interests of the satellite, cable, broadcast, and program content industries. Congress incorporated sunset provisions in SHVERA—and again in STELA—because of its concern that market changes could affect these balances. Indeed, as Congress debated the legislative proposals that were included in, or left out of, STELA, it gave substantial weight to a proposed package of changes in copyright procedures, royalty rates, and other parameters constructed and supported by a wide range of industry players through a process of direct negotiations and compromise.

The statutory provisions distinguish between the retransmission of *local* signals—the broadcast signals of stations located in the same local market (as defined by the 210 designated market areas into which the United States is divided by Nielsen Media Research) as the subscriber—and of *distant* signals. These provisions block or restrict the retransmission of many distant broadcast signals in order to protect local broadcasters from competition from distant signals and to provide them with a stronger negotiating position vis-à-vis the satellite and cable operators. The intent is to foster local programming. But the statutory framework also recognizes that U.S. households benefit from the receipt of certain distant broadcast signals and includes explicit retransmission and copyright rules for these.

The statutory framework for satellite sets the parameters within which industry players must conduct business. It provides answers to four fundamental business questions:

- May—or must—a satellite operator retransmit some or all local broadcast signals?[13]
- May a satellite operator retransmit certain categories of distant (non-local) broadcast signals?
- Is retransmission of those signals contingent on a satellite operator receiving the prior retransmission consent of—and providing compensation to—the broadcaster? and
- Is use of the content on those signals subject to specific copyright license terms?

Satellite operators and broadcasters also must conduct business within the constraints of longstanding industry practice. Broadcast program suppliers—both broadcast networks and owners of non-network, syndicated programming—contractually grant individual broadcast television stations the exclusive broadcast rights to their programming in a geographic area and restrict those broadcast stations from allowing other parties to retransmit the station signals carrying that programming beyond the area of exclusivity. Thus, in some situations where the regulatory framework allows satellite (or cable) operators to retransmit the signals of a distant broadcast station, subject to obtaining the permission of the broadcast station, that station may be—and, in practice, often is—contractually prohibited from granting the MVPD retransmission consent.

Although satellite and cable operators compete directly with one another in most markets, there are significant differences in the regulatory frameworks under which they operate. These differences largely reflect the different origins of the cable and satellite industries—cable beginning as a business with technology focused on serving narrow geographic areas and satellite beginning as a business with technology serving broad geographic areas. To this day, cable network architecture and technology can more efficiently accommodate local programming than can satellite. Some observers have proposed that the retransmission, copyright, and other rules under which these competing multichannel video programming distributors operate should be rationalized to eliminate artificial competitive advantages or disadvantages. For example, the Copyright Office, in a report to Congress required by SHVERA,[14] has proposed that the gross receipts royalty system for cable retransmission of distant broadcast signals in Section 111 of the Copyright Act be replaced by a flat fee per subscriber system of the sort for satellite retransmission of distant broadcast signals in Section 119 of the Copyright Act. The Copyright Office also has proposed[15] that the provisions defining satellite subscriber eligibility for receiving distant signals in Section 119 (the "unserved household" provisions) be replaced by the imposition on satellite operators of the FCC's network non-duplication[16] and syndicated exclusivity rules,[17] which currently are used to limit the retransmission of distant broadcast signals by cable operators. But in the Congressional deliberations leading to passage of STELA, there was little discussion of a major modification of the regulatory framework.

ISSUES ADDRESSED IN STELA

Reauthorization

STELA extended through December 31, 2014, several statutory copyright and communications provisions, required for satellite operators to retransmit distant signals, that would have expired on May 31, 2010. Most significantly:

- Section 119 of the Copyright Act[18] provides satellite operators that retransmit certain "distant" (non-local) broadcast television signals to their subscribers with an efficient, relatively low cost way to license the copyrighted works contained in those broadcast signals—a statutory per subscriber, per signal, per month royalty fee. Had the law expired, it would have been very difficult (and perhaps impossible) for satellite operators to offer the programming of broadcast networks[19] to that subset of subscribers who currently cannot receive that programming from local broadcast stations that are affiliated with those networks.[20] It also would have been difficult for satellite operators to offer their subscribers the signals of distant stations that are not affiliated with broadcast networks, including both "superstations"[21] and other non-network stations.
- In addition, prior to the enactment of STELA, Section 119 provided those satellite operators that retransmit to their subscribers the signals of "significantly viewed" stations—stations that are located outside the local market in which the subscriber is located but have been determined to be "significantly viewed" by those households in the local market that do not subscribe to any MVPD provider—a royalty-free license for the copyrighted works contained in those broadcast signals. Had Section 119 expired, it would have been very difficult (and perhaps impossible) for satellite operators to offer their subscribers the signals of significantly viewed stations. Under STELA, satellite retransmission of significantly viewed stations has been moved from Section 119 to Section 122 of the Copyright Act, under which such retransmission is subject to the royalty-free license in Section 122.
- Section 325(b)(2)(C) of the Communications Act[22] allows a satellite operator to retransmit the signals of distant network stations, without first obtaining the retransmission consent of those distant stations, to those subscribing households that cannot receive the signals of local broadcast television network affiliates. Had it expired, a satellite operator would have had to negotiate compensation terms with those distant network stations whose signals it retransmitted to those "unserved" subscribers.
- Section 325(b)(3)(C)(ii) of the Communications Act[23] prohibits a television broadcast station that provides retransmission consent from engaging in exclusive contracts for carriage or failing to negotiate in good faith. Section 325(b)(3)(iii)[24] prohibits an MVPD from failing to negotiate in good faith for retransmission consent. Had these provisions expired, a broadcaster or an MVPD could have chosen to employ a "take it or leave it" strategy rather than to negotiate retransmission consent terms in good faith, increasing the risk of an impasse that results in subscribers losing access to the broadcast station's programming.

STELA included a provision making the effective date of the act February 27, 2010, in order to protect satellite operators from potential lawsuits for copyright infringement for the brief period of time when the old authorization had expired and Congress had not yet enacted new authorization. At that time, Congress had encouraged the satellite operators not to discontinue retransmission of the distant signals in order to allow satellite subscribers to continue to receive those signals.

Revising Existing Rules That Are Based on Analog Technology

A number of statutory provisions, and many FCC and Copyright Office rules adopted to implement statutory provisions, have been based on the transmission of analog broadcast signals, but during 2009 the transition to digital broadcast signals was largely achieved. As a result, statutes and rules that explicitly referred to analog technology were no longer effective in attaining the objectives for which they were enacted. Thus, Marybeth Peters, Register of Copyrights, proposed five modifications to Section 111 of the Copyright Act and four modifications to Section 119 of the Copyright Act "to accommodate the conversion from analog to digital broadcasting."[25] Analogous changes were proposed for the Communications Act.

STELA included specific changes to language in the Copyright Act and to the Communications Act intended to make them consistent with a digital environment. It also included provisions directing the FCC to develop a predictive model for the reception of digital signals within six months of enactment in order to determine which households are "unserved" and therefore eligible to receive digital network signals. On November 23, 2010, the FCC adopted rules creating measurement standards for digital television signals and establishing a predictive model.[26] STELA also included a provision that provides guidance for the period before the new predictive model has been implemented.

STELA modified the methodology used to determine whether a household is served to reflect the current market and technological environment, including the transition from analog to digital transmission. It is possible that some of the methodological changes may increase the number of households eligible to receive distant network signals.[27] For example, most households now receive their broadcast signals from their cable or satellite service rather than over-the-air and therefore do not use a rooftop antenna. The old definition of unserved household referred to the inability to receive a signal of a specified intensity using a rooftop antenna; STELA changed the definition to refer to *any* antenna. Since indoor antennas, such as "rabbit-ear" antennas, tend to be less effective than rooftop antennas, this may increase the number of households that qualify as unserved.

Fostering Digital Multicasting, Especially Multicasting to Provide Network Programming in Those Markets That Lack a Network Affiliate ("Short Markets")

Although each of the four major broadcast television networks (ABC, CBS, FOX, and NBC) has a local station affiliate in most U.S. markets, 58 of the 210 markets do not have the full complement of four network affiliates.[28] In these short markets, subscribers have been defined as being "unserved" with respect to the missing network and satellite operators have been allowed to retransmit to their subscribers the signals of up to two distant stations that are affiliated with that missing network.[29]

With the transition from analog to digital technology, however, broadcast stations are able to broadcast multiple video streams. Some local television stations in short markets are affiliated with a national network and broadcast that network's programming on their primary video stream, but also have reached agreements with a second national network that lacks an affiliate in the local market to carry the network programming of that second network on a non-primary video stream. This multicasting allows households in the local market to receive the network programming of that second network, although it is unlikely that the local station provides any original local programming on that secondary video stream.

Under STELA, if a local television station broadcasts a non-primary video stream that provides the programming of a national network and was carried by a satellite operator on March 31, 2010, and if the local station continues to carry that network's programming on that video stream, then as of October 1, 2010, that video stream is considered a "qualified multicast video" and households in that local market will be considered served with respect to the broadcast network whose programming is carried on that video stream. Thus, after October 1, 2010, a satellite operator cannot use the statutory distant signal copyright license to retransmit to households in that local market the signal of a distant broadcast station affiliated with that streamed network. Presumably, the satellite operator would have to obtain retransmission consent from the local broadcaster (which probably would entail making a payment to the broadcaster) to retransmit the programming as part of its local-into-local service.

As of January 1, 2011, all non-primary video streams of national network programming offered by a local television station are considered qualified multicast video and households in the local market are considered served with respect to the broadcast network whose programming was carried on those video streams.[30] As a result of this change in treatment of network programming broadcast over non-primary video streams, satellite operators are allowed to retransmit the programming as part of their local-into-local service offering (if they successfully negotiated a retransmission consent agreement with the broadcaster), but are no longer able to retransmit that network programming using a distant broadcast signal.

STELA allowed a satellite subscriber who was lawfully receiving the distant signal of a network station on the day before enactment of the new legislation to receive both that distant signal and the local signal of a network station affiliated with the same network until the subscriber chooses to no longer receive the distant signal from its satellite operator. Thus, if in a short market a local broadcaster began to multicast on a non-primary video stream the programming of the network for which there has been no local affiliate, and the satellite operator chose to retransmit that non- primary video stream, a subscriber who has been receiving the distant network signal could continue to receive that distant signal as well as the local network signal, as long as the subscriber did not discontinue its subscription for that distant signal. A household in that short market would not be allowed to receive a distant network signal, however, if it received from the satellite operator the programming of that same network from the non-primary video stream of a local broadcaster but was not a subscriber lawfully receiving the distant signal on the day before enactment of the new legislation.

Another provision in STELA fostered multicasting in all markets, not just short markets. It encouraged broadcasters to offer programming over multiple digital video streams—both their primary stream and non-primary streams—by clarifying that satellite operators must pay copyright royalty fees for the retransmission of the programming on broadcasters' non-primary as well as primary video streams. Since broadcasters often hold some copyrights for the programming they broadcast, such payments increase their incentive to multicast.

Providing an Incentive for DISH Network to Offer Local-into-Local Service in All Designated Market Areas: Allowing DISH to Use a Statutory License to Retransmit Distant Network Signals into Short Markets

Satellite operators are allowed, but not required, to offer subscribers the signals of all the broadcast stations in their local market. If a satellite operator chooses to retransmit the signal of a local broadcast station and to take advantage of a royalty-free statutory copyright license for the content carried on that signal, it must retransmit the primary signals of all the full power stations in that local market, subject to obtaining local station permission. The satellite operators had chosen not to offer this "local-into-local" service in many small markets, preferring to use their satellite capacity to provide additional high definition and other programming to larger, more lucrative markets than to use the capacity to serve very small numbers of customers. In some cases, those small markets may not generate enough revenues to cover the costs of providing local-into-local service.[31] As a result, approximately 3% of all U.S. households did not have access to any local broadcast signals if they subscribed to satellite video service, unless they could receive those signals directly over-the-air.[32]

Early in the 111th Congress, Representative Stupak had introduced a bill, the Satellite Consumers' Right to Local Channels Act, which, in effect, would have required satellite operators to offer local-into-local service in all markets; if a satellite operator wished to use the royalty-free statutory copyright license to rebroadcast the content on a broadcast signal in *any* local market, it would have had to provide local-into-local service in *every* market. But during markup of the House Energy and Commerce Committee bill, Representative Stupak agreed to withdraw his bill (which he had introduced in the form of an amendment), when DISH Network indicated that it would voluntarily provide local-into-local service in all 210 markets within two years in exchange for statutory relief from a current court injunction prohibiting it from providing its subscribers distant signals using the Section 119 copyright license.[33] That *quid pro quo* was incorporated into STELA.

As a result of repeated violations of Section 119 of the Copyright Act, DISH Network had been subject to a permanent injunction, imposed by the U.S. Court of Appeals for the 11th Circuit,[34] barring it from using the Section 119 statutory license for the copyrighted materials when retransmitting distant signals to its subscribers; it therefore had to employ an arms-length agreement with National Programming Service for that entity to deliver distant signals to its subscribers. Under STELA, the injunction was partially waived if DISH Network provided local-into-local service in all 210 local markets in the United States. Specifically, DISH is allowed to use a Section 119 license for the copyrighted materials when retransmitting to its subscribers in a "short market" the signals of a distant network broadcast station affiliated with a network for which no local broadcaster is providing the network programming over its primary video stream.

Because of DISH's long history of illegally retransmitting distant signals, STELA incorporated a number of safeguards. DISH must demonstrate that it is offering local-into-local service in all 210 local markets in the United States (referred to as designated market areas or DMAs) in order to be deemed qualified by the court for a temporary waiver of the injunction. The Court must select a special master who would make an initial examination and provide on-going monitoring to assure that DISH is serving all 210 DMAs (and if not, make a determination that it is nonetheless acting reasonably and in good faith) and is in compliance with the royalty payment and household eligibility requirements of the license. The initial waiver of the injunction would be temporary, but could be extended for good cause; if DISH lost recognition as a qualified carrier it could not seek to be re-qualified. Also, the Comptroller General was instructed to monitor the degree to which DISH is complying with the special master's examination. DISH would have the burden of proof that it is providing local-into-local service with a good quality satellite signal to at least 90% of the households in each DMA. It would be subject to penalties of between $250,000 and $5 million for failure to provide service, with exceptions for nonwillful violations.

On June 3, 2010, DISH introduced local-into-local service in the 29 DMAs it had not been serving. These markets are: Alpena, MI; Biloxi, MS; Binghamton, NY; Bluefield, WV; Bowling Green, KY; Columbus, GA; Elmira, NY; Eureka, CA; Glendive, MT; Greenwood, MS; Harrisonburg, VA; Hattiesburg, MS; Jackson, TN; Jonesboro, AR; Lafayette, IN; Lake Charles, LA; Mankato, MN; North Platte, NE; Ottumwa, IA; Parkersburg, WV; Presque Isle, ME; Salisbury, MD; Springfield, MA; St. Joseph, MO; Utica, NY; Victoria, TX; Watertown, NY; Wheeling, WV, and Zanesville, OH. On June 30, 2010, DISH filed with the FCC an application for certification as a qualified carrier pursuant to Section 206 of STELA. On September 1, 2010, the FCC adopted an order granting that certification.[35]

STELA also required each satellite carrier to submit a semi-annual report to the FCC setting forth (1) each market in which it offers local-into-local service; (2) detailed information regarding the use of satellite capacity for the provision of local-into-local service; (3) each local market in which it has commenced offering local-into-local service in the six-month period covered by the report; and (4) each local market in which it has ceased to offer local-into-local service in the six-month period.

Reducing the Number of Short Markets by Eliminating the "Grade B Bleed" Problem

Prior to enactment of STELA, in areas where a network-affiliated broadcast station was located near the DMA boundary, so that its signal extended into a portion of a neighboring DMA that did not have a local station affiliated with the same network, households in that neighboring market who could receive that signal at a Grade B level were not considered to be "unserved" for that network. A satellite operator could neither offer that overlapping signal to those households as part of local-into-local service (since it was a distant signal) nor provide to those households the signal of a distant station affiliated with the same network, because those households were not considered unserved. The satellite operators sought to eliminate this so-called "Grade B bleed" problem by modifying the test for a subscriber being unserved to apply only to the strength of the signal from an in-market station or by defining unserved in terms of whether the viewer can get local service from the satellite spot beam, rather than in terms of over-the-air reception.[36]

STELA eliminated the problem by defining as "unserved" those households that do not receive the network programming from an over-the-air signal that originates in the local market, that is the signal of their *local network affiliate.*

Household Eligibility to Receive Distant Signals: Grandfathered Subscribers, Other Subscribers, and Households That Are Not Subscribers When Legislation Is Enacted ("Future Applicability")

The primary mechanism for limiting satellite retransmission of distant network signals has been to restrict such retransmission to "unserved" households that cannot receive the programming of a particular network because either (1) the satellite operator is not offering local-into-local service in that market and the households cannot receive a signal of a threshold quality level over-the-air from the local network affiliate, or (2) there is no local affiliate offering the programming of that network. But both the Copyright Act and the Communications Act include certain grandfathered exceptions to those eligibility restrictions; as a result, many households that are able to receive a network signal from a local broadcast station are allowed to continue to receive the distant signal of a broadcast station affiliated with the same network. STELA retained most of these grandfathered exceptions and in some ways expands on them.

Section 339 of the Communications Act sets the rules for carriage of distant television station signals by satellite operators. Section 339(a)(2) addresses the replacement of distant signals with local signals, enumerating four different sets of rules: for grandfathered subscribers to analog distant signals, for other subscribers to analog distant signals, for households that are not subscribers at the time the legislation is enacted (future applicability), and for subscribers to distant digital signals. STELA:

- Retained Section 339(a)(2)(A), the grandfathering provision that allows certain households that historically had been receiving distant network signals illegally (and therefore otherwise would not have qualified to receive those distant signals) to continue to receive those signals. The language was updated only to reflect the date of enactment of the new legislation and to eliminate reference to analog technology. All these households continue to be grandfathered to receive distant network signals despite being able to receive the signals of local stations with the same network affiliation.
- Eliminated references to analog signals from Section 339(a)(2)(B), but otherwise the two-part provision is retained. Under the first part, if a household's satellite operator had made a local network station available on January 1, 2005, as part of local-into-local service, the operator would nonetheless be allowed to provide to that household a distant signal of a station affiliated with the same network if the operator had submitted to the television network no later than March 1, 2005, a list of households receiving that distant signal that included that household. This continues the grandfathering of households that had been legally receiving a distant network signal

and were allowed to continue to receive that signal when they also had access to the signal of a local broadcast station affiliated with the same network. Under the second part, if the satellite operator had not made available a local network station on January 1, 2005, as part of local-into-local service, the operator would be allowed to offer the household the distant network signal only if (a) the household seeks to subscribe to the distant signal before the date on which the operator begins to offer local-into-local service, and (b) the operator submits to each television network within 60 days of commencing such service the households subscribing to the distant signal. Thus, a household that had legally sought to receive a distant network signal is allowed to continue to receive that signal after the signal of a local broadcast station affiliated with the same network is available.

- Allowed a subscriber who is lawfully receiving the distant signal of a network station from a satellite operator on the day before enactment of STELA to receive both the distant signal and the local signal of the same network until the subscriber chooses to no longer receive the distant signal from the satellite operator (whether or not the subscriber elects to subscribe to local-into-local service). Thus, all the households legally receiving distant network signals under Section 339(a)(2)(B) at the date of enactment of STELA continue to be allowed to receive those distant signals.

- Prohibited a satellite operator from providing a distant network signal to a person who (1) (a) is not a subscriber legally receiving that distant signal on the date STELA is enacted, and (b) at the time the person seeks to receive the distant signal, resides in a local market where the satellite operator offers local-into-local service that includes a local station affiliated to the same network and the person can receive that local-into-local service, *or* (2) (a) is a subscriber legally receiving a distant signal on or after the date STELA is enacted, and (b) subsequent to such subscription the satellite carrier makes available to that subscriber the signal of a local network station affiliated with the same network as the distant signal (and the retransmission of such signal by the carrier can reach the subscriber), unless the person subscribes to the signal of the local network station within 60 days after the signal is made available. The latter is intended to support local stations by requiring the subscriber to obtain local-into- local service in order to continue to receive the distant network signal.

- Defined a subscriber as eligible to receive a distant signal of a network station affiliated with the same network as a local station if, with respect to a local network station: (1) the subscriber's household is not predicted by the model specified in the act to receive the threshold signal intensity; (2) the household is determined, based on a test conducted in accordance with the current model or any successor regulation, not to be able receive the signal of the local station with an intensity that exceeds the standard; or (3) the subscriber is in an unserved household as determined by the definition of an unserved household in Section119(d)(10)(A) of the Copyright Act. The third criterion appears to allow a household that does not meet the signal intensity test for analog service for the signal of a local network station to be grandfathered for the receipt of a distant network signal carrying the same network, even if that household could receive the digital signal of the local network station.

Provisions in Section 119 of the Copyright Act define "unserved households" and set the copyright rules that apply to the secondary transmission of distant signals to those unserved households. STELA modified some of those provisions.

- If a local station is multicasting and offers a second network's programming on one of its non-primary video streams, but a household using an antenna cannot receive that non-primary video stream at the signal intensity specified in FCC rules, then the household is deemed unserved with respect to the network whose programming is being broadcast on that non-primary stream. This took effect on October 1, 2010, for multicast streams that existed on March 31, 2010, and on January 1, 2011, for all other multicast streams.

- References to analog signals were eliminated, but otherwise all the rules covering grandfathered households receiving distant signals currently in Section 119(a)(4)(A) were retained.

- For a subscriber (other than a grandfathered household) who, on the day before enactment of STELA, was lawfully receiving a satellite retransmission of a distant network signal under a statutory license, the statutory license applies for the retransmission of that distant signal. Further, the subscriber's household continues to be considered an unserved household with respect to that network until the subscriber elects to stop receiving that distant signal, whether or not the subscriber has access to the signal of a local network station affiliated with the same network through local-into-local service and whether or not the subscriber elects to subscribe to that local-into-local service. This, in effect, created a new group of grandfathered households.

- The statutory distant signal copyright license in Section 119 of the Copyright Act does not apply to the satellite retransmission of a distant network signal to a person who is not a subscriber lawfully receiving that distant network signal at the date of enactment of STELA if, when that person subsequently seeks to subscribe to a satellite carrier for that distant signal, that person can obtain that network's programming from a local station affiliated with the same network through local-into-local service.

- The statutory distant signal copyright license in Section 119 of the Copyright Act applies to the satellite retransmission of a distant network signal to a person who is a subscriber lawfully receiving that distant network signal on or after the date of enactment of STELA, and the subscriber's household continues to be considered to be an unserved household with respect to that network, until such time as the subscriber elects to terminate such retransmission, but only if the person subscribes to retransmission of a local network station affiliated with the same network (that is, subscribes to local-into-local service) within 60 days of the satellite carrier making local-into-local service available to the subscriber. Thus, a household can be grandfathered for the distant network service only if it subscribes to local-into-local service within 60 days of that service becoming available.

Modified Copyright Treatment of the Satellite Retransmission of Low Power Television Station Signals

Low power television service was created by the FCC in the 1980s to serve small communities (rural or urban) with low cost, limited geographic range facilities that used available spectrum between full power stations. It is a "secondary service" that is not guaranteed protection from interference or displacement by full service stations. Low power stations that produced at least two hours per week of local programming, maintained a production studio within their Grade B contour, and complied with many of the requirements placed on full service stations were given a one-time opportunity to obtain "Class A" status that gave them primary status, that is, protected their channel from interference or displacement.

Historically, satellite retransmission of low power television signals was covered by the statutory distant signal copyright license in Section 119 of the Copyright Act. Satellite operators were allowed to retransmit the signals of low power stations to subscribers within certain geographic limitations—to subscribers within 20 miles of the station transmitter for network-affiliated stations located in the 50 largest markets, within 35 miles of the station transmitter for network- affiliated stations located in other markets, and within the same designated market area as non- network-affiliated stations.[37] Satellite operators had no copyright royalty obligation for retransmission of the low power station content within those same mileage limits; beyond those limits, satellite operators were subject to the statutory copyright license fees for distant signals outlined in Section 119 of the Copyright Act.[38]

Under STELA, if satellite operators seek to use a statutory license for the copyrighted material on the low power television stations whose signals they retransmit, they must use the royalty-free statutory local signal license in Section 122, rather than the Section 119 license. STELA expands the geographic area covered by the royalty-free statutory license to the entire DMA in which the low power station is located.

STELA also explicitly stated that a satellite carrier that retransmits the signal of a low power station under a statutory license is not required to make any other secondary retransmissions. Thus retransmission of a low power station does not trigger the requirement to offer local-into- local service or to retransmit any other low power stations. No local low power station can demand carriage by the satellite operator serving its market area, even if that satellite operator is providing local-into-local service.

Since low power television stations do not have a deadline for their transition from analog to digital transmission, the old, analog-based FCC rules for determining whether a household is eligible to receive distant signals apply with respect to low power television until the station is licensed to broadcast a digital signal.

The statutory local signal copyright license does not apply to satellite retransmission of repeaters or translators.

Satellite Carriage of Noncommercial Educational Television Stations

By statute, providers of direct broadcast satellite service (DirecTV and DISH Network) must reserve between 4% and 7% of their channel capacity exclusively for noncommercial programming of an educational or informational nature.[39] With the digital transition, broadcasters now are able to broadcast high definition signals and multiple digital programming streams over their licensed spectrum, and the public television stations are seeking to expand satellite carriage of their high definition and multicast signals.

At the time STELA was enacted, the public broadcasters had reached retransmission consent agreements with DirecTV, the cable industry (through both the National Cable and Telecommunications Association representing large cable operators and the American Cable Association representing small cable operators), and Verizon for the retransmission of most of their high definition and multicast video streams. The agreement with DirecTV incorporated "creative solutions that recognized [DirecTV's] capacity limitations; ultimately ensuring that subscribers have access to the myriad of content and services provided by the local stations while accommodating their capacity concerns."[40] The public broadcasters had not yet achieved retransmission agreement with DISH Network, but negotiations were continuing.

STELA modified Section 338(a) of the Communications Act, which addresses the carriage of local television signals by satellite carriers, to require any satellite carrier that has not yet negotiated a carriage contract covering at least 30 noncommercial educational television stations by July 27, 2010, (1) to provide subscribers, by the end of 2011, the high definition signals of qualified noncommercial educational television stations in all the local markets in which the carrier currently offers local television broadcasts in high definition (and by the end of 2010 to half of those markets), and (2) when initiating the provision of high definition local broadcast television in a market, to include the high definition signals of all qualified local noncommercial educational television stations. In July 2010, DISH filed suit in the U.S. District Court for Nevada, seeking temporary injunctive relief from FCC enforcement of that provision, claiming the provision infringed on its First Amendment right "to make the editorial judgment whether to carry local PBS stations in HD" and was confiscatory of its property.[41] U.S. District Judge James Mahan declined DISH's request for an injunction.[42] DISH did not reach an agreement with Association of Public Television Stations, which represents all public television stations, in time to meet the July 27, 2010, deadline, but it averted the carriage mandate in STELA by reaching an independent HD carriage agreement with 30 noncommercial stations by July 27, 2010.[43]

At an October 7, 2009, hearing of the Senate Subcommittee on Communications, Technology, and the Internet, public broadcasters identified another problem for which they sought a legislative solution. Most states have developed state public television networks intended to serve the entire state, but in 16 states those networks do not have public stations transmitting signals in each DMA in the state; under current law, satellite carriers are not allowed to use a royalty-free statutory copyright license to retransmit the signals of the in-state, but out-of-market public broadcasting stations to their subscribers in those DMAs.[44] STELA modified the provisions for the royalty-free statutory copyright license in Section 122 of the Communications Act to allow, where there is a public educational network of three or more noncommercial educational broadcast stations in a state, a satellite operator to use the royalty-free license to retransmit the programming on those stations' signals to subscribers in any county in the state whose households are otherwise ineligible to receive retransmissions of those signals.

Satellite Carriage of State Public Affairs Networks

Cable franchise authorities are allowed, by law, to require cable operators to set aside some of their capacity for the carriage of public, educational, and governmental (PEG) programming. This programming is not broadcast to the public; it is sent directly to the cable system's head-end for retransmission. Satellite operators are not required to offer PEG programming, though they have the obligation to allocate between 4 and 7 percent of their channel capacity exclusively to noncommercial programming of an educational or informational nature. In order to foster PEG programming on cable systems, a number of states have created state public affairs networks that produce non-broadcast programming of state-wide interest. Although this programming is available to satellite operators, those operators are not widely offering it to subscribers.

STELA included a provision intended to encourage satellite operators to carry these state public affairs networks. Under the provision, a satellite carrier that provides the retransmission of the state public affairs networks of at least 15 different states, under reasonable prices, terms, and conditions, and does not delete any of the noncommercial educational or informational programming on those networks, would only have to allocate 3.5% of its channel capacity to the retransmission of educational or informational noncommercial programming, rather than 4%. This provision might encourage satellite operators to offer state public affairs networks to subscribers in orphan counties or in short markets.

The Retransmission of In-State, but Non-Local, Broadcast Signals into Counties Assigned to Local Markets in Other States ("Orphan Counties")

The current regulatory frameworks for both satellite and cable distinguish between the retransmission of local and distant signals and require that local markets be defined by the DMAs constructed and published by Nielsen Media Research.[45] The viewing patterns that underlie these Nielsen markets are primarily the result of the physical locations of the various broadcast television stations and the reach of their signals. (They also reflect the boundaries of the exclusive broadcast territories that each of the three original television broadcast networks—ABC, CBS, and NBC—had incorporated into their contracts with their local affiliate stations decades ago.) DMAs do not take into account state boundaries. As a result, under current statutes and rules, a number of counties are assigned to local markets for which the principal city (from which all or most of the local television signals originate) is outside their state.[46] Satellite subscribers (and many cable subscribers) in these "orphan counties" may not be receiving signals from in-state broadcast stations and as a result may not be receiving news, sports, and public affairs programming of interest in their state.

Many residents of orphan counties have proposed that the statutory framework be modified to remove prohibitions or impediments on satellite operators retransmitting to their subscribers in these counties the signals of broadcast stations in in-state, but non-local, markets. (SHVERA selectively removed these impediments through four "exceptions" that allow satellite operators to retransmit to their subscribers in particular orphan counties in New Hampshire, Vermont, Oregon, and Mississippi—but not in other locations—the signals of in-state but out-of-market broadcast stations.[47]) Broadcasters, however, have voiced concern that allowing such retransmission could undermine their financial viability by reducing their audience share and thus reducing their advertising revenues. They also assert such retransmission would weaken the local broadcasters' negotiating position with the satellite and cable operators, who could turn to the programming of an in-state but out-of-market affiliate of a particular network if they failed to reach retransmission consent with the local affiliate of that network. Broadcasters claim this would harm their ability to provide quality local programming, which is expensive to produce.[48]

A number of bills had been introduced in the 111th Congress that directly addressed this issue (either generically or for specific states or geographic areas) by allowing satellite operators to retransmit to subscribers in orphan counties the signals of certain in-state, but non-local broadcast stations.[49] But STELA (reflecting each of the four bills that had been reported out of the House Energy and Commerce, House Judiciary, Senate Commerce, Science, and Transportation, and Senate Judiciary committees, leading to STELA) did not include any provisions that would address this issue directly. During the markup of the Senate Judiciary Committee bill, reportedly several Senators prepared amendments that would have narrowly addressed the orphan county issue in their states, but then agreed to withdraw their amendments when other Senators voiced concern that the provisions would delay passage of the legislation because of unresolved issues among broadcasters and satellite operators. At the markup, reportedly there was discussion of imposing a deadline on the industry to reach a negotiated solution, such as a proposal by Senator Coburn that, if there were no industry agreement by the time the legislation reaches the Senate floor, a trigger provision would be inserted in the bill that would impose a statutory solution for the orphan counties issue if no negotiated compromise is reached after two years.[50] But STELA did not include a trigger provision.

STELA included a provision instructing the FCC to prepare within one year a report containing analysis of (1) the number of households in each state that receive local broadcast signals from stations of license located in a different state; (2) the extent to which consumers have access to in- state broadcast programming; and (3) whether there are alternatives to the use of DMAs to define local markets that would provide more consumers with in-state broadcast programming. The FCC submitted its report to Congress on August 29, 2011.[51] The report provided data, summarized the comment of interested parties, and identified several alternatives to the use of DMAs to define local television markets, but did not provide any conclusions or recommendations.

In addition, a savings clause in STELA—stating that nothing in the legislation, in the Communications Act, or in any FCC regulation shall limit the ability of a satellite operator to retransmit a performance or display of a work pursuant to an authorization granted by the copyright owner—is intended to clarify that a satellite operator always has the opportunity to negotiate a copyright license outside the Section 119 statutory license. This clarification is not likely to result in the satellite retransmission into orphan counties of the sports and network programming on in-state, but out-of-market stations, but could encourage the retransmission of those stations' locally produced news programming.

Changing the Statutory Copyright License Applied to the Content on the Signals of Significantly Viewed and "Exception" Broadcast Stations

The statutory framework for the retransmission of broadcast television signals has been based on a distinction between local and distant signals. The signals of significantly viewed stations and the signals of in-state, out-of-market stations in the four states that satellite operators were allowed to import into orphan counties under the exceptions in SHVERA, originate outside the market into which they are imported; in that regard, they are distant signals and they have been subject to the Section 119 distant signal statutory copyright license. But since significantly viewed stations and the "exception" stations can be presumed to be providing programming of local or state-wide interest to counties in particular local markets, arguably that content could be viewed as local to the counties into which they are imported and should be treated accordingly. STELA modified the Copyright Act to treat those signals as local, moving the relevant provisions from Section 119 to Section 122.

STELA changed language in the heading of Section 122 from "secondary transmission by satellite carriers within local markets" to "secondary transmission of local television programming by satellite." It made satellite retransmission of both significantly viewed stations and the exception stations subject to the local signal statutory copyright license in Section 122 rather than the distant signal statutory license in Section 119, but required the satellite operator to continue to pay the statutory copyright license fees under Section 119 for the retransmission of the exception stations. Since significantly viewed stations already are subject to the royalty-free license in Section 122, effectively there is no change in copyright treatment for the content on the signals of significantly viewed stations. But the statutory change allowed DISH Network, which currently is under a court injunction prohibiting it from using the Section 119 statutory copyright license to retransmit the content of broadcast signals, to use the Section 122 statutory copyright license to do so.

Although STELA changed the statutory license required for satellite retransmission of the signals of significantly viewed and exception stations, it did not affect the retransmission consent requirement or the exemption from the FCC's network non-duplication and syndicated exclusivity rules, as they currently apply to significantly viewed and exception stations. It did, however, include a provision stating that the satellite operator would not be required to carry the significantly viewed stations or exception stations if it offered local-into-local service.

Allowable Signal Formats for the Retransmission of Significantly Viewed Stations

The satellite operators have complained that although both cable and satellite operators may offer significantly viewed stations, only satellite operators have been subject to an "equivalent bandwidth" provision that, as interpreted by the FCC, required the satellite operator to carry the signals of a significantly viewed station that is affiliated to the same network as a local station in the same format as that local station every moment of the day. Thus, for example, if the local station were not transmitting its programming in high definition format, the satellite operator would not be allowed to retransmit into the market the signals of the significantly viewed station in high definition format. According to satellite operators, this was infeasible.

STELA clarified that a significantly viewed signal may only be provided in high definition format if the satellite carrier is passing through all of the high definition programming of the corresponding local station in high definition format as well; if the local station is not providing programming in high definition format, then the satellite operator is not restricted from providing the significantly viewed station's signal in high definition format.

Studying What the Impact Would be If the Statutory Licensing System for Satellite and Cable Retransmission of Distant Broadcast Signals Were Eliminated

The United States Copyright Office has proposed that Congress abolish Sections 111 and 119 of the Copyright Law, arguing that the statutory licensing systems created by these provisions result in lower payments to copyright holders than would be made if compensation were left to market negotiations.[52] According to the Copyright Office, the cable and satellite industries no longer are nascent entities in need of government subsidies, have substantial market power, and are able to negotiate private agreements with copyright owners for programming carried on distant broadcast signals.

One possible way to transition from the current licensing system would be to enact a statutory "trigger" mechanism, by which once a broadcast station successfully demonstrated that it had obtained the rights to negotiate for all the holders of copyrighted materials on its programming, so that a satellite carrier did not have to negotiate with multiple copyright holders, the statutory license for that station would sunset and the satellite operator would have to undertake private negotiations. This is strongly opposed by satellite operators, who question how voluntary licensing arrangements and sublicensing would work in practice.[53] Other parties argue that the current licensing systems are efficient and that the purpose of copyright law is to balance the potentially conflicting goals of fostering the dissemination of copyrighted material and allowing the copyright holder to be compensated by giving the copyright holder a *limited* monopoly over its material; they oppose a rule that allows the copyright holder to fully exploit its monopoly power to receive whatever the market would bear.[54]

STELA instructed the Copyright Office, after consultation with the FCC, to submit to the House and Senate Judiciary Committees, within one year, a report containing proposed mechanisms, methods, and recommendations on how to implement a phase-out of the current statutory license requirements in Sections 111, 119, and 122 of the Copyright Act, including recommendations for legislative or administrative actions. The Copyright Office submitted its report on August 29, 2011.[55] The report included detailed recommendations "to effectuate a phase-out of the statutory licenses."[56]

STELA also instructed the Comptroller General to prepare and submit a report within 12 months that analyzes and evaluates the changes to the cable and satellite carriage requirements in the Communications Act and in FCC rules that would be required if Congress implemented a phase-out of the current Section 111, 119, and 122 statutory licensing requirements in the Copyright Act. It instructed the Comptroller General to consider the impact of such a phase-out on consumer prices and access to programming and to include recommendations for legislative or administrative actions. GAO submitted its report in November 2011.[57]

The Copyright Office report, which focused on how the various industry players would respond to the phase-out of the statutory licenses, was confident that sublicensing and other new business models would develop to replace the statutory licenses. The GAO report, which focused more on the impact on consumers, raised a number of problems that might arise during a phase-out, but also "identified a number of actions to mitigate these problems."[58] The two reports provide a starting point for a policy debate.

Providing Digital Service on a Single Dish

Under Section 338(g) of the Communications Act, satellite operators had been required to provide to their subscribers the analog signals of all broadcast stations on a single roof-top dish. Operators had been allowed to use a second dish for the provision of digital signals, but there was no requirement that all digital signals be provided on the same dish. STELA modified Section 338(g) to require a satellite operator, if it offers local-into-local service in a market, to provide to a subscriber the digital signals of all the local broadcast stations on a single dish.

Modification of the Methodology for Setting Copyright Royalty Rates and of Copyright Administrative Procedures and Requirements

STELA modified the methodology for setting copyright royalty rates as well as copyright administrative procedures and requirements. Among these changes, STELA:

- required satellite operators whose retransmissions of distant broadcast signals are subject to the Section 119 statutory license to pay a filing fee, to be determined by the Register of Copyrights, to help recoup the administrative costs of distributing royalty fees;
- modified the Section 119 statutory royalty fee payable to copyright owners to take into account the non-primary streams of multicasting broadcasters;
- instructed the Register of Copyrights to issue regulations to permit interested parties to verify and audit the statements of account and royalty fees submitted by satellite carriers and cable operators;
- changed the process for adjusting royalty fees. Most significantly, STELA created a proceeding of the Copyright Royalty Judges, which replaced the previously used compulsory arbitration proceeding, to determine royalty rates;
- instructed the Copyright Royalty Judges, when determining royalty rates in those situations where the parties are not able to reach a negotiated agreement, to establish fees that represent the fair market value of the retransmissions, basing their decision on economic, competitive, and programming information presented by the parties;

- required the Copyright Royalty Judges to make an annual adjustment to the royalty fee based on the most recent consumer price index for all consumers and for all items;
- increased the statutory maximum damages to be imposed on satellite operators for violating territorial restrictions on the retransmission of distant broadcast signals from $5 to $250 per subscriber per month during which the violation occurred. It also increased the maximum statutory damages for regional or large- scale violations (that do not trigger a permanent injunction) from $250,000 for each six-month period to $2.5 million for each three-month period. One-half of the statutory damages ordered are to be deposited with the Register of Copyrights and distributed to copyright owners;
- modified the statutory license for retransmission by cable systems in Section 111 of the Copyright Act by increasing the specified percentages of the gross subscriber receipts that cable operators must pay;
- updated the definition of "distant signal equivalent" used to reflect and take into account multicast signals when calculating the cable royalty payment, and set a schedule for when these changes go into effect based on existing contractual agreements;
- clarified that the royalty rates specified in Sections 256.2(c) and (d) of title 37, Code of Federal Regulations, commonly referred to as the "3.65% rate" and the "syndicated exclusivity surcharge," respectively, do not apply to multicast streams;
- clarified that when a cable operator retransmits a distant broadcast signal to a service area comprised of multiple communities, in which some communities are permitted to receive that signal and other communities are prohibited to do so, the royalty calculation does not include payment for the households that are not allowed to receive the signal;[59]
- modified the methodology for determining the maximum and minimum royalty payments for small cable systems; and
- created filing fees for satellite carriers and cable operators filing statements of account for Section 111, 119, and 122 statutory copyright licenses that are reasonable and that do not exceed one-half of the cost incurred by the Copyright Office for the collection and administration of the statements of account and any royalty fees deposited with the statements.

Severability

STELA included a "severability" provision stating that if any provision of the new law, amendment made by the new law, or applications of such provision or amendment is held to be unconstitutional, the remainder of the law, amendments, and applications would not be affected. This provision was included because there has been a long history of litigation in this area and was intended to make sure that the entire law would not be overturned if there were a successful legal challenge to one provision.

End Notes

[1] 124 Stat. 1218.

[2] 17 U.S.C. §§111, 119, and 122.

[3] 47 U.S.C. §§325, 335, 338, 339, 340, and 341.

[4] "Significantly viewed" stations are located outside the local market in which the subscriber is located but have been determined by the Federal Communications Commission to be viewed by a "significant" portion of those households in the local market that do not subscribe to any multichannel video programming distributor (MVPD). The specific threshold viewing level for a significantly viewed station are, for a network affiliate station, a market share of at least 3% of total weekly viewing hours in the market and a net weekly circulation of 25%; for independent stations, 2% of total weekly viewing hours and a net weekly circulation of 5%. The share of viewing hours refers to the total hours that households that do not receive television signals from MVPDs viewed the subject station during the week, expressed as a percentage of the total hours these households viewed all stations during the week. Net weekly circulation refers to the number of households that do not receive television signals from MVPDs that viewed the station for 5 minutes or more during the entire week, expressed as a percentage of the total households that do not receive television signals from MVPDs in the survey area. A satellite operator can retransmit the signals of these significantly viewed stations only with the retransmission consent of the station.

[5] The 2004 Satellite Home Viewer Extension and Reauthorization Act allowed satellite operators to retransmit in-state but non-local broadcast television signals to subscribers located in certain counties in Vermont, New Hampshire, Oregon, and Mississippi that are assigned to local markets (as defined by Nielsen Media Research designated market areas) whose local broadcast stations are located in another state. For convenience, these stations are referred to as statutorily exceptional stations.

[6] The United States Copyright Office submitted its report, *Satellite Television Extension and Localism Act §302 Report* (available at http://www.copyright on August 29, 2011, and the United States Government Accountability Office submitted its report, *Statutory Copyright Licensing: Implications of a Phaseout on Access to Television Programming and Consumer Prices Are Unclear,* GAO-12-75 (available at http://www.gao.gov/new.items/ d1275.pdf), in November 2011.

[7] *In the Matter of In-State Broadcast Programming Report to Congress Pursuant to Section 304 of the Satellite Television Extension and Localism Act of 2010,* MB Docket No. 10-238, Report, adopted August 26, 2011, and released August 29, 2011.

[8] P.L. 100-667.

[9] P.L. 103-369.

[10] P.L. 106-113.

[11] P.L. 108-447, passed as Division J of Title IX of the FY2005 Consolidated Appropriations Act.

[12] For a more detailed discussion of the differences in the rules for cable and satellite providers, see CRS Report R40624, *Reauthorizing the Satellite Home Viewing Provisions in the Communications Act and the Copyright Act: Issues for Congress,* by Charles B. Goldfarb, especially at Table 1, "Current Retransmission and Copyright Rules for Satellite and Cable Operators."

[13] This is formally referred to in the statute as "secondary transmission" of the broadcast signals. The initial transmission of the signals by the broadcast station is the "primary transmission."

[14] *Satellite Home Viewer Extension and Reauthorization Act Section 109 Report,* A Report of the Register of Copyrights, June 2008, at pp. ix-xi and 94-180.

[15] *Satellite Home Viewer Extension and Reauthorization Act Section 109 Report,* A Report of the Register of Copyrights, June 2008, at pp. 167-168.

[16] 47 C.F.R. §§76.92, 76.93, 76.106, 76.120, and 76.122. Commercial television station licensees that have contracted with a broadcast network for the exclusive distribution rights to that network's programming within a specified geographic area are entitled to block a local cable system from carrying any programming of a more distant television broadcast station that duplicates that network programming. Commercial broadcast stations may assert these non- duplication rights regardless of whether or not the network programming is actually being retransmitted by the local cable system and regardless of when, or if, the network programming is scheduled to be broadcast. This rule applies to cable systems with more than 1,000 subscribers. Generally, the zone of protection for such programming cannot exceed 35 miles for broadcast stations licensed to a community in the FCC's list of top 100 television markets or 55 miles for broadcast stations licensed to communities in smaller television markets. The non-duplication rule does not apply when the cable system community falls, in whole or in part, within the distant station's Grade B signal contour. In addition, a cable operator does not have to delete the network programming of any station that the FCC has previously recognized as "significantly viewed" in the cable community. With respect to satellite operators, the network non-duplication rule applies only to network signals transmitted by superstations, not to network signals transmitted by other distant network affiliates.

[17] 47 C.F.R. §§76.101, 76.103, 76.106, 76.120, and 76.123. Cable systems that serve at least 1,000 subscribers may be required, upon proper notification, to provide syndicated protection to broadcasters who have contracted with program suppliers for exclusive exhibition rights to certain programs within specific geographic areas, whether or not the cable system affected is carrying the station requesting this protection. However, no cable system is required to delete a program broadcast by a station that either is significantly viewed in the cable community or places a Grade B or better contour over the community of the cable system. With respect to satellite operators, the syndicated exclusivity rule applies only to syndicated programming transmitted by superstations, not to syndicated programming transmitted by other distant broadcast stations.

[18] 17 U.S.C. §119.

[19] A network is defined as an entity that offers an interconnected program service on a regular basis for 15 or more hours per week to at least 25 affiliated television licensees in 10 or more states. (17 U.S.C. §119(d)(2)(A) and 47 U.S.C. §339(d)(2)(A)) In addition to the four major television networks—ABC, CBS, Fox, and NBC—that provide national news and entertainment programming aimed at a general audience, there are several networks—Univision, Telefutura, and Telemundo—that offer news and entertainment targeted to ethnic communities, as well as smaller networks that provide entertainment or religious programming to their affiliates. Section 119(d)(2)(B) of the Copyright Act defines "network station" to also include noncommercial broadcast stations.

[20] This would include subscribers who are not able to receive network programming because either (1) the satellite operator does not offer the signals of the local broadcast stations and the subscribers are located too far from the transmitter to receive the signals of the local network-affiliated stations over-the-air or (2) there is no network-affiliated station in the local market. The specific household eligibility requirements for receiving distant signals are very complex and include certain grandfathered exceptions, but as a general rule households that can receive the signals of local broadcast television stations either over-the-air or as part of local-into-local satellite service are not eligible to receive distant network signals and would not be affected by the expiration of this provision.

[21] Prior to enactment of STELA, the Copyright Act and the Communications Act both had language referring to "superstations," but that term was defined differently in the two acts, thus creating confusion. The Communications Act identifies a class of "nationally distributed superstations" (47 U.S.C. §339(d)(2)) that is limited to six stations that were in operation prior to May 1, 1991. These are independent broadcast television stations whose broadcast signals are picked up and redistributed by satellite to local cable television operators and to satellite television operators all across the United States. These nationally distributed superstations in effect function like a cable network rather than a local broadcast television station or a broadcast television network. The nationally distributed superstations are WTBS, Atlanta; WOR and WPIX, New York; WSBK, Boston; WGN, Chicago; KTLA, Los Angeles; and KTVT, Dallas. All of these nationally distributed superstations carry the games of professional sports teams. It has become common in FCC proceedings and discussions to refer to these nationally distributed superstations as simply "superstations." In addition to these independent nationally distributed superstations, there also are many independent television stations that are not nationally distributed superstations. This distinction is important because under section 325(b)(2)(B) of the Communications Act, satellite operators may retransmit the signals of "superstations" without obtaining the consent of the stations if they abide by the FCC's network non-duplication and syndicated exclusivity rules (see footnotes 11 and 12 above), but this exemption from the retransmission consent requirement does apply to the retransmission of the signals of other independent stations. On the other hand, until statutory changes were made in STELA, the Copyright Act had defined "superstation" as "a television station, other than a network station, licensed by the Federal Communications Commission, that is secondarily transmitted by a satellite carrier." (17 U.S.C. §119(d)(9)) Thus, under the Copyright Act pre-STELA, all independent stations were considered superstations and the copyright provisions applied the same way to all independent stations. Language in STELA eliminated the definitional inconsistency between the acts by replacing the word "superstation" with "non-network station" throughout the Copyright Act.

[22] 47 U.S.C. §325(b)(2)(C).

[23] 47 U.S.C. §325(b)(3)(C)(ii).

[24] 47 U.S.C. §325(b)(3)(C)(iii).

[25] Marybeth Peters, Register of Copyrights, written statement before the House Judiciary Committee, hearing on "Copyright Licensing in a Digital Age: Competition, Compensation and the Need to Update the Cable and Satellite TV Licenses," at Appendix 1, February 25, 2009. The proposed modifications to section 111 include revising section 111, and its terms and conditions, to expressly address the retransmission of digital broadcast signals; amending the definition of "local service area of a primary transmitter" to include references to digital station "noise limited service contours" for purposes of defining the local/distant status of noncommercial educational stations (and certain UHF stations) for statutory royalty purposes; amending the statutory definition of "distant signal equivalent" (DSE) to clarify that the royalty payment is for the retransmission of the copyrighted content without regard to the transmission format; amending the definitions of "primary transmission" and "secondary transmission," as well as the "station" definitions in section 111(f) so they comport to the amended definition of DSE; and clarifying that each multicast stream of a digital television

station shall be treated as a separate DSE for section 111 royalty purposes. The proposed modifications to section 119 include replacing the existing Grade B analog standard with the new noise-limited digital signal intensity standard; adopting the Individual Location Longley Rice (ILLR) predictive digital methodology for predicting whether a household can receive an acceptable digital signal from a local digital network station; mandating that the FCC adopt digital signal testing procedures for purposes of determining whether a household is actually unserved by a local digital signal; and deleting various references in section 119 to "analog" unless that reference is to low power television stations that have not yet converted to digital broadcasting.

[26] *In the Matter of Measurement Standards for Digital Television Signals Pursuant to the Satellite Home Viewer Extension and Reauthorization Act of 2004*, ET Docket No. 06-94, Report and Order, adopted November 22, 2010, and released November 23, 2010, and *In the Matter of Establishment of a Model for Predicting Digital Broadcast Television Field Strength Receive at Individual Locations*, ET Docket No. 10-152, Report and Order and Further Notice of Proposed Rulemaking, adopted November 22, 2010, and released November 23, 2010.

[27] See, for example, Lauren Lynch Flick and Scott R. Flick, "Congress Passes Satellite Television Extension and Localism Act of 2010," Pillsbury Winthrop Shaw Pittman LLP Client Alert, May 14, 2010, available at http://www.ilba.org/downloads/~mo~FCC/Congress_Passes_STELA.pdf, viewed on June 2, 2010. Pillsbury is a law firm with many broadcaster clients.

[28] Warren Communications, *Television & Cable Factbook 2010*, Station Volume 2, "Affiliations by Market for TV Stations, as of October 1, 2009," at pp. C-5 – C-8.

[29] 47 U.S.C. §339. This provision applies to all network stations, but in practice it primarily involves the retransmission of distant signals into short markets that do not have local broadcast stations affiliated with each of the four major national broadcast networks.

[30] There remains a brief transition period, October 1, 2010, to January 1, 2011, during which if a local broadcaster were to begin multicasting another broadcast network signal, the signal would not be deemed a qualified multicast video and a satellite carrier could import into the local market the signal of a broadcaster affiliated with the same network.

[31] Paul Gallant, an analyst with Stanford Washington Research Group, reportedly stated that mandatory provision of local-into-local service in all markets "would impose significant new costs on Dish Network and DirecTV and generate virtually no new revenue" because the markets in question are so small. See Todd Shields, "DirecTV, Dish May Face Requirement for More Local TV (Update1)," Bloomberg.com, February 23, 2009, available at http://www.bloomberg.com/apps/news?pid=newsarchive&sid=ayQ_vo3nJImo, viewed on April 27, 2009.

[32] According to the written testimony of Charles W. Ergen, chairman, president, and chief executive officer of DISH Network Corporation, submitted for the hearing on "Reauthorization of the Satellite Home Viewer Extension and Reauthorization Act," before the Subcommittee on Communications, Technology, and the Internet, Committee on Energy and Commerce, U.S. House of Representatives, February 24, 2009, at p. 2, "DISH provides local service in 178 markets today, reaching 97 percent of households nationwide." According to the written testimony of Bob Gabrielli, senior vice president, broadcasting operations and distribution, DIRECTV, Inc., before the House Judiciary Committee, February 25, 2009, at p. 10, "DIRECTV today offers local television stations by satellite in 150 of the 210 local markets in the United States, serving 95 percent of American households. (Along with DISH Network, we offer local service to 98 percent of American households.)"

[33] See John Eggerton, "DISH: Local Into Local Within Two Years—No. 2 DBS Provider Said It Will Deliver Local TV Stations to All 210 DMAs During that Time Frame," *Multichannel News*, October 15, 2009.

[34] *CBS Broad. Inc. v. Echostar Comm. Corp.*, 11th Cir. Docket No. 03-13671 (May 23, 2006).

[35] *In the Matter of Application of DISH Network, LLC for Qualified Carrier Certification*, MB Docket No. 10-124, adopted on September 1, 2010, and released on September 2, 2010.

[36] See, for example, the written testimony of Derek Chang, executive vice president, content strategy and development, DirecTV, Inc., before the House Committee on Energy and Commerce, Subcommittee on Communication Technology, and the Internet, June 16, 2009, at pp. 5-6.

[37] 17 U.S.C. §119(a)(15)(B).

[38] 17 U.S.C. §119(a)(15)(D).

[39] 47 U.S.C. §335(b)(1).

[40] Written Testimony of Bill Acker, Director of Broadcasting and Technology, West Virginia Public Broadcasting, before the Senate Committee on Commerce, Science, and Transportation, Subcommittee on Communications, Technology and the Internet, October 7, 2009, at p. 3.

[41] Kamala Lane, "Dish Sues FCC Over 'PBS HD Mandate,'" *Satellite Week*, July 12, 2010.

[42] "Satellite TV," *Satellite Week*, July 26, 2010.

[43] John Eggerton, "Dish Averts STELA Carriage Mandate," *Multichannel News*, August 2, 2010.

[44] Ibid. at pp. 8-10.

[45] The statutory provisions for satellite explicitly require the use of Nielsen's DMAs. (17 U.S.C. §122(j)(2)(A) and (C).) The statutory provisions for cable instructed the FCC to make market determinations "using, where

available, commercial publications which delineate television markets based on viewing patterns." (47 U.S.C. §534(h)(1)(C).) Nielsen had already delineated such television markets, assigning geographic areas to markets based on predominant viewing patterns in order to construct ratings data for advertisers, and the FCC therefore adopted Nielsen's market delineations.

[46] For a complete state-by-state list of these counties, their populations, and the full power television stations located in the counties, see the Appendix to CRS Report R40624, *Reauthorizing the Satellite Home Viewing Provisions in the Communications Act and the Copyright Act: Issues for Congress*, by Charles B. Goldfarb.

[47] 17 U.S.C. §§119(a)(2)(c)(i)-(iv) and 47 U.S.C. §341.

[48] See, for example, John Eggerton, "Affiliate Associations Warn Legislators Against Allowing Imported Signals from In-State, Distant Markets," *Broadcasting & Cable*, March 30, 2009.

[49] Representative Ross had introduced the Local Television Freedom Act of 2009, which would have allowed multichannel video programming distributors (MVPDs)—satellite operators and cable operators (including telephone companies)—serving an orphan county to retransmit to their subscribers in that county the signals of television broadcast stations located in an adjacent in-state market. In addition, the Four Corners Television Access Act of 2009 had been introduced in both the House (by Representatives Salazar and Coffman) and the Senate (by Senators Bennet and Udall) to allow satellite operators to retransmit the signals of certain in-state broadcast stations to subscribers located in two Colorado counties that are assigned to the Albuquerque, NM, local market and to allow cable operators located in those counties to retransmit the signals of certain in-state stations without having to obtain retransmission consent from the stations. Also, Representative Boren had introduced a bill which would have allowed satellite operators to retransmit to any subscriber in the state of Oklahoma—not just those in orphan counties—the signals of any broadcast station located in that state.

[50] See Anandashankar Mazundar, "Senate Judiciary Committee Votes Out Satellite Television Reauthorization Bill," *BNA Daily Report for Executives*, September 25, 2009.

[51] See footnote 7 above.

[52] *Satellite Home Viewer Extension and Reauthorization Act Section 109 Report*, A Report of the Register of Copyrights, June 2008, at p. xiv.

[53] See, for example, the Written Testimony of Robert Gabrielli, senior vice president for program operations, DirecTV, Inc., before the Senate Committee on Commerce, Science, and Transportation, October 7, 2009, at p. 8.

[54] See, for example, the website of Public Knowledge at http://www.publicknowledge.org/issues

[55] *Satellite Television Extension and Localism Act §302 Report*, August 29, 2011.

[56] Ibid. at pp. 139-140.

[57] *Statutory Copyright Licensing: Implications of a Phaseout on Access to Television Programming and Consumer Prices Are Unclear*, GAO-12-75, November 2011.

[58] Ibid. at unpaginated section entitled "What GAO Found."

[59] Prior to this clarification, there had been situations in which a cable operator has been required to make a copyright payment as if it were retransmitting a distant signal to all the communities in a service area, but in fact was not allowed to retransmit the signal to certain communities in the service area. Cable operators have referred to the signals that they were not allowed to retransmit, but for which they had to make copyright payments, as "phantom signals."

INDEX

C

E

F

G

H

W

waiver, 89, 96, 99, 113, 262
wants and needs, 22
var, 229
Washington, 18, 20, 119, 126, 175, 177, 198, 223, 233, 276
water, 35, 74, 81
web, 48, 51, 166
web pages, 48
websites, 47, 89, 121, 163, 175, 176, 192, 193, 250
welfare, 25, 77
White House, 40, 177, 248, 250
White Paper, 181, 195, 197
wholesale, 57
Wi-Fi, 33, 83, 85
wildlife, 81
wireless broadband deployment, vii, 1, 23

wireless devices, xiv, 72, 80, 81, 201, 213, 216, 221
wireless networks, 6, 24, 36, 51, 79, 84, 244
wireless systems, 16
wireless technology, xiv, 66, 74, 80, 201, 217, 221, 222
wires, 13
Wisconsin, 18, 20
witnesses, 83, 165, 225
word processing, 80
workers, 202
workload, 237
World Wide Web, 157
worldwide, 164, 173, 187, 217, 228
worms, 52

Y

yield, 95